The Science of AI in Environmental Engineering

This book explores the integration of artificial intelligence (AI) in environmental engineering, emphasizing the unique challenges and approaches required for the accurate modeling of physical phenomena. It clearly explains how AI should be developed and applied specifically in this field, offering definitions, examples, and practical guidance. It is designed to be accessible, featuring tables, figures, and illustrations to simplify complex topics like water hydraulics, air pollution, waste management, and more. Suitable for professionals in the field and students, this book explains the benefits of AI in environmental engineering and discusses the latest developments and environmental concerns.

This book:

- Explains the nexus between artificial intelligence and environmental engineering.
- Includes illustrative problems and solutions commonly used in current environmental practices.
- Covers the latest AI developments and how they can be effectively applied to solve modern engineering challenges.

I0894925

The Science of AI in Environmental Engineering

Frank R. Spellman

CRC Press
Taylor & Francis Group
Boca Raton London New York

CRC Press is an imprint of the
Taylor & Francis Group, an **informa** business

Designed cover image: Shutterstock

First edition published 2025
by CRC Press
2385 NW Executive Center Drive, Suite 320, Boca Raton FL 33431

and by CRC Press
4 Park Square, Milton Park, Abingdon, Oxon, OX14 4RN

CRC Press is an imprint of Taylor & Francis Group, LLC

ISBN: 978-1-032-98950-1 (hbk)
ISBN: 978-1-032-98951-8 (pbk)
ISBN: 978-1-003-60156-2 (ebk)

DOI: 10.1201/9781003601562

Typeset in Times
by codeMantra

Contents

Preface

This book, *The Science of AI in Environmental Engineering*, is the fifteenth volume in the acclaimed series that includes *The Science of Fluid Mechanics: Basic Concepts and Applications in Water and Wastewater Treatment, The Science of Land Subsidence, The Science of Green Energy, The Science of Carbon Capture and Storage (CCS) (in production), The Science of Ocean Pollution), The Science of Lithium (Li), The Science of Electric Vehicles (EVs): Concepts and Applications, The Science of Rare Earth Elements: Concepts and Applications, The Science of Water, The Science of Air, The Science of Environmental Pollution, The Science of Renewable Energy, The Science of Waste,* and *The Science of Wind Power* all of which bring this highly successful series fully into the twenty-first century. This book continues the mantra series based on sound and not feel-good science. It also continues to be presented in the author's trademark real-world conversational style—making sure communication is certain—not a failure. I like to ensure that the reader sees the forest, the trees, and the bushes within, maybe because I want to converse with the reader. I aim to be comprehensive in coverage, understandable in what I deliver, and void of excess verbiage so that you, the reader, can feel fully informed and knowledgeable. Based on 50 years of experience in environmental concerns, interests, and personal involvements, I am also opinionated to provide you with a wealth of expertise and knowledge in a style that engages and connects with you.

Following the successful format of the other editions in this series, this no-holds-barred book, based on science, operation, and experience, aims to provide an understanding of the current science underpinning artificial intelligence (AI) as it is related to and interfaced with environmental science and environmental engineering.

Moreover, the purpose of this book, which I refer to as an elementary primer (the keyword is "elementary"), is to introduce and explain the nexus between artificial intelligence and environmental engineering. This book is not just about theory presented in the abstract; it is about practical application. This book is not a mathematics book; it is an information source. It is designed to prepare you for real-world scenarios and give you the confidence to apply your knowledge effectively.

One may ask: "Why is the focus on artificial intelligence (AI) in environmental engineering?" The answer is simple: based on personal and professional experience, almost 70% of environmental practice is involved with, computed by, explained by, and shown through computer models. Models provide a system description or describe a theory or phenomenon that accounts for its known or inferred properties, and they are commonly used for further analysis of their characteristics. Following this, most models today are based on a mathematical structure that can describe and analyze an actual environmental proposal, question, problem, or concern.

When Dylan stated that "things they are changing," I would like to slightly modify this celebrated and legendary statement by saying, "times they are changing." Anyone who does not recognize this is unfamiliar with the computer revolution and the tools that computing affords the environmental engineer (and many other professionals and others, as well) that are needed to delve into "things" as "time" advances and changes. Engineers and their associates must leverage AI to be competitive and competent—decisions must be data-driven.

With times changing, Earth faces various problems and challenges, such as natural disasters, a shortage of resources—oil, coal, gas, food, water, and others—environmental pollution, and safety in the workplace—all mixed with pollution issues and hazards.

Safety in the workplace? What does safety in the workplace have to do with environmental engineering?

When teaching full-time college undergraduate and graduate classes in environmental health and engineering, as well as a few nursing program courses, I was often asked: Why industrial hygiene? Why do we have to learn that? Is AI modeling used in industrial hygiene, and if so, for what?

After hearing these statements from various students for a few years (since about 2019), I explained to them that industrial hygienists use environmental modeling, monitoring, and analytical methods to detect the extent of worker exposure to hazards within the workplace environment. I explain that the keyword in the preceding is "environment." We all live, work, and play in an environment—some environment. Again, I point out that industrial hygienists utilize engineering models, monitoring, work practice controls, and other methods to control potential "environmental" health risks in the workplace.

This book is directed at those involved in environmental engineering, environmental health, and general environmental concerns, as well as for the general reader who wants to add to their knowledge of artificial intelligence and its potential impact on the environment. It can also serve as an introduction to environmental engineering for environmental undergraduate students, professionals, and scientists. It is an essential reference that provides a broad multidisciplinary view of one of the most critical engineering disciplines/technologies in use today. The goal of this manual is not to make environmental engineers or scientists out of anyone but to provide the "basic" principles required to understand the nexus between artificial intelligence and environmental technology. Illustrative problems regarding commonly used environmental practices are provided with artificial drivers to ensure correlation to modern practice and design.

The bottom line: in the evolving landscape of environmental studies, research, and practice, transformative technologies—artificial intelligence, machine learning, deep learning, and the Internet of Things—are steering industries toward unparalleled sustainability, growth, and efficiency. It can be said that AI stands at the helm, steering toward and boosting environmental sustainability, refining engineering practices, and optimizing supply chains worldwide. However, there are limitations because the helm must be cautiously steered, navigating a guided, unobstructed course. AI is revolutionizing environmental engineering (and just about everything else), enabling advanced monitoring, management, and optimization of all aspects of technology.

Frank R. Spellman,
Norfolk, VA

About the Author

Frank R. Spellman, Ph.D., is a retired assistant professor of environmental health at Old Dominion University, Norfolk, Virginia, and the author of more than 160 books covering topics ranging from concentrated animal feeding operations (CAFOs) to all areas of environmental science and occupational health. Many of his texts are readily available online, and several have been adopted for classroom use at major universities throughout the United States, Canada, China, Europe, South Korea, and Russia; two have been translated into Spanish for South American markets. He has been cited in more than 1,150 publications. He is a professional expert witness for three law groups and an incident/accident investigator for the U.S. Department of Justice and a northern Virginia law firm. In addition, he consults on homeland security vulnerability assessments for critical infrastructures, including water/wastewater facilities nationwide. He conducts nationwide pre-Occupational Safety and Health Administration (OSHA)/Environmental Protection Agency EPA audits. He receives frequent requests to co-author with well-recognized experts in several scientific fields; for example, he is a contributing author of the prestigious text *The Engineering Handbook,* 2nd ed. (CRC Press). He lectures on sewage treatment, water treatment, homeland security lectures, and safety topics throughout the country and teaches water/wastewater operator short courses at Virginia Tech (Blacksburg, Virginia). He holds a BA in Public Administration, a BS in Business Management, an MBA, and an MS and a PhD in Environmental Engineering (Figure 0.1).

FIGURE 0.1 Author in Badlands of South Dakota.

Part I

The Foundation

1 The Digital Enlightenment

Headline: This book defines AI as the simulation of human intelligence by mimicking the human brain for analysis, environmental system modeling, and decision-making. The U.S. Department of Energy (2008) defines AI as:

> ...a machine-based system that can, for a given set of human-defined objectives, make predictions, recommendations, or decisions influencing real or virtual environments. Artificial intelligence systems use machine- and human-based inputs to perceive real and virtual environments; abstract such perceptions into models through analysis in an automated manner; and use mode inference to formulate options for information or action.

THE FOURTH INDUSTRIAL REVOLUTION

Humans love to label. We label everything, every place, event, and many other things. This tendency to label makes sense whenever we need to distinguish one thing or time from the other, to classify, distinguish, and sometimes to judge. The list of labels is long. Some of our labels are benign—when we label something as delicious, the majority of us get the message, and some are harmful—when we label something hazardous, the majority of us get the message meant by this label. Let us cut to the chase: we label something or anything to distinguish, to express, to be descriptive, or to use euphemisms. The truth be told, we often label, which is about all many of us know or want to know about something or anything.

Let us talk about labeling time. We label historical times as Prehistory (the beginning to 600 B.C.), the Classical Era (600 B.C.–A.D. 476), the Middle Ages (A.D. 476–A.D. 1450), and the Early Modern Era (A.D. 1450–A.D. 1750). The present age, era, period, or time is labeled the Modern Era (A.D. 1750–Present).

We also label human generations. Consider the following generation names:

- The Lost Generation: Born 1883–1910
- The Greatest Generation: Born 1901–1927
- The Silent Generation: Born 1928–1945
- The Baby Boomer Generation: 1946–1964
- Generation X: Born 1964–1980
- Xennial Generation: Born 1977–1983
- Millennial Generation: Born 1981–1996
- Gen Z: Born 1997–2010
- Gen Alpha: Born after 2010

DOI: 10.1201/9781003601562-2

We also label revolutions; the French Revolution, the Russian Revolution, the Chinese Communist Revolution, and the Cuban Revolution probably come to mind. However, another type of revolution is pertinent to this text and discussion: the Industrial Revolution. In the initial Industrial Revolution (IR) (commonly labeled the First Industrial Revolution), arguably the best-known revolution, humans employed water, steam power, and other inventions to mechanize production. This was when we shifted from an almost total agrarian life to urbanization—working in factories, railroads, steamships, and other occupations remote from the farm. In 1870, we entered the Second Industrial Revolution, where electricity, mass production, and the division of labor came to the forefront. In 1969, we entered the Third Industrial Revolution, where the mighty transistor and other electronic components and machines came into use with the development of Artificial Intelligence and automated production systems. Today, we are living during the Fourth Industrial Revolution (aka 4th IR; the term was popularized in 2016 by Klaus Schwab)—what is often also known as the cyber-physical era or the age of digitization, which is certainly appropriate.

As with the previous Industrial Revolutions, the 4th IR punctually has changed the world. It is driven and powered by the cloud, social media, mobile, the Internet of Things (IoT), cybersecurity, cognitive computing, advanced robotics, 3D printing, machine-to-machine communication, and artificial intelligence (AI)—all of which are being pushed to more fantastic applications by increased computing power and data.

Robotics, AI, IoT, and augmented reality (AR) are some of the 4IR technologies changing the world and how Environmental Engineering has progressed nonlinearly, utilizing cyber technologies to model and mitigate environmental issues in the environment and the workplace at warp speed.

DID YOU KNOW?

The cloud was created because of organizations' storage needs. Based on rapid growth and experience, organizations needed more computing and storage space, which was too expensive. The cloud turned out to be the white knight that removed this burden. Keep in mind that cloud computing is vast—maybe measureless in extent.

NOTHING WILL BE LEFT UNTOUCHED

What exactly is AI? How exactly does AI work?

Many of us have asked or are currently asking these questions.

Why?

The reason is that AI has emerged as a transformative force in various industries, and environmental engineering is no exception. Moreover, AI is growing in use and applications, so you can hardly avoid hearing about AI today. For example, most of us have heard about or have used Alexa or Siri, the voice-controlled virtual assistants. However, not everyone using these virtual assistants understands their source, what wake word detection is all about (i.e., the special word or phrase that tells a voice assistant that a command is about to be spoken), how voice processing works, or how machine learning models use custom triggers and generative AI. However, the user proceeds without asking any questions. Whatever system is being used, the user finds that it either operates or does not operate, which is about the extent of most users' AI knowledge. This mindset, a way of thinking and operating, is a typical modus operandi in our daily ride through life. For example, most people have no idea how a vehicle motor powers their vehicle. Still, they turn the ignition key or push the start button anyway, start the car, and move along without knowing anything about vehicle engines. Today, many of those using AI, machine learning (ML), deep learning (DL), or IoT do so without thinking about it; it's just like breathing, eating, sleeping, or sky diving, just not thought about until we run out of electrical power in the first case, or in the other cases, when we need oxygen, food, the ability to sleep, or realize when we jump that we forgot our parachute.

Again, during this developing age of technological advancement, also known as the Fourth Industrial Revolution, the real-world applications of AI are expanding. AI is a work in progress, and we have only reached the edge of implementing AI technology. Figure 1.1 illustrates where we stand today (2024) in AI applications. Note that it is essential to point out that Figure 1.1, illustrates, again, where we are at this moment, and it is likely only the beginning of the applications of AI in the future.

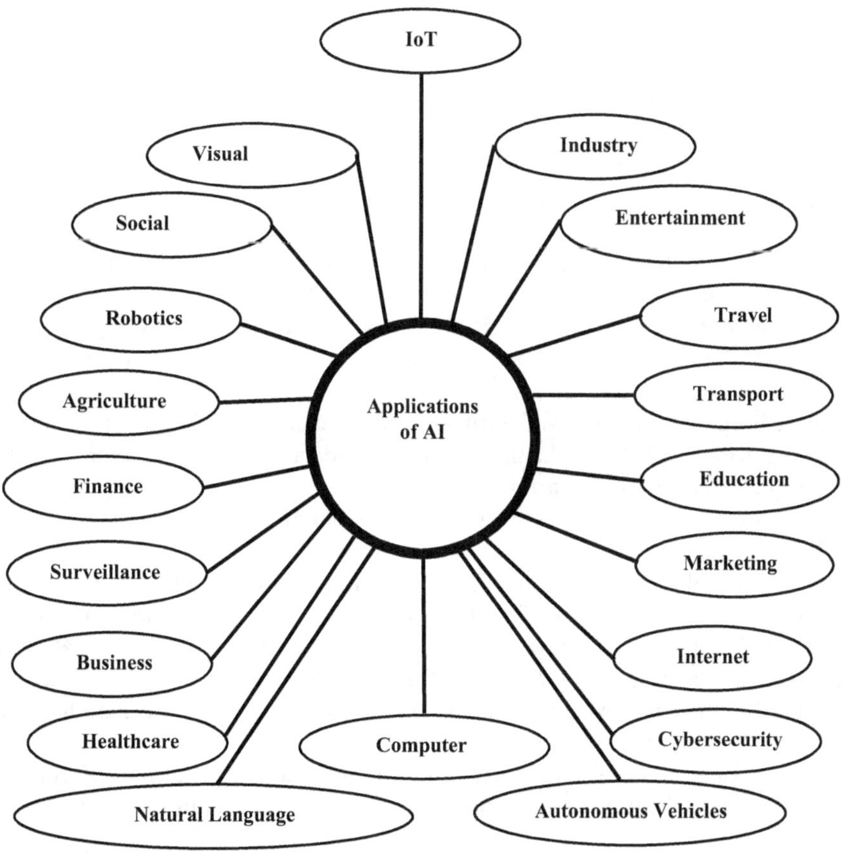

FIGURE 1.1 Real-world AI applications.

Although we are only at the outer edge of using AI applications, one thing is sure: AI is shaping the future of almost every sector and every human on the globe. So far, AI has acted as the driving force behind developing technologies for industrial automation, agriculture, IoT applications, cybersecurity services, medical applications, smart cities, business and finance, computer vision, mobile applications, etc.

DID YOU KNOW?

Another factor that users of AI must be aware of is that much of the data contained in many applications is in the "Black Box Mode," meaning that various sources contain data that is obscure and difficult to analyze. In contrast, in the "White Box Mode," data is transparent and can be analyzed by the user.

In addition to AI-assisted triggers used in virtual assistants, AI is everywhere, reaching further into our homes, cars, and phones. For example, AI makes social media content recommendations, detects financial fraud, and is employed in transportation, manufacturing, healthcare, and education.

THE NEED TO ADAPT, TO LEARN

Environmental engineers should (must) learn AI.

Why?

It is the future, like the IoT,

like ML,

like DL.

All sorts of digitization are occurring now and even more so in the future.

It is the future. This is the future. This is now!

Get aboard the digital train, and do not worry about a caboose—riding at the end of the train is not an option. No, not for anyone, especially not for technology-driven fields like environmental engineering.

I have been lucky. I have lived during the advancement of the digital age, and the passage of time is all about adapting; learning to adapt is how we move on and survive. Moreover, adapting is learning.

How about you? Are you ready for the AI revolution?

Can you adapt?

For me, the digital enlightenment, adaptation, and switchover from typewriter to personal computer began in 1984. I remember this year well for two reasons: my favorite NFL team, the Washington Redskins got skinned 38–9 by the Los Angeles Raiders, and secondly, during the Super Bowl broadcast, the viewing audience was introduced to the Apple Macintosh computer. The Redskins' loss was hard to swallow, but the advent of an affordable personal computer available for use in preparing my college environmental engineering, health, and a few medical nurses' courses I taught was beneficial, especially in the fashioning of my lecture notes, handouts, and examinations.

As the years passed and my lecture attendance continued to grow, the students felt I was somewhat of a new hat (a bit different) and quickly advertised that I was a 'professor easy' in favor of a wide range of students. Another item I noticed is that by 2020, almost every student possessed a laptop computer, smartphone, and smartwatch.

This 2020 awakening, for me, also was my first exposure to AI available at my fingertips. That Super Bowl ad was what some called a "resource" for humanity from the computer giant IBM. The ad conveyed that there was now a computer for the rest of us. I quickly became a fan of the Mac but did not express my satisfaction with Mac tattoos, a "MacQuarium," an Apple log haircut, paper Macs, Mac-inspired much is, and more—no, for me, the old hat was the call of the age, to a degree, at least.

One drastic change I noticed in the classroom, delivered via AI, is that the quality of students' reports, papers, theses, and dissertations is primarily well-written and grammatically correct—AI-assisted work.

Does this AI-assisted student work bother me? No, it is smooth sailing when I have to read their work. And besides, who wants to ride in the caboose in this digital age?

By the way, I currently use Microsoft Word and other associated applications in my work—it is just a personal choice—but Apple hooked me into the computer age and Microsoft into the AI age … and hopefully more to come.

THE BOTTOM LINE

The future is occurring faster than expected. Artificial intelligence and computer applications, both those we use daily and those yet to be discovered and applied, are part of the changing times. We can ride the AI train or get left behind, ride the caboose of technology, or stand in a forest without trees.

REFERENCE

USDOE. (2008). *The Smart Grid: An Introduction.* Accessed 9/30/24 @ https://www.energy.gov/smart-grid.

2 The Digital Professional

This book defines artificial intelligence (AI) as the simulation of human intelligence through mimicking the human brain for analysis, environmental system modeling, and decision-making.

DIGITALIZATION OF ENVIRONMENTAL PRACTICE

In September 2003, at a meeting in a 1936 hurricane-proof building in Newport News, Virginia, during Hurricane Isabel's arrival and short-term unwanted stay, I was sitting in an empty classroom with a headache. No, it was not Isabel's fury that was causing my headache. The sound of the wind striking the windows at my back seemed more soothing than the headache-knocker.

No, the headache-knocker was the training session I had just attended. I had listened to experts on the rapidly evolving digital world and its potential water/wastewater technology applications.

Note: Some might think that the presentation in this chapter is a personal monologue about entering and adapting to digital-age applications in the industry. And so it may be, but this chapter is essential and relevant to the discussion, worthy of presentation.

The information I received that day during the storm outside initiated a paradigm shift in my thinking about environmental engineering, managing and operating water and wastewater treatment plants, renewable energy, fracking, pollution, and much more—well, this was the case because water/wastewater and the others have always been close to the top areas of interest. First, concerning water and wastewater engineering and treatment, one of the ingredients in the mix causing my headache was a thought about my future, my fellow water and wastewater management professionals, operators, laboratory technicians, water quality technicians, and others associated with waterworks and wastewater treatment plants.

The key to thinking about personal job security—the ol' self-preservation motive—and that of others in what I call the fluid mechanic's field was initiated, punctuated, and aggravated by the stark realization that my livelihood might be threatened. However, these feelings of dread did not stay with me for long for several reasons. The first reason I recognized (with some internal pushing and recognition) was that it was time for me to retire at the ripe old age of 66 (do we ever really retire?). And shortly after this awakening to reality, I did retire. Well, not entirely, because I retired from official work to the bounty of teaching college students environmental health and engineering and even a few nursing subjects at two different universities. Also, the U.S. Department of Justice called upon me to investigate on-the-job fatalities in water/wastewater treatment plants, along with U.S. military operations where personnel were accidentally contaminated with methicillin-resistance *Staphylococcus aureus* (MRSA). But I have found all those enterprises to be more

DOI: 10.1201/9781003601562-3

adventure than work and have (to this day) the same feelings toward writing the books that I felt were and are necessary for various technical fields.

Anyway, after calming my self-preservation fears, my focus quickly turned to the job security of those in water/wastewater operations who were not eligible to retire. These were the operators (the fluid mechanics), trained, experienced, and licensed after years of schooling and apprenticeship service in water/wastewater operations.

After some thought and a couple of aspirins, I realized I was *not* jumping overboard from a sinking ship. My reliable life jacket (the computer world) saved me and allowed me to metamorphose, leading to modernization and more efficiency. Computerized operations or not, the fact struck me that the operation's personnel were still needed. In particular, computers require operators (well, at least then and at present), and all the unit *processes*, machinery, distribution systems, pumping stations, wastewater outfalls, and laboratory analysis units need someone to operate, monitor, repair, analyze lab samples, and adjust the flow within the system as required.

The point is that present-day water/wastewater operations personnel—and several other working technological and other professionals—who are not computer literate need training in computer operations to be digitized quickly.

Adapt, adapt, or otherwise.

The point is that digitization "greases" many skids. We need to apply the grease.

After absorbing this paradigm shift and its possible impact, including worries about self-preservation and other trained operators, maintenance personnel, laboratory technicians, and upper management personnel, I realized the only thing needed was a realignment in plant operations, including training in computerized, digitized systems.

But the headache did not depart. Instead, my focus shifted to another headache driver: the almost thirty pages of written notes I had taken during that day's training session.

I remember reviewing the notes and the deep learning (DL) indoctrination I had received earlier. The official title of the training was "*Deep Learning in Water and Wastewater Treatment*." It was exciting. More importantly, even though I was a member of the "old school"—known as a dinosaur-type to many folks—one filled with the knowledge of the past and pretty new to anything digital other than the word processing I used every day, even so, I understood the implications—the headache drivers from several different angles. These angles ranged from the advantages and disadvantages of incorporating artificial intelligence (AI), machine learning (ML), and DL into water and wastewater operations—this is not to ignore the Internet of Things (IoT) systems that several treatment facilities were and are using. In light of this ongoing incorporation of learning digitization, evolving technology, and its implications, there are many disadvantages for one who does not know, read, and understand the language—the "new" language.

I always recall what Voltaire said so long ago, "If you wish to converse with me, please define your terms."

AI and its associates have created a new language we must define, understand, grasp, and use. This language is in addition to the language associated with environmental engineering.

New language?

Yes, indeed—for sure. Let me present a real-world digital-based employment example. Let us say that you are the chief engineer of a large engineering firm.

You need to hire a new engineer-in-training (EIT). You have sorted through half a dozen job applicants' resumes and company job application forms. You are impressed with the two candidates' resumes and other information. You schedule personal interviews with them. In the morning, during your interview with Candidate 1, you were pleased with her extensive education and experience and her answers to all your queries—especially when she could correctly answer and define all the digitization terms you threw at her.

So, after the Candidate 1 interview, you were convinced that if her references were positive, she would be the one to hire.

Later in the day, during the Candidate 2 interview, you were just as impressed with this young woman's background, presentation, appearance, attitude, and the rest. You could feel the onset of a puzzling question: Which candidate should I hire?

Before the interview ended, Candidate 2 answered this question for you. After preliminary greetings, a review of Candidate 2's background, resume, college studies, course grades, and some general small talk, you decided it was time to present technical questions—the same questions you posed to Candidate 1. The difference between the experiences is that it took only one question and the response to it to tell you all you needed to know about Candidate 2 and your determination of whom to hire.

You stated that in filling the open position, it was important that the hired person be utterly literate in computer operations and have a strong background in AI, robotics, and digital systems, all of which are increasingly used in environmental engineering.

So, your first and only query was as follows:

"Okay, in a database management system, better known as DBMS, what is the difference between indexing and hashing?"

There was a moment of silence.

There was more than a moment of silence.

You looked at Candidate 2 and noticed that her forehead was lined, her eyes wide, and her head slightly turned to the left. She tried to answer your question but stumbled, her voice changing. Spasmodic dysphonia (a shaky voice) was apparent. She fumbled for an answer and instead kept looking away from you.

After a short period, you stood up, reached out your hand to shake hers, and waited for her to stand, shake hands, and depart.

She slowly got up from the chair and gave you a weak handshake, often described as limp. Then she turned and left the room, not to be seen by you again.

You hired Candidate 1 the following day and have not looked back since.

DID YOU KNOW?

The difference between indexing and hashing is as follows:

Indexing is a technique that quickly retrieves records from database files and is generally used to optimize database performance.

Hashing, on the other hand, is a faster and more flexible technique that allows searching for the desired data location on a disk without relying on an index structure.

ADAPT, ADAPT, OR OTHERWISE …

Many humans are under extreme emotional stress when their job environment or potential new job environment shifts from "old school" operations to the latest digital wave. They have trouble with the wave and cannot surf it. The simple truth is that, in this digital age, we must quickly adapt, get laid off, or change careers.

Not all the news is terrible. We must accept that we are living in a time of exciting (hopefully for most of us) technological innovations, with the digital sector driving transformative change. Business paradigms are changing and shifting. Digital technologies are reshaping the traditional workplace, enabling us to work from home instead of at the office or another work site. The latest developments in AI, generative AI, and related advances are expanding the frontiers of the digital revolution. You might think or suppose that the COVID-19 pandemic has slowed innovation and transformation, but the opposite is true; digital conversion is accelerating. The future is here, and you are either riding the wave or left riding the caboose in this digital age—you must adapt, adapt, or otherwise….

THE DIGITAL METAMORPHOSIS

To ride the digital wave (aka the digital train), we must, as stated, adapt, adapt, or otherwise. In this adaptation process, we need to learn the nuts and bolts and applications of AI and its subsets, including ML, DL, natural language processing (NLP), expert systems, computer vision, and the Internet of Things (IoT). While traditional web technologies (i.e., programming languages, databases, server administrations, servers, markup language, style rules, and cloud-side codes) remain at the top of the ladder in the majority of usage at present, the engine driving digital services is a new wave of cloud-based services. Following the engine (aka the flagship part of the digital train at present) in the digital train are the mobile internet, Big Data, design thinking, AI, robotics, advanced neural machine learning, augmented-reality technology, and additive manufacturing, in that order, are moving (the movers) up the train of usage, rails of popularity and innovation.

DID YOU KNOW?

NLP focuses on understanding and generating human language. Expert systems use rules and knowledge to make decisions in specific domains. Computer vision involves analyzing visual information to understand and interpret images and videos.

Let us look at the current components that make up the digital train and list their functions:

- **Cloud-Based Services:** Traditionally, organizations had to purchase and maintain their physical servers. Now, cloud-based servers run in the cloud on infrastructure owned by the cloud service provider; they remove the need to buy, run, and manage physical servers.

- **Mobile Internet:** Also known as the *mobile web*, it refers to an internet connection for handheld devices, such as smartphones or feature phones (i.e., a pre-wireless smartphone), through a wireless mobile network.
- **Big Data Architecture**: This refers a systemic design and structure of a data management system.
- **The Internet of Things (IoT)**: This is a network of physical objects embedded with sensors, software, and other technologies to connect and exchange data with other devices and systems via the Internet.
- **Digital Thinking (DT)**: This is a non-linear, iterative aspect or process that challenges assumptions, redefines complications, and creates groundbreaking models to test.
- **AI Technology**: This enables computers and machines to simulate human learning, understanding, problem-solving, decision-making, originality, and independence.
- **Robotics**: This follows the digital train. Within environmental engineering and science, robotics focuses on robotic automation algorithms used in environmental functions such as performing vital tasks related to air pollution monitoring, water, soil, and wildlife monitoring, and waste cleanup operations.
- **Advanced Neural ML**: The term "neural" is based on a model of how the human brain functions and comprises a collection of processing units called *nodes*. Advanced neural ML involves learning without definite instructions and is primarily used in deep learning.
- **Augmented-Reality Technology (AR)**: Augmented-reality technology (AR) is following in the digital train. AR is popular because it presents an enhanced version of the real-world via computer-generated digital information, including visual, sound, and other sensory elements.
- **Additive Manufacturing**: The final section of the digital train, not the caboose presented in this discussion, is known as additive manufacturing (aka 3D printing). In operation, additive manufacturing constructs a three-dimensional object from a CAD or digital 3D model.

Note: To be successful in AI-based environmental engineering, you must know the terminology, live it, write it, and use it in speech. Therefore, the keywords and abbreviations used in this book are presented and defined in the following section, in addition to the glossary at the end of the book.

KEY TERMS AND ABBREVIATIONS

Back to sitting in that empty classroom and looking over my notes, I noticed that the first part of the notes was all about AI nomenclature. However, the terms that were included and related to water and wastewater unit operations, such as anaerobic digestion, activated sludge, filtration systems, disinfection types and operation, chemical oxygen demand, biological oxygen demand, dissolved oxygen, mixed liquor suspended solids, nitrous oxide, suspended solids, and typical waterworks and wastewater treatment plant nomenclature, were not new to me and were quickly reviewed.

Terms and Definitions

Algorithm: A specification of how to solve a class of problems by performing calculations, data processing, and automated reasoning tasks.

Artificial Intelligence (AI): Intelligence generated by machines. AI is applied when a machine mimics cognitive functions that humans associate with learning and problem-solving.

Chatbot: An artificial intelligence that conducts a conversation via auditory or word-based methods.

Generative AI: Refers to models or algorithms that generate new outputs, such as text, photos, videos, data code, and 3D representations, from vast amounts of data they were prepared on. Understanding the differences between AI and Generative AI is crucial in today's technology-driven world, as these two concepts, while married at the hip, so to speak, offer unique capabilities and applications in several different fields. Traditional AI systems are designed to understand, interpret, and make decisions based on the data they have been supplied. They can answer questions, recommend purchasing products, and drive motor vehicles autonomously. This is accomplished through machine learning algorithms that allow these systems to gain knowledge from experience and improve over time. On the other hand, Generative AI goes beyond understanding and interpreting data. It generates entirely new content from the very beginning. It can create human-like text through advanced deep learning, compose music, and produce images indistinguishable from those made by humans.

Generative Pre-trained Transformers (GPT): Denotes a family of AI models. These models, primarily large language models such as those developed by OpenAI (founded in 2015 in San Francisco by Elon Musk and others), are based on a transformer architecture capable of understanding and generating human-like language.

Machine Learning (ML): The scientific study of statistical models and algorithms that computer systems use to perform specific tasks.

Yes, water and wastewater nomenclature were simply an old hat and trite to me; it was nothing new that I did not already know and feel comfortable with.

However, the following AI nomenclature and the numerous industry abbreviations were new to me—as foreign as hieroglyphics—and were head-scratchers.

As I looked down my list of AI jargon, the terms and brief definitions I had written next to each abbreviation were all new to me, like rocket science. My list began with and continued with:

AE: Auto Recorder
AI: Artificial Intelligence
ANN: Artificial Neural Network
CFD: Computational Fluid Dynamics
CUSUM: Shewhart Cumulative Sum
DBN: Deep Belief Network
DL: Deep Learning
DRL: Deep Reinforcement Learning
GPU: Graphical Processing Unit
GRU: Gated Recurrent Unit

LSTM: Long Short-Term Memory
MLP: Multi-Layer Perception
ML: Machine Learning
NN: Neural Network
TL: Transfer Learning

It took me a few months to understand the meaning of all these abbreviations and their impact and place in water and wastewater treatment management and operations. It took longer to escape my tendency toward linear and binary thinking and move on to the divergent and far-thinking mode. For example, I understood AI's Deep Learning technique for creating a model of a complex non-linear system by learning from examples of the system's inputs and outputs. Deep learning extracts essential features or representations directly from the raw input data by organizing its computational processing units (aka artificial neurons) into a hierarchy of layers.

Hierarchy of layers?

Yes.

As a subset of machine learning, methods based on neural networks with representation learning are what the hierarchy of layers is all about. I learned that "deep" means using multiple layers in the network. The organization in these layers provides the view that the computational depth of the network inspires the name "deep learning." When trained for a particular task, whatever that task might be, the model can be used (and is used) to generate outputs. The function of these outputs is based on the neural network's architecture.

DID YOU KNOW?

What exactly is AI? Artificial intelligence allows a computer program to think and learn independently. It is synthetic because it simulates human intelligence in machines that do things humans usually do. This technological marvel extends beyond mere automation, incorporating a broad spectrum of AI skills, including enabling machines to understand, reason, learn, and interact in a human-like manner. AI is designed to complement humans and not replace them—at least for now.

THE MACHINE LEARNING INTERFACE

In 2022, I learned a lot while watching the installation of computerized automatic sampling at water and wastewater facilities. Moreover, I watched the integration of AI, the IoT, and ML into large plant operations, which was (and is) revolutionizing water and wastewater treatment (as well as the other sectors of environmental engineering).

The critical facets of ML and its co-technologies that I observed and learned include the following points:

- **Real-time Monitoring**: Devices, such as IoT sensors, are embedded wholesale, blanket-fashion, in waterworks and wastewater treatment plants. Real-time monitoring delivers updated data about events, processes, and systems. This capability is essential because it provides real-time data on water quality parameters. Another advantage of this type of monitoring is that it enables continuous monitoring and immediate recognition of glitches and abnormalities.

- **Predictive Modeling**: This technique uses statistics to predict outcomes (most commonly future outcomes). The algorithms analyze the data collected in ML to predict potential issues and optimize treatment processes. In a technical sense, predictive modeling begins with data processing, data splitting, and model tuning (essentially meaning that deep learning models can be trained with limited data). From my perspective, the huge benefit is that data-driven decisions enhance efficiency and reduce operational costs—especially in energy conservation and water quality management.
- **Computerization and Manipulation (Automation and Control):** When IoT and ML are combined, they enable an automated control system that dynamically adjusts treatment procedures based on real-time data. This ensures that the treatment unit processes are continuously operating at peak levels.
- **Viability:** These technologies contribute to more sustainable wastewater management by improving the accuracy and efficiency of treatment processes, minimizing environmental impact, and reducing energy consumption. Water, and especially wastewater treatment processes, are highly energy-dependent. Finding cost savings in energy usage is often tricky, but it is possible with planning and competent supervision.
- **Blockchain Integration**: Blockchain is an emerging technology that creates a secure and transparent framework for managing water/wastewater treatment systems data and works to monitor water quality (SmartWater Magazine, 2024).

Note: I discuss AI and water/wastewater treatment in the early sections of this book but remember that the following section also applies to all environmental engineering sectors.

GOOD AND BAD OF AI IN ENVIRONMENTAL ENGINEERING

After training in AI, ML, and IoT and realizing that instead of being a job killer in water treatment and distribution, as well as wastewater treatment and discharge or reuse, I began to look at the potential advantages of incorporating AI into water/wastewater operations. So, what I did was first ease my headache by treating it with medicine, and then before leaving the empty classroom, I began listing the advantages, at least from my point of view. What I came up with is based on Duggal (2024) and is listed as follows:

- **Reduction of Human Error**: AI can reduce errors to null when appropriately programmed.
- **Human Risk Reduction**: The robotics involved in AI, ML, and IoT have revolutionized the field and allow hazardous processes to be accomplished without human hands-on involvement. Typical applications of AI in risk

management include threat intelligence analysis, event management, security information, fraud detection, risk reduction in the workplace, and data monitoring and classification.

- **On-Call 24 × 7**: Twenty-four by seven service means AI, ML, and IoT are productive on a nearly continuous basis, enhancing plant productivity, and can handle repetitive jobs easily, with near seamless operations around the clock.
- **Digital Assistants**: To deliver operational content upon user request.
- **New Discoveries**: AI, ML, and IoT are driving forces behind numerous innovations and discoveries that are aiding humans in resolving the most challenging issues.
- **Unemotional Operation**: AI, ML, and IoT are devoid of emotions and are highly practical and rational in their approach. Whenever human bias is removed from operations, it is a plus.
- **Repetition, Repetition, Repetition, and More Repetition**: This is the old 'here we go again' repetition. Based on experience and observation, little is as dull as repetition. The problem is that we all do a lot of repetitive tasks as part of our daily work, our work-related tasks. AI, ML, and IoT may efficiently automate these menial chores and even remove the "boring" tasks for people. When we are not bored, we are more creative.
- **Daily Uses:** While teaching upper-grad and graduate students one of my college topics on the Internet and Homeland Security, I always injected the following in my lectures: "Where do you think we would be today without the Internet?" We utilize a variety of apps, including Alexa, Windows, Google, Siri, and Google Maps, to make calls, take selfies, and respond to emails. With all this available technology, we give little thought to the enormous technological advancements for our daily use.
- **AI in Perilous Situations:** The most common uses of AI, ML, and IoT to support risk management and mitigation processes are essential. Having a robot perform perilous tasks lets us avoid many dangerous situations.
- **Medical Uses**: AI-powered tools have applications ranging from diagnosis and treatment to drug discovery and clinical trials. They help medical professionals keep employees and patients in top condition.

The Flip Side

We all know a coin has two sides. AI, ML, and IoT have the advantages listed above. Still, like the coin, these critical technologies have another side or downside, depending on how you view them or their applications.

Let us now look at artificial intelligence's main disadvantages, again depending on one's point of view.

- **Significant Costs**: "A significant cost" because having the ability to create a robot, a machine that can simulate human intelligence, is no small feat. This achievement is not cheap because enormous amounts of time and resources can cost a lot of money. Also, AI is constantly changing, adapting, and advancing with the latest hardware and software; staying updated is a critical requirement, thus making it quite expensive.

- **Imagination Restrained**: This points to immobilized or held-back creativity. AI cannot learn to think outside the digital box. However, AI can learn over time with pre-fed data and past experiences but cannot be imaginative or creative in its approach. AI can only do what it is programmed or told to do.
- **Makes Humans Lethargic and Lazy:** AI is addictive and has rapidly integrated into many aspects of our daily lives. Whether used as personal assistants on our smartphones or the algorithms underpinning AI and its messaging, AI does the work humans used to do. AI technologies allow us to use our brains less and less, and future generations are at risk.
- **There is No Right from Wrong:** In this situation, ethics and morality are essential to humans, but ethics and morality in AI are a different story. These human features are challenging to incorporate into AI. With no concern for ethics and morality, this moment is known as AI singularity, referring to when AI intelligence surpasses human intelligence, leading to machines that can outthink and outnumber us. AI can grow uncontrollably and eventually take over or wipe out humanity.
- **Lack of Feelings**: We have been taught or advised that machines, including computers, do not have feelings. Even though robots are superior to humans when functioning effectively, humans function as a team and cannot be replaced by digital technology.
- **Machines are Programmed**: AI, ML, and IoT machines can only be used to complete tasks they have been programmed for; they cannot do anything else. If adjustments are needed, we must manually alter the codes.
- **Robot It!** The most obvious (perceived) disadvantage of AI, ML, and IoT is their use in robots and chatbots that take human jobs. Unemployment is a situation most of us attempt to avoid. The bottom line for workers worrying about digital processes or systems taking their jobs is to join the chorus. In this digitized world, we must adapt and become computer literate. Moreover, while disruption caused by advanced technology is a fact of life, AI, ML, and IoT do not cause widespread loss of jobs (presently). However, they change the demand for specific skills (e.g., plant maintenance operators and samplers) and contribute to more flexible contingent work arrangements.

Throwing one's hands in the air and accepting defeat and loss of employment due to chatbots and robots in this digital age is not the answer. The answer is, again, adapting to change by working with and on implementing and maintaining the changes, not being against them.

An offshoot of the robot is the Internet bot. One of the questions I often hear is, "What is a bot?" Now, my students do not ask this question because I have found that they have become experts at or with bots—the question usually arises from those who are older and from a different time and place and not all that up on computers, the Internet, bots, or anything digital other than their TV remotes.

Different types of bots, including *chatbots*, simulate human conversation and provide customer service and information. *Web crawlers* index content for search engines. *Social bots* operate on social media platforms to respond to prompts and interact with users. Finally, there are the *malicious bots* we all need to be aware of, along with their potential to conduct cyber-attacks.

THE BOTTOM LINE

We need humans behind the digital revolution. To qualify and succeed in engineering or any other technology field (even in business, finance, banking, and so forth), we need to adapt to the skills required for the job. Times are changing fast, and roles are being redefined and recast. There is a definite need to board the digital train and hold on while the industry advances to levels not yet identified.

REFERENCES

Duggal, N. (2024). *Advantages and Disadvantages of Artificial Intelligence (AI)*. Accessed 8/11/24 @ https://muchrach.com.

SmartWaterMagazine. (2024). *How AI Technologies Are Revolutionizing Wastewater Treatment.* Accessed 8/11/2024 @ https://smartwatermagazine.com/news/smart-water-magaine/how-ai-technolgies-are-revolutionizing-wastewater-treatment.

3 Fundamentals of Environmental Engineering

AI represents a transformative branch of computer science dedicated to creating intelligent agents capable of representing human-like thought functions.

INTRODUCTION

From the air we breathe to the water we drink and the food we eat, the environment plays an important and crucial role in determining our well-being. As you might expect, environmental and human health concerns are growing globally. Environmental engineering is a subdiscipline of civil engineering, where, in practice, environmental engineers focus on what factors are prevalent and how they impair our day-to-day lives (see Figure 3.1). Understanding these connections is essential for

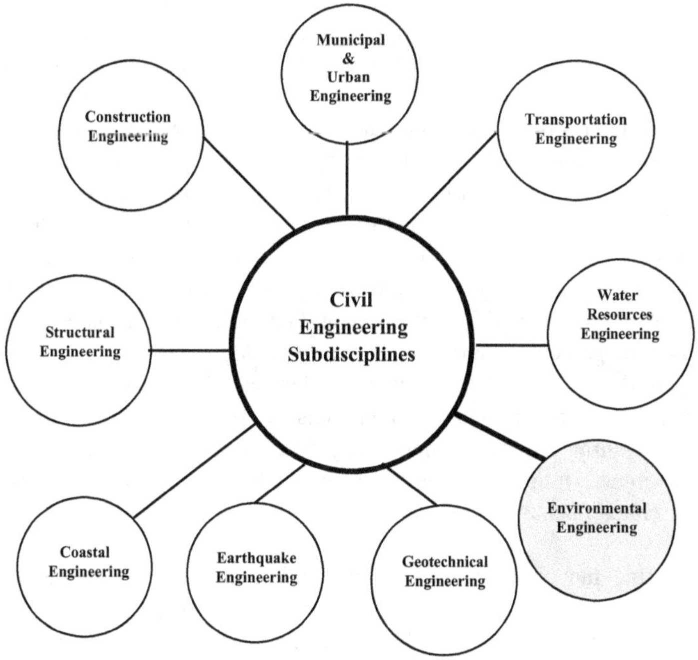

FIGURE 3.1 Subdisciplines of civil engineering.

DOI: 10.1201/9781003601562-4

developing effective strategies to mitigate environmental health risks and promote viable and sustainable life for all.

Is using AI an essential strategy?

Artificial intelligence (AI) has surfaced as a transformative influence in all branches of engineering, including environmental engineering. In the realm of practice, the projects environmental engineers work on include air pollution, water pollution, wastewater treatment, agriculture (soil pollution), recycling, waste disposal, public health, clean energy, sustainability, climate change, rainwater harvesting, green energy, environmental impact studies, and dam design.

Environmental engineers work to sustain human existence by balancing human needs with the environment's natural state per Nature's grand design. It goes without over-emphasizing the obvious truth that sustaining human existence and natural processes in harmony is no easy undertaking. In the face of global pollution, diminishing natural resources, increased population growth (especially in disadvantaged countries), geopolitical warfare, terrorism, global climate change (cyclical and human-caused), and other environmental problems, it is essential to the argument or thesis, of *The Science of AI in Environmental Engineering* that we live in a world undergoing rapid ecological and technological transformation.

It has been made clear through various personal observations, including an exponential increase in graduate and undergraduate student populations filling my classes in environmental engineering, environmental health, and environmental science, that the scope and importance of these environmental disciplines continue to grow. Because of this rapid recognition of pressing environmental changes and concerns, the role of environmental engineering has become increasingly prominent. Moreover, technological advances have created many modern environmental issues and professional opportunities for well-trained practitioners. To mitigate these issues, we must capitalize on the environmental protection and remediation opportunities presented by the technology that created many of the problems confronting us today.

To find and implement solutions to create a balance between the environment and those of us who occupy this planet (who seek the so-called good life), we must have trained personnel who know the issues and how to solve them. Familiarity with all environmental engineering specialties recognizes the importance of obtaining a solid quantitative background in environmental engineering, computer applications, science, and management principles that govern ecological processes. Of course, all the training in the world will never suffice unless we blend it with that uncommon trait known as common sense tempered with loads of practical on-the-job experience.

Regarding training for environmental engineers, students and practitioners in the field need to obtain an up-to-date level of knowledge and understanding in the following interrelated fields:

- Knowledge and understanding of physical, chemical, and biological processes, measurement techniques, and environmental issues that illustrate the application of certain fundamental principles of engineering and science required for any quantitative treatment of environmental problems fundamental to understanding the environment's fate and engineered treatment of environmental contaminants

- Knowledge and understanding of the sources and nature of waste materials that contribute to air, oil, and water pollution and relevant management and control technologies
- Knowledge and understanding of science, impact mitigation, adaptation, and policy relevant to climate change and global environmental sustainability, energy planning, alternative energy technologies (e.g., hydraulic fracking), sustainable development, and next-generation processes
- Knowledge and understanding of the transport and transformation of contaminants through environmental pathways
- Knowledge and understanding of pollution prevention technologies and designs associated with the treatment and disposal of waste materials
- Knowledge and understanding of the connection between the engineering and scientific aspects of environmental problems and decision-making processes

In environmental engineering, AI comes into play with knowledge and understanding related to applications in various environmental sectors, including energy management, wildlife protection, pollution control, agriculture, clean energy, natural resources conservation, and waste management.

Understanding these areas is achieved through quantitative educational training programs and years of on-the-job experience built around the common theme of engineering, science, and AI supporting environmental decision-making and management.

Currently, assorted critics view technological advances as the culprits humankind has employed to harm Nature's Way and the environment. Environmental engineers respond to society's needs with these technological innovations; they use technology to solve technologically-driven problems.

The question is: How do environmental engineers employ and put to use AI technology to correct or mitigate environmental problems? What are the simple and compound answers? The simple answer first: Environmental engineers use their tools. Tools? Yes, absolutely. Environmental engineers have a toolbox complete with tools. What are these tools? Good question. The short list of these tools includes knowledge of AI, metrics, constants, parameters, fluid mechanics, ecology, principles of toxicology, risk assessment, management principles, hydrogeology, modeling contaminant migration through multimedia systems, aquatic chemistry, environmental microbiology, applied statistical analyses, open channel hydraulics, field methods in habitat analysis and wetland delineation, principles of estuarine environments, hydrology, resources modeling, environmental sampling, sediment transport and river mechanics, geomorphic and ecological foundations of stream restoration, atmospheric chemistry, environmental chemicals, economic foundations for public decision-making, business law for engineers, environmental impact assessment, geographic information systems (GIS), global environmental sustainability, water resources management, sustainable development, green engineering, energy planning, renewable energy, innovative growth strategies for sustainable urban development and revitalization, environmental safety and health, and many other specialties. One essential specialty that is often overlooked, not studied, or not even thought about in college-level environmental engineering curriculums is industrial hygiene. Usually, it is not until the graduate environmental engineer enters the working world that they recognize that industrial hygiene is an essential subsector of environmental engineering. Because of this shortcoming in present-day textbooks and many college-level training programs, and the need for environmental engineers to be well-versed in the critical aspects of industrial hygiene, one part of this book is devoted to industrial hygiene.

Okay, now for the compound answer: From the preceding list, the well-trained environmental engineer must have an extensive toolbox; that is, they must be a generalist—they must know a little bit about a whole bunch. This whole bunch includes broad scientific topics like physics, ecology, chemistry, biology, hydraulics, hydrology, microbiology, and mathematics. This is the point the author is making. This is the compound part of the answer to the question: How do environmental engineers employ and put to use technology to correct or mitigate environmental problems? One way in which environmental engineers do this is by employing AI in practice.

A NOTE TO THE READER: PARDON MY REDUNDANCY, PLEASE!

Note: The following passage has been used by me several times over the years while lecturing, in writing, in public speaking, and in general thinking (on my part). I do not apologize for inserting it here in this text mainly because the critics have overwhelmingly (at last count in 5-log 99.999 fashion) responded positively to the passage, with many suggestions and constructive ideas. Learning is about broadening one's thinking, opening one's mind, and thinking for oneself. If this passage accomplishes any of this for you, please reread it—redundancy is good.

*****

I remember the moment well. It was a few minutes before I entered my classroom for the first time that semester to present my opening lecture on An Introduction to Environmental Engineering to a room full of college students (a large mix of juniors, seniors, and graduate students). At that exact moment, I stood at my office window looking out onto the campus grounds. I just stood there, bringing on one of those thinking-too-hard headaches. Of course, because I had developed the tendency to think on two wavelengths simultaneously (i.e., thinking about general thoughts and subliminally about what I was viewing at that moment—I call this compound thinking), a super headache manifested as always.

Anyway, back to that day and my compound thinking. On the general wavelength, I was thinking about my pending lecture presentation. More specifically, more to the point...I was thinking about how to grab the students' attention, that is, how to hook them. To deliver the message, you MUST hook the students' attention. I knew I NEEDED to hook them; otherwise, I would lose them to one of those annoying Twitter tweets, songs, or daydreams (boredom was not an option and impossible in my presence). So, I had the choice to take the milk toast presentation method: yes, I could point out that the environment is in trouble; yes, that they should be careful about that glass of water they might drink; yes, they should think about the air they are breathing; yes, they should think about the ozone hole; yes, they should think about Arctic ice melt; and, yes, they should think about their current practice as a contributor to the world's foremost "throw-away society" (this was an American campus, of course). And no, it is not hopeless. No, we are not doomed. No, we are not headed to the Sixth or Seventh Extinction, nor are we aware of how many there have been. The fact is I knew I could present all this. I did not believe (and still do not) that environmental professionals should focus on reactionary actions only to mitigate any of them. This type of presentation is a milk toast presentation. A milk toast presentation is just that: milk toast with no body or substance, a timid, unassertive, spineless (jellyfish) approach lacking boldness and vigor. One thing is sure: if you set out to save the world from human-derived destruction, you best leave the milk toast approach at home. Why only fix the problem after it occurs?

At that moment, I remember rubbing that throbbing forehead of mine as I contributed to that other compound ingredient of my headache: the visual aspect. I contributed to it in a cursory way... subliminally, you understand...well, sort of...by staring through my office window at the campus surroundings. I looked at what should have been a glorious early Fall campus panorama; instead, I took in a disgusting trash-laden panorama of the campus and then directly at a large tree with its full-leafed branches loaded with candy wrappers, remnants of this or that...hallmarks of the throw-away society; just a bunch of those fluttering banners of debris, those candy wrappers and lost plastic bags, scraps of newspaper, the rings of recently, and not so recently, gulped six packs, and to my right, on a singular branch, hung a signature of misguided passion or ignorance, an exclamation point: a battered political sign...all hanging around and just blowing in the wind. Then my glance shifted to the flower bed and the beauty of fully bloomed flowers, interrupted here and there by plastic and paper wraps of previous uses...and, of course, those trail marker beer cans and bottles here, there, and everywhere. This borders and bisects one of the numerous campus parking lots, and the debris marks those pathways most often taken. It was funny that the cars and trucks parked there did not register in my brain cells...only the discarded paper, plastic, beer cans/bottles, cigarettes, and other smoking debris registered on the lid and margin of my aching brain cells.

I remembered then that my thought process had shifted gears to my preferred hardtack approach (i.e., I would change to scaring the bejesus out of those bright-eyed students). More specifically, I had decided to present the obvious or that which was not so obvious. Simply, I would point out the need for environmental studies because of a litany of environmental issues, harmful aspects of human activity on the biophysical environment, and, as such, how they relate to the anthropogenic effects on the natural environment, which were combining to kill us all off eventually. I planned to loosely divide causes, effects, and mitigation, pointing out that effects are interconnected and can cause new effects. To accomplish all this, I initially decided to present, in one form or another, a semester of presentations that would include the following causes and effects of environmental problems:

Causes:

- Human overpopulation
- Hydrological issues

- Intensive farming
- Land use
- Nanotechnology
- Nuclear issues
- Biodiversity issues
- Ecophagy

Effects:

- Climate change (I think this is a natural phenomenon, not human-caused, but is undoubtedly exasperated by humans)
- Environmental degradation
- Environmental health
- Environmental issues with energy
- Environmental issues with war
- Overpopulation
- Genetic engineering
- Pollution—Air, Water and Soil
- Resource depletion
- Toxicants
- Waste

Of course, I had also planned to point out (and did point out) to the students that all of the above causes and effects were about to be their focus in life, their mantra, their burden, and their passion (hopefully). As prospective environmental professionals, there were, in this class and those to follow, students to learn not only about the causes and effects of environmental issues but also how to mitigate them, not only by practicing various mitigation measures and complying with environmental laws but, more importantly, by performing practices to prevent the causes and thus prevent environmental damage in the first place. It was necessary to point out the environmental issues and imprint them on those young minds while expressing the importance of engineering out a potential problem before it manifests itself. Prevention is always better than mitigation (mitigation consists of actions that include recycling, treatment, and disposal). Proactive is better than reactive. Building one's box to think outside of makes better sense than thinking outside someone else's creation.

So, that day, I faced that same never-ending dilemma: How could I harness the students' attention and hook them to preserve our environment? How was I to make the lasting point that our environment was (is) in trouble? Intuitively, I knew scaring the bejesus out of them was the straightforward approach to make them tune into my lectures, but I also knew that words only have so much weight—whether they are scary or not. Remember what they say about being safe? It is all about fear. We act safely because we fear being hurt or worse. If I could instill this thought in my students—protecting our environment because we fear the potential negative consequences—I would have found the solution to pollution.

I rubbed my head and indirectly the headache, and then I had my Eureka moment. Right there in front of me...you know the old saying, a picture is worth a thousand words, and it often provides the cement needed to imprint to memory the sights beheld...all that was visible outside my window was all I needed to show them. And I did. I ushered them into my office, single-filed them to the window to glance out and onto the campus grounds, and then sent them back to their desks in the classroom. Of course, I had directed them to write a 150-word description of what they had observed through the window (I always make my students write, write, and write some more. Knowing environmental engineering without knowing how to write is like knowing dentistry without knowing how to pull teeth). They were instructed to email their papers to me. I have found that whenever I put a student on their computer, I have my foot in the door to gain their attention and advance their learning. By the way, 77 out of 79 of the digital responses I received related observations of precisely what I wanted them to observe outside that window. What they saw was this textbook: a ready reference and guidebook of information and instruction on environmental issues and how to engineer out a problem before it occurs.

In this note to the reader, a final word about this textbook. This is not your Granddad's textbook. This textbook presents the facts, as it must, but it is opinionated with statements based on the author's observations, experiences, mistakes, misjudgments, and a few triumphs. You may disagree with the opinions stated within. But if this is the case, you will think independently. And this is what I call the hardtack approach; that is, to make you think! Is there any other way?

By the way, if my students do not learn a thing from me other than the following statement, then MISSION ACCOMPLISHED!

The first written warning regarding the need for accident prevention is in Hammurabi's Code, about 1750 B.C. It states, "If a builder constructs a house for a person and does not make it firm and the house collapses and causes the death of the owner, the builder shall be put to death."

WHAT THEY DO

Before I retired, I was often asked what I did for a living. I answered that I worked in environmental engineering. Then, the next question would be, "What is environmental engineering?" After briefly explaining environmental engineering, some would ask what an environmental engineer does. The answer to this question takes me more time to explain, but usually, I list the typical job duties of an environmental engineer, including:

- Preventing harmful biological and chemical contaminants from being released into the air, water, and soil.
- Controlling pollution and ensuring good air, water, and soil quality.
- Detecting pollutants and their sources.
- Obtaining permits and ensuring companies follow standard operating procedures (SOPs).

- Inspecting facilities to ensure environmental regulations are followed.
- Determining the environmental impacts of planned industrial, commercial, and residential projects.
- Conducting hazardous waste management studies and finding ways to clean contaminated areas.
- Assessing the safety, quality, and effectiveness of existing equipment (this is where Industrial Hygiene comes into play)
- Overseeing industries' and companies' environmental management systems.
- Investigating environmental complaints and creating reports on them.
- Reworking existing industrial and agricultural processes to reduce air, water, or soil pollution.
- Taking air, water, and soil samples for analysis to determine the extent of contamination or level of pollutants.
- Designing equipment and processes necessary for water reclamation, waste management, air pollution control, wastewater treatment, and other systems and facilities.
- Staying current on environmental improvement programs, new technologies, regulations, and scientific data.
- Working with government agencies to draft new regulations.
- Working with businesses to find ways to reduce their waste and pollution.
- Creating and reviewing environmental reports.
- Studying the effects of specific technologies like AI and environmental advancements.
- Predicting weather conditions.
- Conserving water and fighting forest fires.
- Creating low-carbon materials.
- Monitoring deforestation.

ENGINEER OUT THE PROBLEM

Although this text takes the proactive approach (engineer out the problem of pollution before it occurs), the reality is that environmental engineers also work to clean up pollution sites (the reactive mode, for sure). I include environmental law and sustainability in the components of environmental engineering. These two specialties have become buzzwords in today's environmental world and necessary skill sets for practitioners in environmental engineers' toolboxes to succeed. We could say (and I do) that to be well-rounded in practical knowledge and practice, the environmental engineering practitioner must know environmental law and understand resource sustainability.

Environmental engineers are also concerned with finding plausible solutions to public health problems, such as vector-borne (arthropod-borne) diseases, and implementing laws promoting adequate sanitation in urban, rural, and recreational areas.

To demonstrate, explain, and illustrate what environmental engineers do is to provide an example. Example 3.1 is just one hypothetical example of environmental professionals working in the real world.

Example 3.1: Salmon and the Rachel River[1]

The Rachel River, a hypothetical river system in the northwestern United States, courses its way through an area that includes a Native American reservation. The river system outfalls to the Pacific Ocean, and its headwaters begin deep and high within the Cascade Mountain Range of Washington State. For centuries, this river system provided a natural spawning area for salmon. The salmon fry thrived in the river and eventually grew the characteristic dark blotches on their bodies, transforming from fry to parr. When the time came to make their way to the sea, their bodies larger and covered with silver pigment, the salmon, now called smolt, inevitably migrated to the ocean, where they thrived until it was time to return to the river and spawn (about 4 years later). During spawning season, the salmon instinctively homed their way toward the odor generated by the Rachel River (their homing signal) and up the river to their home waters, as their life cycle instincts demanded.

Before non-Native Americans (settlers) arrived in this pristine wilderness region, nature, humans, and salmon lived in harmony and provided for each other. Nature gave the salmon the perfect habitat; the salmon provided Native Americans with sustenance. Native Americans provided their natural world and the salmon with the respect they deserved.

After the settlers came to the Rachel River Valley, changes began. The salmon still ran the river, and humans still fed on the salmon, but circumstances quickly altered. The settlers wanted more land, and Native Americans were forced to give way; they were destroyed, diseased, and died, or forcibly moved to other places, to reservations, where the settlers did all they could to erase Native American beliefs and cultural inheritance.

The salmon still ran the river.

After the settlers drove out the Native Americans, the salmon continued to run for a while. But more non-Native Americans poured into the area. As the area became more crowded, the salmon still ran, but now their home, their habitat, the Rachel River, started to show the effects of modern civilization's influence. The "civilized" practice and philosophy were, "If I don't want it anymore, it's trash. Throw it away," and the river provided a seemingly endless dump—out of the way, out of sight, out of mind. And they threw their trash away, all the mountains of trash they could manufacture, into the river.

The salmon still ran the river.

More time passed. More people moved in, and the more people, the bigger their demands. In its natural course, sometimes the river flooded, creating problems for the settler populations. Besides, everyone wanted power to maintain modern lifestyles—and hydropower poured down the Rachel River to the ocean constantly. So, they built flood control systems and a dam to convert hydropower to hydroelectric power. (Funny. The Native Americans did not have a problem with flood control. When the river rose, they broke camp and moved to higher ground. Hydroelectric power? If you do not build your life around things, you don't need electricity to make them work. Who would want hydroelectric power with the sun, the moon, the stars, and their healthy, vital land at hand? Not the Native Americans.)

The salmon still ran.

Building dams and flood control systems takes time, but humans, though impatient, can conquer and use time (and anything else that gets in the way) to accomplish their tasks, goals, objectives—and construction projects. As the years passed, the construction moved on to completion and finally ended. The salmon still ran—but in reduced numbers and size. Soon, local inhabitants could not catch the quantity and quality of salmon they had in the past. When the inconvenience finally struck home, they began to ask, "Where are the salmon?"

But no one seemed to know. The time had come to call in the scientists, the experts. So, the inhabitants' governing officials formed a committee, funded a study, and hired some scientists to tell them what was wrong. "The scientists will know the answer. They will know what to do," they said, and that was partly true. Notice they did not try to ask the Native Americans. They also would have known what to do. The salmon had already told them.

The scientists came and studied the situation, conducted testing, tested their tests, and decided that the salmon population needed to increase. They determined that an increased salmon population could be achieved by building a fish hatchery, which would take the eggs from spawning salmon, raise the eggs to fingerling-sized fish, release them into specially built basins, and later release them to restock the river.

A lot of science goes into operating a fish hatchery. It cannot operate successfully independently (though Mother Nature never has a severe problem with it when left alone). Still, it must be run by trained scientists and technicians following a proven protocol based on biological studies of salmon life cycles and employing AI.

Note: AI is making significant strides in the fish hatchery industry, enhancing efficiency and sustainability. Some key ways AI is utilized include monitoring and analysis, feeding management, environmental control, and predictive analytics.

When the time was right, the salmon were released into the river. Meanwhile, other scientists and engineers realized that some mechanism had to be installed in the dam to allow the salmon to swim downstream to the ocean and then to reverse. In salmon lives (since they are anadromous species—they spend their adult lives at sea but return to freshwater to spawn), what goes down must go up (upstream). Those salmon would eventually need some way of getting back up past the dam and into home water, their spawning grounds. So, the scientists and engineers devised, designed, built, and installed fish ladders in the dam so that the salmon could climb the ladders, scale the dam, and return to their native waters to spawn and die.

The salmon again ran strong in the Rachel River in a few seasons. The scientists had temporarily—and at a high financial expenditure—solved the problem. Nothing in life or Nature is static or permanent. All things change. They shift from static to dynamic in natural cycles that defy human intervention relatively quickly, without notice—like a dormant volcano or the Pacific Rim tectonic plates. In a few years, local Rachel River residents noticed an alarming trend. Studies over 5 years showed that no matter how many salmon were released into the river, fewer and fewer returned to spawn each season.

So, they called in the scientists again. "Don't worry. The scientists will know. They'll tell us what to do."

The scientists came in, analyzed the problem, and came up with five conclusions:

1. The Rachel River is highly polluted from point and nonpoint sources.
2. The Rachel River Dam has radically reduced the number of salmon returning to the spawning grounds.

3. Foreign fishing fleets off the Pacific Coast are depleting the salmon.
4. Native Americans were removing salmon downstream before getting close to the fish ladder at Rachel River Dam.
5. A large percentage of water is withdrawn yearly from the river for cooling machinery in local factories. Large rivers with rapid flow rates usually can dissipate heat rapidly and suffer minor ecological damage unless seasonal fluctuations sharply reduce their flow rates. This was not the case, of course, with the Rachel River. The significant input of heated water from Rachel River area factories back into the slow-moving Rachel River created an adverse effect called *thermal pollution***.** Thermal pollution and salmon do not mix. In the first place, increased water temperatures lower the dissolved oxygen (DO) content by decreasing the solubility of oxygen in the river water, and warmer river water also causes aquatic organisms to increase their respiration rates and consume oxygen faster, increasing their susceptibility to disease, parasites, and toxic chemicals. Although salmon can survive in heated water, many other fish (the salmon's food supply) cannot. Heated discharge water from the factories also disrupts spawning and kills the young fry.

The scientists prepared their written findings and presented them to city officials, who read them and were (at first) pleased. "Ah!" they said. "Now we know why we have fewer salmon!"

But their short-lived pleasure faded. They did indeed have the causal factors defined—but what was the solution? The scientists looked at each other and shrugged. "That's not my job," they said. "Call in the environmental engineers."

The salmon still ran, not up the Rachel River to its headwaters.

Within days, the city officials hired an environmental engineering firm to study the salmon depletion problem. The environmental engineers came up with the same causal conclusions as the scientists (also related to the city officials). Still, they also related the political, economic, and philosophical implications to the city powers. Based on AI's processing of large volumes of data, the environmental engineers explained that most pollution pouring into the Rachel River would soon be eliminated when the city's new wastewater treatment plant came online. That specific *point source pollution* would be eliminated. They explained that the state agricultural department and their environmental staff worked with farmers along the lower river course to modify their farming practices and pesticide treatment regimens to help control the most destructive types of *nonpoint source pollution*. The environmental engineers explained that the Rachel River dam's present fish ladder was incorrectly configured but could be modified with minor retrofitting.

The over-fishing by foreign fleets off the Pacific Coast is a problem that the Federal Government is working to resolve with the involved governments. The State of Washington and the Federal Government are addressing the issue with the Native Americans who fish the down-river locations. These local tribes are not just passive participants in the negotiations but active stakeholders with pending litigation against the state and the federal government over fishing rights to the Rachel River and the salmon.

For the final problem, thermal pollution from the factories, making the Rachel River unfavorable for spawning, decreasing salmon food supply, and killing off the young salmon fry, the environmental engineers explained that to correct this problem, the outfalls from the factories would have to be relocated. The environmental engineers also recommended constructing a channel basin whereby the ready-to-release salmon fry could be released in a favorable environment at ambient stream temperatures. They would have a controlled one-way route to safe downstream locations where they could thrive until it was time to migrate to the sea.

After many debates and newspaper editorials, city officials decided to fund the projects needed to solve the salmon problem in the Rachel River. It is important to note that some short-term projects are already showing positive signs of change, long-term projects are underway, and the Rachel River is recovering.

In essence, environmental engineers seek to solve problems through scientific analysis and study, often with the assistance of AI-processed data. They focus on pure science and can reach the same causal conclusions as scientists. However, they also consider socio-economic, political, and cultural influences and are at the forefront of incorporating the latest advancements in AI into their work.

But wait! It still is not over. Concerns over the wild salmon gene pool disruption by hatchery trout are drawing attention from environmentalists, conservationists, and wildlife biologists. Hatchery—or farm-raised stock of any kind—is susceptible to problems caused by, among other things, a lack of free genetic mixing, the spread of disease, infection, and parasites, and reinforcement of negative characteristics—when escaped hatchery salmon breed with wild salmon, the genetic strain is changed, diseases can be transmitted…, and many problems arise.

So, the environmental engineers recommended helping mitigate these issues by upgrading area computer systems and incorporating the latest advancements in AI, borrowed from the significant strides in aquaculture, into hatchery management. The engineers explained to local officials that AI would help solve their fish hatchery issues, including optimizing feeding using machine learning to monitor fish behavior and hunger levels. This step works to dispense the right amount of food, improving growth rates and reducing waste. Another benefit of installing AI is that it can analyze data from sensors and cameras to observe fish behavior and physiology and detect early signs of disease or stress in fish, allowing for early intervention and better overall health management. By the way, this allows for remote monitoring of these issues and others. To ensure optimal living conditions, installing AI in the incubator can allow control and monitoring of water quality parameters such as oxygen levels, pH, and temperature. AI also helps in selecting broodstock by assessing genetic traits simultaneously.

DID YOU KNOW?

Genetic algorithms (GA) are used in various fish hatcheries to find high-quality solutions efficiently. GA allows AI to analyze genetic data to create breeding programs that produce healthier and more resilient fish. Experience has demonstrated that GA used in AI is one of the essential technologies that fish hatcheries can use to be more efficient and sustainable.

Another important contribution of AI in hatcheries is using it to optimize the nutrient content of culture media for growing microalgae, which is essential for feeding larval fish. Food nutrition is crucial to guarantee the healthy growth of aquatic animal offspring. Microalgae are ubiquitous photosynthetic microorganisms, with their high nutritional value and suitable cell size (3–50 μm) and are indispensable food sources for fish larvae and juveniles.

DID YOU KNOW?

The microalgae commonly used as live prey in fish hatcheries include *Chaetoceros muller, Skeletonema costatum, Phaeotactylum tricornutum, Thalassiosira* sp., *Amphora* sp., *Isochronic zhaangiiangensis, Dunaliella salina*, and others.

The environmental engineers' final recommendation to city officials was to install biomass cameras and devices like Innovasea's BiomassPro™ biomass cameras, which use AI to estimate fish stocks' size and weight in real time (BiomassPro, 2024).

The city officials adopted all of the environmental engineers' recommendations. After funding the changes, Rachel's Creek became a work in progress, a pure example of far-thinking in AI.

AI SUBSETS MACHINE AND DEEP LEARNING

To say that technology is a dynamic (ever-changing and advancing) discipline and that it is challenging to keep up with the daily, maybe hourly, advances we are making in the discipline is a gross understatement, but it is also a truism. The point is that if the current environmental engineering professor, environmental engineering professional, environmental engineer, or engineering student is to stay tuned in to and current with what is 'happening' in the real world of technological advancement, then he or she must practice like the doctor who performs while medical advancements are occurring all the time; like the lawyer who must stay tuned into legal advances, reversals, case law, and the whims and wants of state legislatures—those who love to initiate and pass new laws, regulations, rules, and so forth, especially if they are named for or after the originator.

An advancing technical area that is presently making its way into the field of environmental engineering (and several other disciplines) is AI, with its subset of machine learning and its branch of deep learning (DL). Deep learning deals with the development and application of modern neural networks. A neural network refers to or relates to a network of electronic components designed to mimic the operation of the human brain. One example of a neural network is the feed-forward neural network, an artificial neural network wherein connections between the nodes do not form a cycle (Zell, 1994). In this feed-forward neural network, information always moves in one direction; it never goes backward.

In addition, with the need for environmental engineers and programs in environmental learning to get on board in their dynamic field(s), so to speak, with AI, machine learning, deep learning, and feed-forward neural network information as it applies to engineering and the environment, current personnel within the environmental engineering field must also become familiar with Python programming, algorithms, TensorFlow/PyTorch, linear algebra, optimization/probability theory, GPU hardware, Convolutional Neural Networks,

Recurrent Neural Networks, Generative Adversarial Networks/Variational Autoencoders, Physics-informed Learning, Reinforcement Learning/Active Learning/Transfer Learning, and other machine learning techniques along with terms such as *Perceptron.*

THE 411 ON ALGORITHMS: WHAT ARE THEY?

An *algorithm* is a specific mathematical calculation procedure, a computable set of steps to achieve a desired result. According to Cormen et al. (2002), "an algorithm is any well-defined computational procedure that takes some value, or set of values, as input and produces some value, or set of values, as output." In other words, an algorithm is a recipe for an automated solution to a problem. A computer model may contain several algorithms. According to Knuth, the word "algorithm" is derived from the name "al-Khowarizmi," a ninth-century Persian mathematician.

Algorithms should not be confused with computations. While an algorithm is a systematic method for solving problems, and computer science is the study of algorithms (although algorithms were developed and used long before any device resembling a modern computer was available), the act of executing an algorithm—that is, systematically manipulating data—is called *computation.*

For example, the following algorithm (attributed to Euclid circa 300 B.C. and thus known for millennia) for finding the greatest common divisor of two given whole numbers may be stated as follows:

- Set a and b to the values A and B, respectively.
- Repeat the following sequence of operations until b has a value of 0:
  - let r take the value of a mod b;
  - let a take the value of b;
  - let b take the value of r.
- The greatest common divisor of A and B is a's final value.

Note: The operation a mod b gives the 'remainder' obtained by dividing a by b.

The problem of finding the greatest common divisor of two numbers is specified by stating what is to be computed; the problem statement itself does not require that any particular algorithm be used to calculate the value. Such method-independent specifications can be used to define the meaning of algorithms: the meaning of an algorithm is the value that it computes.

Several methods can be used to compute the required value; Euclid's method is just one. The chosen method assumes a set of standard operations (such as basic operations on whole numbers and a means to repeat an operation). It combines these operations to form an operation that computes the required value. Also, it is not apparent to most people that the proposed algorithm computes the required value. That is one reason why a study of algorithms is essential—to develop methods that can be used to establish what a proposed algorithm achieves.

Expressing Algorithms

Although an in-depth discussion is beyond the scope of this text, we point out that analyzing algorithms often requires drawing upon a body of mathematical operations. Some of these operations are as simple as high-school algebra, but others may be less familiar to the average environmental engineer. Along with learning to manipulate asymptotic notations and solve recurrences, several other concepts and methods must be understood to analyze algorithms.

For example, methods for evaluating bounding summations (e.g., when an algorithm contains an iterative control construct such as a *while* or *for* loop, its running time can be expressed as the sum of the times spent on each execution of the body of the loop), which frequently occur in the analysis of algorithms, are essential. Many formulas commonly used in analyzing algorithms can be found in any calculus text. In addition, to explore many algorithms, we must be familiar with the basic definitions and notations for sets, relations, functions, graphs, and trees. A basic understanding of elementary counting principles (permutations, combinations, and the like) is also essential. Most algorithms used in environmental engineering require no probability for their analysis; however, familiarity with these operations can be helpful.

Because mathematical and scientific analysis (and many environmental engineering functions) is so heavily based on numbers, computation has tended to be associated with numbers. However, this need not be the case: algorithms can be expressed using any formal manipulation system, that is, any system that defines a set of entities and a set of unambiguous rules for manipulating those entities. For example, the SKI calculus system consists of three combinators (entities): coincidentally, S, K, and I. The computation rules for the calculus are:

- Sfgx = fx (gx)
- Kxy = x
- Ix = x

where f, g, x, y are strings of the three entities. The SKI calculus is computationally complete; that is, any computation that can be performed using any formal system can be performed using the SKI calculus (equivalently, all algorithms can be expressed using the SKI calculus). Not all computation systems are equally powerful, though—some problems that can be solved using one system cannot be solved using another. Further, it is known that there are problems that cannot be solved using any formal computation system.

DID YOU KNOW?

The SKI combinator calculus is a combinative logic and computational system that can be considered a computer programming language but is not practical for writing software.

GENERAL ALGORITHM APPLICATIONS

Practical applications of algorithms are ubiquitous. All computer programs are expressions of algorithms, where the instructions are expressed in the computer language used to develop the program. Computer programs are described as expressions of algorithms because an algorithm is a general technique for achieving some purpose, which can be expressed in several ways. Algorithms exist for many purposes and are described in many different ways. Examples of algorithms include recipes in cookery books, servicing instructions in a computer's hardware manual, knitting patterns, digital instructions to a welding robot for where each weld should be made, or cyber-speak to any system working in cyberspace.

Algorithms can be used in sorting operations to reorder a list into some defined sequence. It is possible to express the same algorithms as instructions to a human who has a similar requirement to reorder some list, for example, to sort a list of tax record reports into a sequence determined by the date of birth on the record. These instructions could employ the *insertion sort algorithm*, the bubble sort algorithm, or one of many other available algorithms. Thus, an algorithm is a general technique for expressing the process of completing a defined task, independent of the precise manner in which it is expressed.

Sorting is not the only application for which algorithms have been developed. Practical applications of algorithms include the following examples:

- Internet routing: single-source shortest paths
- Search engine: string matching
- Public-key cryptography and digital signatures: number-theoretic algorithms
- Allocating scarce resources in the most beneficial way: linear programming
- Algorithms are at the core of most technologies used in contemporary computers
    - Hardware design uses algorithms
    - The design of any GUI relies on algorithms
    - Routing in networks relies heavily on algorithms
    - Compilers, interpreters, or assemblers make extensive use of algorithms

A few classic algorithms are commonly used to illustrate their function, purpose, and applicability. One of these classics is known as the ***Byzantine Generals***. Briefly, this algorithm is about the problem of reaching a consensus among distributed units if some of them give misleading answers. To be memorable, the problem is couched in terms of generals deciding on a joint plan of attack. Some traitorous generals may lie about whether they will support a particular strategy and what other generals tell them. When exchanging only messages, what decision-making algorithm should generals use to reach a consensus? What percentage of liars can the algorithm tolerate and still correctly determine a consensus (Black, 2012)?

In the following, we use another classic algorithm to illustrate (because of its general usefulness and because it is easy to explain to just about anyone) how an algorithm can be applied to real-world situations, known as the ***Traveling Salesman*** problem. The Traveling Salesman problem is the most notorious NP-complete problem (that is, NP-complete means that no polynomial-time algorithm has yet been discovered for an NP-complete problem, nor has anyone yet been able to prove that no polynomial-time algorithm can exist for any of them).

The Problem

We have pointed out some of the functions that algorithms can perform. The question arises: "Can every problem be solved algorithmically?" The simple and complex answer is NO. For example, there are problems for which no generalized algorithmic solution can exist (unsolvable). Also, there are problems for which no efficient solution is known: *NP-Complete Problems;* that is, it is unknown if efficient algorithms exist for NP-complete problems. If an efficient algorithm exists for any one of them, then efficient algorithms exist for all of them (e.g., Traveling Salesman Problem). Finally, there are problems that we do not know how to solve algorithmically (Figure 3.2).

From the discussion above, it should be apparent that computer science is NOT about word processing and spreadsheets; it can include *applications,* not just *software applications.*

ENVIRONMENTAL PRACTICE ALGORITHM APPLICATIONS

Although algorithms can be used in transportation applications (traveling salesperson problems), many critical applications are applied to environmental engineering functions. For example, consider a robot arm assigned to weld all the metal parts on an automobile in an assembly line. The shortest tour that visits each weld point exactly once defines the most efficient path for the robot. A similar application arises in minimizing the time a design engineer or draftsperson takes to draw a given structure.

Environmental engineering tackles complex issues from natural environments, including water, soil, air, organisms, and biological data. In practice, environmental engineers collect considerable amounts of data. Thanks to current AI technologies, such large amounts of data can be efficiently managed and assessed. Data from various sources, including energy usage, traffic trends, demographic information, and socioeconomic factors, can provide valuable awareness and understanding and help predict environmental conditions.

Algorithms have found widespread application in all branches of environmental practice. For example, the USEPA uses computer models and various algorithms to monitor chemical spills and ultimate fate data in environmental engineering. In the following, we provide selected model summary descriptions of applications used in

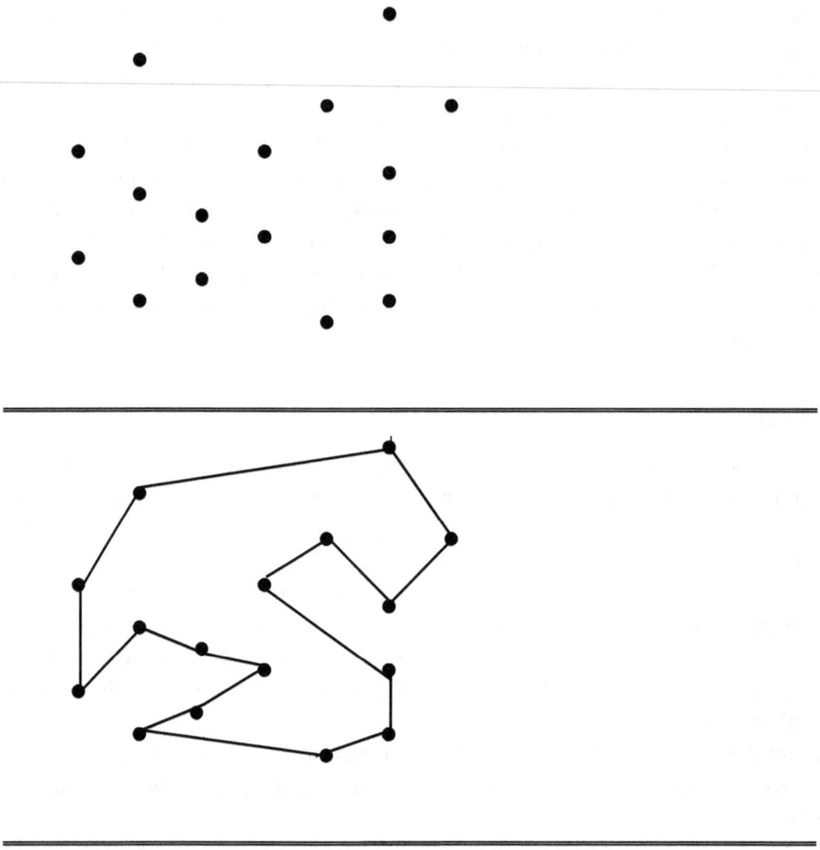

FIGURE 3.2 Imagine a traveling salesperson who has to visit each of a given set of cities by car. Using (a) below, shows how to find the minimum cost cycle (b), visiting all the vertices exactly once.

Dispersion Modeling. Specifically, we discuss how the USEPA (and others) employ preferred/recommended models (i.e., refined models recommended for a specific type of regulatory application) in monitoring air quality, that is, ambient pollutant concentrations and their temporal and spatial distribution. Further information on this critical topic can be accessed at the USEPA's Technology Transfer Network Support Center for Regulatory Air Models at www.epa.gov/scram001/tt22.htm.

MODELS

The definition of an environmental model used in this book is:

> A simplification of reality that is constructed to gain insights into select attributes of a particular physical, biological, economic, or social system.
>
> *(NRC, 2007)*

Dispersion Models

The following algorithm-based models described below are currently listed in Appendix A of the *Guidelines on Air Quality Models* (published as Appendix W of 40 CFR Part 51):

Dispersion Models:

- **BLP** *Buoyant Line and Point Source Model* is a Gaussian plume dispersion model designed to handle unique modeling problems associated with aluminum reduction plants and other industrial sources where plume rise and downwash effects from stationary line sources are essential.
- **CALINE3** A steady-state Gaussian dispersion model designed to determine air pollution concentrations at receptor locations downwind of "at-grade," "fill," "bridge," and "cut section" highways located in relatively uncomplicated terrain.
- **CALPUFF** A multi-layer, multi-species, non-steady-state puff dispersion model that simulates the effects of time- and space-varying meteorological conditions on pollution transport, transformation, and removal. CALPUFF can be applied on scales of tens to hundreds of kilometers. It includes algorithms for subgrid-scale effects (terrain impingement) and longer-range effects (such as pollutant removal due to wet scavenging and dry deposition, chemical transformation, and visibility effects of particulate matter concentrations).
- **CTDMPLUS** Complex Terrain Dispersion Model Plus Algorithms for Unstable Situations is a refined point source Gaussian air quality model for use in all stability conditions for complex terrain (i.e., terrain exceeding the height of the stack being modeled as contrasted with *simple terrain* which is defined as an area where terrain features are all lower in elevation than the top of the stack of the source). The model contains, in its entirety, the technology of CTDM for stable and neutral conditions.
- **ISC3** Industrial Source Complex Model is a steady-state Gaussian plume model that can be used to assess pollutant concentrations from various sources associated with an industrial complex. This model can account for the following: settling and dry deposition of particles; downwash; point, area, line, and volume sources; plume rise as a function of downwind distance; separation of point sources; and limited terrain adjustment. ISC3 operates in both long-term and short-term modes.
- **OCD** Offshore and Coastal Dispersion Model is a straight-line Gaussian model developed to determine the impact of offshore emissions from point, area, or line sources on the air quality of coastal regions. OCD incorporates overwater plume transport and dispersion and changes that occur as the plume crosses the shoreline. Hourly meteorological data are needed from both offshore and onshore locations.

SCREENING TOOLS

Note: Screening tools are relatively simple analysis techniques used to determine if a given source is likely to threaten air quality. Concentration estimates from screening techniques precede a refined modeling analysis and are conservative.

- **CAL3QHC/CAL3QHCR** CALINE3 with queuing and hot spot calculations is a CALINE3-based CO model with a traffic model to calculate delays and queues that occur at signalized intersections; CAL3QHCR requires local meteorological data.
- **COMPLEX 1** A multiple-point source screening technique with terrain adjustment that incorporates the plume impaction algorithm of the VALLEY model.
- **CTSCREEN** Complex Terrain Screening model is a Gaussian plume dispersion model designed as a screening technique for regulatory application to plume impaction assessments in complex terrain. CTSCREEN is a screening version of the CTDMPLUS model.
- **LONGZ** A steady-state Gaussian plume formulation for both urban and rural areas in flat or complex terrain to calculate long-term (seasonal and annual) ground-level ambient air concentrations attributable to emissions from up to 14,000 arbitrarily placed sources (stack, buildings, and area sources).
- **SCREEN3** A single-source Gaussian plume model that provides maximum ground-level concentrations for point, area, flare, and volume sources, as well as concentrations in the cavity zone and concentrations due to inversion break-up and shoreline fumigation. SCREEN3 is a screening version of the ISC3 model.
- **SHORTZ** A steady-state bivariate Gaussian plume formulation for urban and rural areas in flat or complex terrain to calculate ground-level ambient air concentrations. It can calculate 1-, 2-, 3-hours, etc., average concentrations due to emissions from stacks, buildings, and area sources for up to 300 arbitrarily placed sources.
- **VALLEY** A steady-state, complex terrain, univariate Gaussian plume dispersion algorithm designed for estimating 24-hour annual concentrations resulting from emissions from up to 50 (total) point and area sources.
- **VISCREEN** Calculates the potential impact of a plume of specified emissions for specific transport and dispersion conditions.

Note: Environmental modeling is discussed in detail later in this book.

THE BOTTOM LINE

When you throw a stone into a pool of quiet water, the ensuing ripples move out in concentric circles from the point of impact. Eventually, those ripples, much dissipated, reach the pond's edge, where they break, disturbing the shore environment. When we alter our environment, similar repercussions affect the world around us—and some of these actions can—or will be—felt worldwide. We use technology to alter our environment to suit our needs. That same technology can be implemented to protect our environment from unrecoverable losses. Environmental engineers must maintain an acute awareness of the global repercussions of problems we create for the environment—to extend the boundaries of the problem beyond our backyard.

Historically, as long as human population numbers remained small, the environment could adequately process human pollutants. However, as population numbers increased, human waste overwhelmed natural systems. As societies become technologically sophisticated, humans produce large numbers and volumes of new substances yearly. The environment's ability to absorb and process these substances (to self-purify) has natural limitations. Our environment can only handle so much. Complicating the situation even further, consumption of resources per person in the developed world increases daily—sometimes dramatically. A prevailing attitude of "consumerism in a throw-away society" leads to resource abuse, resulting in ever-increasing pollution released into the environment. Consequently, pollution has become an increasing source of environmental problems.

Human progress has led inexorably to air, water, and soil pollution. Is this trend likely to continue? Maybe, maybe not. We do not know the answer to this vital question. Does anyone know the answer? We hope so. However, some things we do know. We understand that the total de-industrialization of the developed world is improbable and, barring some major catastrophic event, unlikely—out of the question. A more practical view of pollution is required, allowing sustainable development of healthy and vigorous industrial activity to provide the goods and services needed by modern civilization.

We also know we must do all this with a sense of balance.

What kind of balance? Another good question.

The balance we are referring to is the tradeoff we are willing to accept for the things we do. More specifically, if we are to build a new factory that produces widgets which, during the production phase, will contribute an unreasonable level of pollutants to our air, water, and soil, we have to ask ourselves: Is it worth it? Is the tradeoff between creating jobs and enhancing the material livelihood of several individuals worth the potential damage to the environment?

We need to find a middle ground. We must balance our desires against possible or potential results. Thus, what we have is a balancing act… a delicate balance—one we must work hard to maintain.

So, what is the bottom line of environmental engineering, pollution, and its effects on us? This book provides facts and data, explains, and illustrates the bottom line. We consistently point to two overriding and connected themes in this text. First, when it comes to environmental pollution and degradation and their potentially harmful effects on our lives, one thing is sure: we do not know the extent of the problem—we do not know what we do not know (scary when you think about it). Second, the tragedy that underlies politics with the "pollution thing" (remember, in the real world, everything revolves around politics) can be summed up by the words of philosopher Georg Hegel: "The nature of tragedy is not the conflict between right and wrong but between right and right."

For now, read this book with a mind open to the facts. Further, when you hear someone state that what we need to solve all of Earth's problems (social and otherwise) is innovation, innovation, and innovation, you can accept this view, or you can settle for what this book preaches: what it is, discovery, invention, and then innovation to the nth degree. Make your own choices, decisions, and judgments because the natural bottom line is that, when defining environmental pollution, you must make a judgment—pollution is a judgment call. When pollution is identified, call on computer-literate environmental engineers to study and mitigate the issue—whatever the problem.

Computer-literate or computer-savvy environmental engineers are a must in addressing ever-changing environmental issues; the mitigation procedures and tools needed to stay attuned to today's and tomorrow's problems are aided exponentially by employing AI tools.

So, solving environmental problems is greatly assisted by computer-savvy environmental engineers. That is undoubtedly the case, but what is a computer-savvy environmental engineer? A better question is: What does a computer-savvy environmental engineer need to know or use to succeed in this digital world?

The digital environmental engineer needs to go way beyond the simple use of word processing. Knowing how to hook up a computer and accessories (you might be surprised at how many people have problems hooking up a computer system), use search engines, spreadsheets, browser basics, familiar keyboard commands, virus and malware scanning, simple networking diagnosis, security, and basic hardware terminology. Suppose you want to digitize yourself (i.e., you can use AI to assist in planning and solving problems). In that case, you must know, understand, and speak the language and use the terminology correctly.

NOTE

1 This section is modified from F.R. Spellman's (2020). *Environmental Engineering: Concepts and Applications.* Boca Raton, FL: CRC Press.

REFERENCES

Biomass Pro. (2024). *Biomass Estimation for Aquaculture—Innovasea.* Accessed 5/5/24 @ https://www.innovasea.com-biomass-estimation.

Black, P.E. (2012). "Byzantine generals," in *Dictionary of Algorithms and Data Structures.* The US. National Institute of Standards and Technology. Accessed 12/21/12 @ https://www.nist.gov/dada/HTML/by zaintine.html.

Cormen et al. (2002). *Introduction to Algorithms*, 3rd ed. Cambridge, MA: The MIT Press.

NRC. (2007). Models in environmental regulating environmental models. Washington, DC: National Academies Press.

Zell, A. (1994). *Artificial Neural Network Simulation Neuronal. Algorithm Netze.* Boston, MA: Addison-Wesley Publishing Co.

4 The Environmental Engineer's Toolbox

MASS BALANCE

(Mass at time $t + \Delta t$) = (mass at time t) + (mass that entered from t to t = Δt) – Mass that exited from t to $t + \Delta t$ + net mass of pollutant produced from other compounds by chemical reactions between t and $t + \Delta t$)

ENERGY BALANCE

(Change in internal + external energy per unit time) = (Energy flux in) – (Energy flux out)

Note: This chapter presents only basic information and does not include higher mathematics. Instead, it is intended to provide explanations and examples of basic measurements. This book focuses on AI and its applications in environmental engineering, not higher mathematics.

INTRODUCTION

Units, standards of measurement, conversions, and mass and energy transfer are foundational to environmental engineering, which is essential because an environmental engineering book that does not include units and conversions is like a book without a table of contents or an index. For example, converting units from grams to slugs, centistokes to square feet per second, or pounds per million gallons to milligrams per liter can be accomplished automatically, right away, if you have a mind containing a library of facts, figures, conversions, data, and all kinds of stuff. However, if you are normal, there are times when even the most adept, confident, competent, and brilliant engineer or engineer-to-be must refer to some reference to find certain facts. I always hand out a printed copy of units and conversions in my college environmental classes. Most students welcome the handout; however, I have had a few who shunned it at first, saying they knew how to convert units and could do it from memory. I like to counter with, "Well, that means you know how to convert abamperes to statamperes, right?" As former Colorado Springs police Lt Joe Kenda (*Homicide Hunter*) often says when surprised or having a eureka moment, "Oh my! My, my, my!" The look on their faces is always priceless. Students who do not know everything are my kind of students; they are teachable, they have open minds, and their brain chips are clear but ready to be filled. Because not many of us are human computers with digital memories, this textbook includes a discussion of units, standards of measurement, and conversions. Moreover, because this is the

DOI: 10.1201/9781003601562-5

age of energy consumption, renewable energy production, and hydraulic fracturing to capture more energy, energy conversion calculations, energy transfer, and mass transfer are also included. By the way, does the reader know the difference, in gallons, between a barrel of U.S. liquid and a barrel of petroleum? Do not worry; you will find out if you read this chapter.

UNITS OF MEASUREMENTS: THE BASICS

A basic knowledge of units of measurement and how to use and convert them is essential. Environmental engineers should be familiar with the U.S. Customary System (USCS), the English System, and the International System of Units (SI). Table 4.1 summarizes some necessary units; these are a few of the primary SI and USCS units of measurement that will be encountered in practice.

In studying environmental engineering math operations (and in actual practice), it is quite common to encounter very large and very small quantities. The concentrations of some toxic substances may be measured in parts per million, billion, or trillion (ppm, ppb, or ppt), for example.

Note: PPM may be roughly described as an amount contained in a full shot glass at the bottom of a standard-sized swimming pool. The entire shot glass is 1 ppm relative to the rest of the water in the pool. PPT may be roughly described as an amount included in one drop from an eye-dropper—the one drop is 1 ppt relative to the rest of the water contained in the pool. Because most concentrations of contaminants are minimal, 1 L of the mixture has a mass that is essentially 1,000 g, so for all practical purposes:

$$1\,\text{mg/L} = 1\,\text{g/m}^3 = 1\,\text{ppm}\left(\text{by weight}\right) \tag{4.1}$$

$$1\,\mu\text{g/L} = 1\,\text{mg/m}^3 = 1\,\text{ppb}\left(\text{by weight}\right) \tag{4.2}$$

TABLE 4.1
Commonly Used Units

Quantity	SI Units	USCS Units
Length	meter	feet (ft)
Mass	kilogram	pound (lb)
Temperature	Celsius	Fahrenheit (F)
Area	square meter	square foot (ft^2)
Volume	cubic meter	cubic foot (ft^3)
Energy	kilojoule	British thermal unit (Btu)
Power	watt	Btu/hr
Velocity	meter/second	mile/hour (mi/hr)

On occasion, when the concentration of liquid wastes may be so high that the specific gravity of the mixture is affected, corrections to Equations 4.1 and 4.2 may be required:

$$\text{mg/L} = \text{ppm}\left(\text{by weight}\right) \times \text{specific gravity of mixture} \tag{4.3}$$

Example 4.1: Fluoridation of Drinking Water

Problem: The optimum dose of fluoride in drinking water to prevent mottling of the teeth is 0.7 mg/L. If 0.7 mg/L of fluoride water is contained in 1 glass (12 oz), what is the mg of fluoride per glass?

1 glass = 12 oz of water = 0.3548 L

0.7 mg/L is the fluoride concentration.

Solution:

0.7 mg/L of fluoridated water × [0.3548 L/1 glass of water (12 oz)] = 0.248 mg of fluoride per 1 glass (12 oz).

DID YOU KNOW?

A pea-sized (0.25 g) amount of toothpaste contains approximately the same amount of fluoride as a 12 oz glass of water.

To describe quantities that may take on such large or small values, it is helpful to have a system of prefixes that accompany the units. Some of the more critical prefixes are presented in Table 4.2.

TABLE 4.2
Common Prefixes

Quantity	Prefix	Symbol
10^{-12}	pico	p
10^{-9}	nano	n
10^{-6}	micro	μ
10^{-3}	milli	m
10^{-2}	centi	c
10^{-1}	deci	d
10	deca	da
10^{2}	hecto	h
10^{3}	kilo	k
10^{3}	mega	M

CONVERSION FACTORS AND CONVERSIONS

Sometimes, it is necessary to convert between different units. Suppose that a 60-inch piece of wood is attached, end-to-end, to a 6-ft piece of wood. How long are they together? We cannot find the answer to this question by adding 60 to 6. The reason is that the two figures are given in different units. Before we can add the two numbers, we have to convert one of them to the units of the other. Then, when we have two numbers in the same units, we can add them. To perform this conversion, we need a *conversion factor.* In this case, we must know how many inches make up a foot; 12 inches equals 1 ft. Knowing this, we can perform the calculation in two steps as follows:

a. 60 inches is 60/12 = 5 ft
b. 5 ft + 6 ft = 11 ft

In the example above, a conversion factor changes known quantities in one unit of measure to an equivalent amount in another.

In making the conversion from one unit to another, we must know two things:

1. The exact number that relates the two units
2. Whether to multiply or divide by that number

When making conversions, confusion over whether to multiply or divide is common; on the other hand, the number relating the two units is usually known and, thus, is not a problem. Understanding the proper methodology—the "mechanics"—to use for various operations requires practice and common sense.

Using the proper "mechanics" (and practice and common sense) to convert units is probably the easiest and fastest method of using a conversion table.

The most straightforward conversions require the measurement to be multiplied or divided by a constant value. For instance, in wastewater treatment, if the depth of biosolids on a drying bed is 0.85 ft, multiplying by 12 inches per foot converts the measured depth to inches (10.2 inches). Likewise, if the depth of the solids blanket in the secondary clarifier is 16 inches, dividing by 12 inches per foot converts the depth measurement to feet (1.33 ft).

Table 4.3 lists many of the conversion factors used in environmental engineering. Note that Table 4.3 is designed with a unit of measure in the left and right columns and a constant (the conversion factor) in the center column (Table 4.4, a more extensive conversion table, is presented later).

- **Key Point:** To convert in the opposite direction (i.e., inches to feet), divide by the factor rather than multiply.

WEIGHT, CONCENTRATION, FLOW CONVERSIONS

Using Table 4.3 to convert from one unit expression to another and vice versa is good practice. However, in making conversions to solve process computations in water/wastewater treatment, we must be familiar with conversion calculations based on a relationship between weight, flow, or volume, and concentration. The basic relationship is

TABLE 4.3
Conversion Table

To Convert	Multiply By	To Get
Feet	12	Inches
Yards	3	Feet
Yards	36	Inches
Inches	2.54	Centimeters
Meters	3.3	Feet
Meters	100	Centimeters
Meters	1,000	Millimeters
Square yards	9	Square feet
Square feet	144	Square inches
Acres	43,560	Square feet
Cubic yards	27	Cubic feet
Cubic feet	1,728	Cubic inches
Cubic feet (water)	7.48	Gallons
Cubic feet (water)	62.4	Pounds
Acre-feet	43,560	Cubic feet
Gallons (water)	8.34	Pounds
Gallons (water)	3.785	Liters
Gallons (water)	3,785	Milliliters
Gallons (water)	3,785	Cubic centimeters
Gallons (water)	3,785	Grams
Liters	1,000	Milliliters
Days	24	Hours
Days	1,440	Minutes
Days	86,400	Seconds
Million gallons/day	1,000,000	Gallons/day
Million gallons/day	1.55	Cubic feet/second
Million gallons/day	3.069	Acre-feet/day
Million gallons/day	36.8	Acre-inches/day
Million gallons/day	3,785	Cubic meters/day
Gallons/minute	1,440	Gallons/day
Gallons/minute	63.08	Liters/minute
Pounds	454	Grams
Grams	1,000	Milligrams
Pressure, psi	2.31	Head, ft (water)
Horsepower	33,000	Foot-pounds/minute
Horsepower	0.746	Kilowatts
To Get	Divide By	To Convert

TABLE 4.4
Weight, Volume, and Concentration Calculations

To Calculate	Formula
Pounds	Concentration, mg/L × Tank Vol. MG × 8.34 lb/MG/mg/L
Pounds/day	Concentration, mg/L × Flow, MGD × 8.34 lb/MG/mg/L
Million gallons/day	$\frac{\text{Quantity, lb/day}}{(\text{Conc., mg/L} \times 8.34\ \text{lb/mg/L/MG})}$
Milligrams/liter	$\frac{\text{Quantity, lb}}{(\text{Tank Volume, MG} \times 8.34\ \text{lb/mg/L/MG})}$
Kilograms/liter	Conc., mg/L × Volume, MG × 3.785 lb/MG/mg/L
Kilograms/day	Conc., mg/L × Flow, MGD × 3.785 lb/MG/mg/L
Pounds/dry ton	Conc. mg/kg × 0.002 lb/d.t./mg/kg

$$\text{Weight} = \text{Concentration} \times \text{Flow or Volume} \times \text{Factor} \tag{4.4}$$

Table 4.4 summarizes weight, volume, and concentration calculations. With practice, many of these calculations become second nature to operators. These calculations are essential relationships and are often used in water/wastewater treatment process control calculations, so on-the-job practice is possible.

The following conversion factors are used extensively in water/wastewater and air pollution operations and are commonly needed to solve problems on licensure examinations. The environmental engineer should keep them accessible to all plant operators:

- 7.48 gal/ft^3
- 3.785 L/gal
- 454 g/lb
- 1,000 mL/L
- 1,000 mg/g
- 1 ft^3/s (cfs) = 0.6465 MGD
- **Key Point:** Density (specific weight) is mass per unit volume and may be registered as lb/ft^3, lb/gal, g/mL, or g/m^3. If we take a fixed-volume container, fill it with a fluid, and weigh it, we can determine the density of the fluid (after subtracting the weight of the container).
- 8.34 lb/gal (water)—(density = 8.34 lb/gal)
- 1 mL of water weighs 1 g—(density = 1 g/mL)
- 62.4 lb/ft^3 (water)—(density = 8.34 lb/gal)
- 8.34 lb/gal = mg/L (converts dosage in mg/L into lb/day/MGD)
  Example: 1 mg/L × 10 MGD × 8.3 = 83.4 lb/day
  - 1 psi = 2.31 ft of water (head)
  - 1 ft head = 0.433 psi

- °F = 9/5(°C + 32)
- °C = 5/9(°F – 32)
- Average water usage: 100 gal/capita/day (gpcd)
- Persons per single-family residence: 3.7

Important Constants

$$\pi = 3.14(\text{pi})$$

$$\text{e} = 2.7183(\text{Napierian})$$

Important Gas/Vapor Conversions

For most air pollution work, by custom, we express pollutant concentrations in volumetric terms. For example, the concentration of a gaseous pollutant in parts per million (ppm) is the volume of pollutants per million volumes of the air mixture.

$$\text{ppm} = \frac{\text{parts of contaminant}}{\text{million parts of air}}$$

For example, the concentration of a gaseous pollutant in parts per million (ppm) is the volume of contaminants per million volumes of the air mixture. Note that it is expected to add a "v" to the ppm, giving ppmv, as shown in Equation 4.3:

$$\frac{1 \text{ volume of gaseous pollutant}}{10^6 \text{ volumes of air}} = 1 \text{ ppm}(\text{by volumc}) = 1 \text{ ppmv} \tag{4.6}$$

In some situations, concentrations are expressed as mass per unit volume, such as mg/m³ or µg/m³. The association between ppmv and mg/m³ depends on the contaminant's pressure, temperature, and molecular weight. The ideal gas law establishes this relationship, and calculations for gas and vapor concentrations are based on the gas laws:

- The volume of gas under constant temperature is inversely proportional to the pressure.
- The volume of a gas under constant pressure is directly proportional to the Kelvin temperature. The Kelvin temperature scale is based on absolute zero (0°C = 273.15 K).
- The pressure of a constant-volume gas is directly proportional to the Kelvin temperature.
- Pressure can be expressed in atmospheres. One pressure atmosphere equals 101.325 kPa (Pa is the abbreviation for Pascals).
- One atmosphere equals 14.7 pounds per square inch (psi), so 1 psi = 6.89 kPa.
- 100 kPa is called a bar, and 100 Pa is a millibar.

The gas laws help us establish the ideal gas law:

$$PV = nRT \tag{4.7}$$

where
P = absolute pressure (atm)
V = volume (m^3)
n = mass (mol)
R = ideal gas constant = 0.082056 L/atm^{-1}/K/mol
T = absolute temperature (K)

When measuring contaminant concentrations, you must know the atmospheric temperature and pressure under which the samples were taken. At standard temperature and pressure (STP), 1 gm-mol of an ideal gas occupies 22.4 liters (L). The STP is 0°C and 760 mmHg. If the temperature is increased to 25°C (room temperature) and the pressure remains the same, 1 g-mol of gas occupies 24.45 L.

Sometimes, you will need to convert milligrams per cubic meter (mg/m^3) (weight-per-volume ratio) into a volume-per-unit-volume ratio. If 1 g-mole of an ideal gas at 25°C occupies 24.45 L, the following relationships can be calculated:

$$\text{ppm} = \frac{24.45\ \text{mg/m}^3}{\text{molecular wt}} \tag{4.8}$$

$$\text{mg/m}^3 = \frac{\text{molecular wt}}{24.45}\ \text{ppm} \tag{4.9}$$

Standard Units and Different Quantities

Before illustrating a few sample conversion problems, it is essential to note that some standard units do not represent the exact quantities. This is important because U.S. engineers unfamiliar with SI or MKS systems are sometimes confused by the various unit comparisons. To illustrate this point, for example, as shown in Figure 4.1, the pound and slug and the newton and pound do not represent the same quantity of matter and force units, respectively.

Conversion Problems

Use Tables 4.3 and 4.4 to make the conversions indicated in the following example problems. Other conversions are presented in appropriate sections of the text. These conversions are used in water/wastewater and air pollution operations.

Note: Many of the computed answers are rounded.

Example 4.2

Convert cubic feet to gallons.

$$\text{Gallons} = \text{Cubic Feet}\left(\text{ft}^3\right) \times 7.48\ \text{gal/ft}^3$$

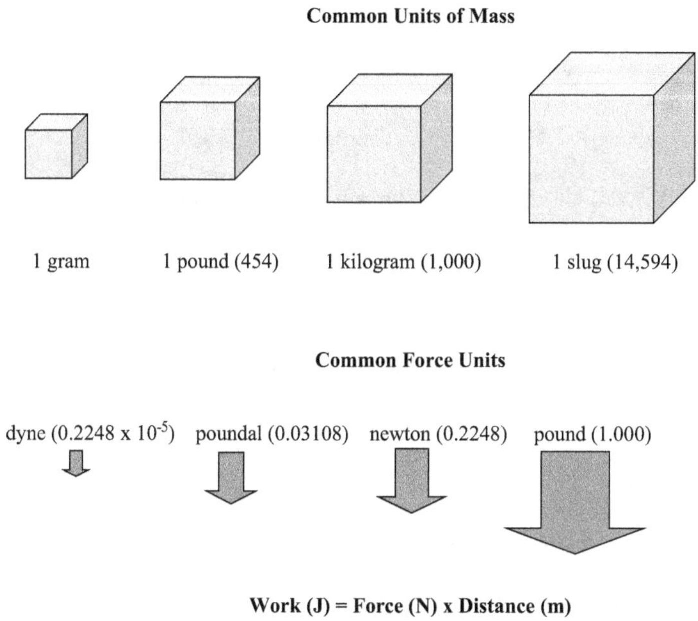

FIGURE 4.1 Common units of mass and force. (Adapted from Lindeburg, M.R. (2009). *Environmental Engineering Reference Manual*, 2nd ed. Belmont, CA: Power to Pass.)

Problem: How many gallons of biosolids can be pumped to a digester with 3,000 ft^3 of volume available?

Solution:

$$\text{Gallons} = 3,000\ \text{ft}^3 \times 7.48\ \text{gal/ft}^3 = 22,440\ \text{gal}$$

Example 4.3

Convert gallons to cubic feet.

$$\text{Cubic Feet} = \frac{\text{Gallons}}{7.48\ \text{gal/ft}^3}$$

Problem: How many cubic feet of biosolids are removed when 16,200 gal are withdrawn?

Solution:

$$\text{Cubic Feet} = \frac{16,200\ \text{gal}}{7.48\ \text{gal/ft}^3} = 2,166\ \text{ft}^3$$

Example 4.4

Convert gallons to pounds.

$$\text{Pounds (lb)} = \text{Gallons} \times 8.34 \text{ lb/gal}$$

Problem: If 1,450 gal of solids are removed from the primary settling tank, how many pounds of solids are removed?

Solution:

$$\text{Pounds} = 1{,}450 \text{ gal} \times 8.34 / \text{gal} = 12{,}093 \text{ lb}$$

Example 4.5

Convert milligrams per liter to pounds per day.

$$\text{Pounds/Day} = \text{Concentration (mg/L)} \times \text{Flow, MGD} \times 8.34 \text{ lb/mg/L/MG}$$

Problem: How many pounds of solids are discharged daily when the plant effluent flow rate is 4.70 MGD, and the effluent solids concentration is 21 mg/L?

Solution:

$$\text{Pounds/Day} = 21 \text{ mg/L} \times 4.70 \text{ MGD} \times 8.34 \text{ lb/mg/L/MG} = 823 \text{ lb/day}$$

Example 4.6

Convert pounds to gallons.

$$\text{Gallons} = \frac{\text{Pounds}}{8.34 \text{ lb/gal}}$$

Problem: How many gallons of water are required to fill a tank with 7,500 lbs of water?

Solution:

$$\text{Gallons} = \frac{7{,}500 \text{ lb}}{8.34 \text{ lb/gal}} = 899 \text{ gal}$$

Example 4.7

Convert milligrams/liter to pounds.

Note: Concentrations in milligrams per liter or parts per million determined by laboratory testing are typically converted to quantities of pounds, kilograms, pounds per day, or kilograms per day.

$$\text{Pounds} = \text{Concentration, mg/L} \times \text{Volume, MG} \times 8.34\ \text{lb/mg/L/MG}$$

Problem: The solids concentration in the aeration tank is 2,500 mg/L. The aeration tank volume is 0.95 MG. How many pounds of solids are in the tank?

Solution:

$$\text{Pounds} = 2{,}500\ \text{mg/L} \times 0.95\ \text{MG} \times 8.34\ \text{lb/mg/L/MG} = 19{,}808\ \text{lb}$$

Example 4.8

Convert milligrams per liter to kilograms per day.

$$\text{Kilograms/Day} = \text{Concentration, mg/L} \times \text{Volume, MG} \times 3.785\ \text{kg/mg/L/MG}$$

Problem: The effluent contains 25 mg/L of BOD_5. How many kilograms of BOD5 are discharged daily when the effluent flow rate is 9.0 MGD?

Solution:

$$\text{kg/Day} = 25\ \text{mg/L} \times 9.0\ \text{MG} \times 3.785\ \text{kg/mg/L/MG} = 852\ \text{kg/day}$$

Example 4.9

Convert pounds to milligrams per liter.

$$\text{Concentration, mg/L} = \frac{\text{Quantity (lb)}}{\text{Volume, MG} \times 8.34\ \text{lb/mg/L/MG}}$$

Problem: The aeration tank contains 89,900 lbs of solids. The volume of the aeration tank is 4.4 MG. What is the concentration of solids in the aeration tank in mg/L?

Solution:

$$\text{Concentration, mg/L} = \frac{89{,}900\ \text{lb}}{4.40\ \text{MG} \times 8.34\ \text{lb/mg/L/MG}} = 2{,}450\ \text{mg/L}$$

Example 4.10

Convert gallons per minute (gpm) to million gallons per day (MGD).

$$\text{Flow(MGD)} = \frac{\text{Flow(gpm)} \times 1{,}440\,\text{min/day}}{1{,}000{,}000\,\text{gal/MG}}$$

Problem: The flow meter indicates the current flow rate is 1,400 gpm. What is the flow rate in MGD?

Solution:

$$\text{Flow, MGD} = \frac{1{,}400\,\text{gpm} \times 1{,}440\,\text{min/day}}{1{,}000{,}000\,\text{gal/MG}} = 2.02\,\text{MGD}$$

Example 4.11

Find the volume of an ideal gas.

Problem: Find the volume of 1 mole of an ideal gas at standard temperature and pressure (STP) conditions of 2 atmospheres of pressure and 0°C temperature. Repeat the calculation for 2 atm and 25°C.

Solution:

Using Equation 2.4 at a temperature of 0°C (273.15 K) gives:

$$V = \frac{1\,\text{mol} \times 0.082056\,\text{L / atm}^{-1}\text{ / K / mol} \times 273.15\,\text{K}}{2\,\text{atm}} = 11.207\text{L}$$

And at 25°C (298.15 K):

$$V = \frac{1\,\text{mol} \times 0.082056\,\text{L/atm}^{-1}\text{/K/mol} \times 298.15\,\text{K}}{2\,\text{atm}} = 12.232\,\text{L}$$

TEMPERATURE CONVERSIONS

Environmental engineers are familiar with the formulas used for Fahrenheit and Celsius temperature conversions:

- °C = 5/9 (°F – 32)
- °F = 9/5 (°C) + 32

The difficulty arises when one tries to recall these formulas from memory. Probably the easiest way to recall these essential formulas is to remember three basic steps for both Fahrenheit and Celsius conversions:

1. Add 40°
2. Multiply by the appropriate fraction (5/9 or 9/5)
3. Subtract 40°

The only variable in this method is the choice of 5/9 or 9/5 in the multiplication step. You must be familiar with the two scales to make the proper choice. The freezing point of water is 32° on the Fahrenheit scale and 0° on the Celsius scale. The boiling point of water is 212° on the Fahrenheit scale and 100° on the Celsius scale.

Note: At the same temperature, higher numbers are associated with the Fahrenheit scale and lower numbers with the Celsius scale. This important relationship helps you decide whether to multiply by 5/9 or 9/5. Let us look at a few conversion problems to see how the three-step process works.

Problem: Suppose that we wish to convert 240°F to Celsius.

Solution: Using the three-step process, we proceed as follows:

1. Add 40°

$$240° + 40° = 280°$$

2. 280° must be multiplied by either 5/9 or 9/5. Because the conversion is to the Celsius scale, we will move to a number *smaller* than 280. Through reason and observation, if 280 were multiplied by 9/5, the result would be almost the same as multiplying by 2, which would double 280 rather than make it smaller. If we multiply by 5/9, the result will be about the same as multiplying by ½, which would cut 280 in half. Because in this problem, we wish to move to a smaller number, we should multiply by 5/9:

$$(5/9)(280°) = 156.0°\text{C}$$

3. Now subtract 40°

$$156.0°\text{C} - 40.0°\text{C} = 116.0°\text{C}$$

Therefore, 240°F = 116.0°C

Example 4.12

Problem: Convert 22°C to Fahrenheit.

1. Add 40°

$$22° + 40° = 62°$$

Because we are converting from Celsius to Fahrenheit, we are moving from a smaller to a larger number, and 9/5 should be used in the multiplication:

2. (9/5) (62°) = 112°
3. Subtract 40:

$$112° - 40° = 72°$$

Thus, 22°C = 72°F

Knowing how to make these temperature conversion calculations is useful. However, in practical *in situ* or non-*in situ* operations, you may wish to use a temperature conversion table.

DID YOU KNOW?

U.S. environmental engineers find identifying certain quantities used in the U.S. can be confusing. For example, at the beginning of this chapter, the question was asked if you knew how many gallons are contained in a barrel of U.S. liquid compared to a barrel of petroleum. Along with answering this question and to clear the air on other quantities, several are listed below:

- Atmosphere standard atmosphere
- Barrel 31.5 gal (U.S., liquid)
- Barrel 42 gal (petroleum)
- Btu traditional (thermochemical) value
- Calories thermochemical
- Gallons U.S., liquid
- Horsepower U.S. or mechanical
- Ounces avoirdupois
- Pints U.S., liquid
- Pounds avoirdupois
- Quarts U.S., liquid
- Tons short tons of 2,000 lbs

MKS System (SI Units)

The *MKS system* (so named because it uses the meter, kilogram, and second as base units) comprises *SI units*. All other units are derived from the base units listed in Table 4.5. This is an entirely consistent system; there is only one recognized unit for each variable (physical quantity).

TABLE 4.5
SI Base Units

Quantity	Name	Symbol
Length	meter	m
Mass	kilogram	kg
Time	seconds	
Electric current	ampere	A
Temperature	kelvin	K
Amount of substance	mole	mol
Luminous intensity	Candela	cd

DID YOU KNOW?

An abampere, in electricity, is a centimeter-gram-second unit of electromagnetic current, equivalent to 10 A. A statampere is the electric unit of current equal to the current produced by an electromotive force of one statvolt acting through a resistance of one statohm. An abampere is multiplied by 2.99793×10^{10} to obtain a statampere.

MASS BALANCE AND MEASURING PLANT PERFORMANCE

The simplest way to express the fundamental engineering principle of *mass balance* is to say, "Everything has to go somewhere." More precisely, the *law of conservation of mass* states that when chemical reactions take place, matter is neither created nor destroyed. What this important concept allows us to do is track materials, that is, pollutants, microorganisms, chemicals, and other materials from one place to another. The concept of mass balance plays an important role in treatment plant operations (especially wastewater treatment), where we assume a balance exists between the material entering and leaving the treatment plant or a treatment process: "what comes in must equal what goes out." The concept is very helpful in evaluating biological systems, sampling and testing procedures, and many other unit processes within the treatment system.

MASS BALANCE AND MEASURING PLANT PERFORMANCE

The environmental engineer's toolbox includes two commonly used tools: mass balance and energy balance. As stated above, the most crucial concept in explaining and understanding mass balance is the law of conservation of mass, which states that mass can neither be produced nor destroyed; everything has to go somewhere (the law of conservation of matter).

The principle of conservation of mass means that if the amount of a substance (e.g., a chemical) increases somewhere (e.g., in a river or steam), then that increase cannot be the result of hocus pocus or some other mysterious configuration. The substance must have either been carried into the river from elsewhere or produced by a biological or chemical reaction from substances already in the river. Similarly, if reactions produced the mass increase of this substance, they must also have caused a corresponding decrease in the mass of some other compound(s).

Conservation of mass provides a basis for compiling a budget for any substance's mass. In the case of a river, this budget keeps track of the amounts of substance entering and leaving the river and the amounts formed or destroyed by chemical reactions. This budget can be balanced over a given period, just as a checking account register is balanced.

In the following sections, we illustrate how, in wastewater treatment operations, the mass balance concept is used to determine the quantity of solids entering and leaving settling tanks and mass balance using Biological Oxygen Demand (BOD) removal.

Mass Balance for Settling Tanks

The mass balance for the settling tank calculates the quantity of solids entering and leaving the unit.

Note: The two numbers (influent and effluent) must be within 10%–15% of each other to be considered acceptable. More significant discrepancies may indicate sampling errors increasing solids levels in the unit or undetected solids discharge in the tank effluent.

Consider the steps used in the example computation below to better understand how the mass balance for the settling tanks procedure is formatted for actual use.

Example 4.13

Step 1: Solids In = Pounds of Influent Suspended Solids
Step 2: Pounds of Effluent Suspended Solids
Step 3: Biosolids Solids Out = Pounds of Biosolids Solids Pumped Per Day
Step 4: Solids In - (Solids Out + Biosolids Solids Pumped)

Problem: The settling tank receives a daily flow of 4.20 MGD. The influent contains 252 mg/L suspended solids, and the unit effluent contains 140 mg/L suspended solids. The biosolids pump operates at 10 min/h and removes biosolids at 40 gpm. The biosolids are 4.2% solids. Determine if the mass balance for solids removal is within the acceptable 10%–15% range.

Solution:

Step 1: Solids In = 252 mg/L × 4.20 MGD × 8.34 = 8,827 lb/day
Step 2: Solids Out = 140 mg/L × 4.20 MGD × 8.34 = 4,904 lb/day
Step 3:

$$\text{Biosolids Solids} = 10\frac{\text{min}}{\text{h}} \times 24\frac{\text{h}}{\text{day}} \times 40\,\text{gpm} \times 8.34 \times 0.042$$

$$= 3,363\,\text{lb/day}$$

Step 4:

$$\text{Balance} = 8,827\,\text{lb/day} - \left(4,904\,\text{lbs/day} + 3,363\,\text{lb/day}\right)$$

$$= 560\,\text{lb or } 6.3\%$$

TABLE 4.6
General Conversion Rates

Process Type	Conversion Factor (lb Solids/lb BOD Removal)
Primary treatment	1.7
Trickling filters	1.0
Rotating biological contactors	1.0
Activated biosolids w/primary	0.7
Activated biosolids without/primary conventional	0.85
Extended air	0.65
Contact stabilization	1.0
Step feed	0.85
Oxidation ditch	0.65

Mass Balance Using BOD Removal

The amount of BOD removed by a treatment process is directly related to the quantity of solids the process will generate. Because the actual number of solids generated will vary with operational conditions and design, exact figures must be determined on a case-by-case basis. However, research has produced general conversion rates for many standard treatment processes. These values are given in Table 4.6 and can be used if plant-specific information is unavailable. The mass balance procedure uses these factors to determine the number of solids the process will produce. This is compared with the actual biosolids production to determine the sampling accuracy and/or the potential for solids buildup in the system or unrecorded solids discharges.

Step 1: $BOD_{in} = \text{Influent BOD} \times \text{Flow} \times 8.34$
Step 2: $BOD_{Out} = \text{Effluent BOD} \times \text{Flow} \times 8.34.$
Step 3: $\text{BOD Pounds Removed} = BOD_{in} - BOD_{out}$
Step 4: Solids Generated, lb – BOD Removed, lb × Factor
Step 5: Removed Solids = Sludge Pumped, gpd × % Solids × 8.34
Step 6: Effluent Solids, mg/L × Flow, MGD × 8.34

Example 4.14

Problem: A conventional activated biosolids system with primary treatment operates at the levels listed below. Does the mass balance for the activated biosolids system indicate a problem exists?

Plant Influent BOD 250 mg/L
Primary Effluent BOD 166 mg/L
Activated Biosolids System Effluent BOD 25 mg/L
Activated Biosolids System Effluent TSS 19 mg/L
Plant Flow 11.40 MGD
Waste Concentration 6,795 mg/L
Waste Flow 0.15 MGD

Solution:

BOD_{in} = 166 mg/L × 11.40 MGD × 8.34 = 15,783 lb/day
BOD_{out} = 25 mg/L × 11.40 MGD × 8.34 = 2,377 lb/day
BOD Removed = 15,783 lb/day – 2,377 lb/day = 13,406 lb/day
Solids Produced = 13,406 lb/day × 0.7 lb Solids/lb BOD = 9,384 lb Solids/day
Solids Removed = 6,795 mg/L × 0.15 MGD × 8.34 = 8,501 lb/day
Difference = 9,384 lb/day – 8,501 lb/day = 883 lb/day or 9.4%
These results are within the acceptable range.
Note: Two ways of using mass balance have been demonstrated. However, the mass balance concept can be used for all aspects of wastewater and solid treatment. In each case, the calculations must consider all the sources of material entering the process and the methods available for solids removal.

Measuring Plant Performance

To evaluate how well a plant or unit process is performing, **performance efficiency** or **percent (%) removal** is used. The results can be compared with those listed in the plant's operation and maintenance (O&M) manual to determine if the facility is performing as expected. This section presents sample calculations often used to measure plant performance/efficiency. The *efficiency* of a unit process is its effectiveness in removing various constituents from the wastewater or water. Therefore, suspended solids and BOD removal are the most common calculations of unit process efficiency. In wastewater treatment, the efficiency of a sedimentation basin may be affected by such factors as the types of solids in the wastewater, the temperature of the wastewater, and the age of the solids. Typical removal efficiencies for a primary sedimentation basin are as follows:

- Settleable Solids----------90%–99%
- Suspended Solids---------40%–60%
- Total Solids----------------10%–15%
- BOD-----------------------20%–50%

AND THERE IS MORE

In addition to the basic units and conversions used in water, air, and other media presented above, environmental engineers have several tools to perform their duties properly in these areas and others. For example, the AI tools available to environmental engineers are substantial, growing in availability and importance, and have proven to be handy. For example, the platforms Grammarly, Otter.ai, GPS, and Search Summaries.

Let us examine these platforms and how they apply to environmental engineering, aiding environmental engineers in performing their responsibilities—helping them perform their duties more correctly and professionally.

Grammarly

I remember a couple of occasions when I was asked to provide a reference for some of my students applying for employment in the private sector, civilian government sector, and those wanting to join the officer ranks in the military. Over 25 years, one of the many questions I was asked to comment on was the prospective candidate's ability to write and perform public speaking—two fundamental abilities required in any engineering branch. As far as their ability with the written word, in reading and writing, it has commonly been a sort of side joke that it is relatively easy to find a well-trained candidate who can perform the engineering duties required, but it is (or was—AI to the rescue) common knowledge that engineers are poor public speakers and even worse at writing.

Well, welcome to AI's Grammarly. This AI-powered spelling and grammar checker can be integrated with your browser to check your writing automatically. In teaching my students in my college courses, I have found that almost every student has and uses a laptop in class (taking lecture notes, hopefully), and many of my students get so used to using Grammarly that I notice they get anxious when they stop or when their equipment fails.

What Grammarly does for engineers is help ensure that clarity is the bottom line in their writing and that precision in technical documents is unquestioned. This AI platform also aids in crafting error-free emails. So, it comes down to more than just wishing to converse with me; not only define your terms but also make your presentation clear and understandable. Of course, this makes communication with colleagues and clients not quite a breeze but undoubtedly easier, especially for colleagues and clients.

So, does it bother me that my students use AI-assisted tools like Grammarly? I used to be asked that question often by my teaching colleagues. My answer: "Oh, absolutely not … I no longer get headaches trying to read or decipher their work." The next question I am often asked is, "Do you use AI assistance in your work, writing, and so forth?" Answer: "Does a healthy cat MEOW?"

OTTER.AI

Otter.ai is an American-made transcription tool that allows engineers (and others) to record meetings and interviews. What is very helpful is that this tool provides an automatic summary of these sessions and text that can be searched through to locate the parts of a lecture, meeting, or interview you want to revisit. I recommend this system to my students who record lectures.

GPS

GPS? What does that have to do with AI-assisted environmental engineering?

Well, GPS is so widespread, so standard now that including it here in this book might seem weird to some folks. But truth be told, without the help of AI-powered GPS apps in your phone, you wouldn't be able to make it to half of your meetings because you'd get lost trying to find the meeting locations.

Recall in the "old days," we had to use MapQuest to print out directions, or we used a roadmap to find where we needed to go.

AI Summarizers

AI search summaries are not new. Microsoft's GPT-powered Copilot has provided them for quite some time. AI speeds up the search. For example, an AI research paper summarizer is a tool that utilizes artificial intelligence algorithms to summarize lengthy discussions, research articles, and research papers. This tool helps engineers and others quickly extract essential information from long-form content such as reports, papers, or articles.

THE BOTTOM LINE

Empowering AI-enabled tools assists the workforce with data. Not only do environmental engineers use the precision of engineering, soil science, biology, and chemistry to develop solutions to environmental problems—such as improving recycling, waste disposal, public health, and water and air pollution control—but they also potentially help solve environmental issues. AI also assists with addressing environmental questions, issues, and problems, such as predicting weather and tracking icebergs. For example, NOAA and other organizations leverage AI to track icebergs more efficiently.

The real bottom line is that we have only taken baby steps into AI's future applications in engineering and environmental science.

Part II

Applications and Impacts

5 Water Supply and Treatment

INTRODUCTION

Where do we get our drinking water? What is the source? Our answer would most likely turn to one of two possibilities: our public water is provided by a groundwater or surface water source—simple enough, because these two sources are the primary sources of most water supplies. We know from the hydrologic or water cycle that from whichever sources we obtain our drinking water, the source is constantly replenished (we hope) with a freshwater supply. Heraclitus of Ephesus best summed up this water cycle phenomenon when he said, "You could not step twice into the same rivers; for other waters are ever flowing on to you." In this chapter, one of the primary duties of the drinking water practitioner is discussed—to find and secure a source of potable water for human use.

DID YOU KNOW?

If water has a bad taste or odor, unless they are dying of thirst, people will not usually drink it.

If water is turbid (i.e., it is muddy or cloudy) and has a color, unless they are dying of thirst, people will not usually drink it. If water makes people sick, unless they are dying of thirst, people will not drink it.

WATER SOURCES[1]

In the real estate business, location is everything. The same can be said for sources of water. The presence of water defines "location" for communities. Although communities differ widely in character and size, all are concerned about finding water for industrial, commercial, and residential use. Freshwater sources that can provide stable and plentiful supplies for a community do not always occur where we wish. The availability of a regular potable water supply is the most important factor affecting many lifeforms' presence—or absence. A map of the world immediately shows us that surface waters are not uniformly distributed over the Earth's surface. U.S. land holds rivers, lakes, and streams on only about 4% of its surface. The heaviest populations of any lifeforms, including humans, are found in regions of the U.S. (and the rest of the world) where potable water is readily available because lands barren of water will not support large populations.

***Note*:** The volume of freshwater sources depends on geographic, landscape, and temporal variations and the impact of human activities.

DOI: 10.1201/9781003601562-7

Just How Readily Available Is Potable Water?

Approximately 326 million cubic miles of water comprise Earth's entire water supply. Of this massive amount of water, while indirectly providing us with fresh water through ocean evaporation, only about 3% is fresh. Even most of the minute percentage of fresh water Earth holds is locked up in polar ice caps and glaciers. The rest is held in lakes, inflows through the soil, and in river and stream systems. Only 0.027% of Earth's freshwater is available for human consumption (see Table 5.1 for the distribution percentages of Earth's water supply). We see from Table 5.1 that the significant drinking water sources are surface water, groundwater, and groundwater under the direct influence of surface water (i.e., springs or shallow wells).

DID YOU KNOW?

Of all the freshwater on Earth, only about 0.3% is contained in rivers and lakes—yet rivers and lakes are not only the water we are most familiar with, but they are also where most of the water we use in our everyday lives exists.

Surface Water Supplies

Most surface water originates directly from precipitation—rainfall or snow. To appreciate the impact of this runoff on surface water supplies, let us look at the water balance in the United States. Over the U.S. mainland, rainfall averages about 4,250 billion gallons per day. Of this massive amount, about 66% returns to the atmosphere through evaporation directly from the surface of lakes and rivers and transpiration from plants. This leaves about 1,250 billion gallons daily to flow across or through

TABLE 5.1
World Waste Distribution

World Water Distribution	
Location	**Percent of Total**
Land Areas	
Freshwater lakes	0.009
Saline lakes and inland seas	0.008
Rivers (average instantaneous volume)	0.0001
Soil moisture	0.005
Groundwater (above depth of 4,000 m)	0.61
Ice caps and glaciers	2.14
Total: Land areas	**2.8**
Atmosphere (water vapor)	0.001
Oceans	97.3
Total all locations (rounded)	**100**

Source: Adapted from Peavy et al. (1985), p. 12.

TABLE 5.2
Water Balance in the U.S.

Water Balance in the U.S.	(bgd)
Precipitation	4,250
Evaporation and transpiration	3,000
Runoff	1,250
Withdrawal	310
Irrigation	142
Industry (principally utility cooling)	142
Municipal	26
Consumed (principally irrigation loss)	90
Returned to streams	220

Source: National Academy of Sciences—National Research Council Publ. 1000-B, 2002.

the earth to return to the sea (see Table 5.2). While municipal water usage is only a small fraction of this significant volume, the per capita water consumption in the U.S. is relatively high—about 150 gallons per person per day, probably because public water is relatively inexpensive in the U.S. In areas where water supplies are less readily available and thus more costly, per capita consumption is much lower due to financial and conservation concerns.

The 1,250 billion gallons per day of surface runoff water, exposed and open to the atmosphere, result from the movement of water on and just beneath the Earth's surface (overland flow). Overland flow and surface runoff are the same—water flow that has not yet reached a definite stream channel. This occurs when the precipitation rate exceeds either the rate of interception and evapotranspiration or the amount of rainfall readily absorbed by the Earth's surface. The land area that contributes runoff to a stream or river is called a *watershed, drainage basin,* or *catchment area.*

Specific sources of surface water include:

- Rivers
- Streams
- Lakes
- Impoundments (man-made lakes made by damming a river or stream)
- Very shallow wells that receive input via precipitation
- Springs affected by precipitation (flow or quantity directly dependent upon precipitation)
- Rain catchments (drainage basins)
- Tundra ponds or muskegs (peat bogs)

Surface water has advantages as a source of potable water. Surface water sources are usually easy to locate; unlike groundwater, finding surface water does not require a geologist or hydrologist. Usually, surface water is not tainted with minerals precipitated from the Earth's strata. Ease of discovery aside, surface water also presents some disadvantages: surface water sources are easily contaminated (polluted) with microorganisms that can cause waterborne diseases and chemicals that enter from surrounding runoff and upstream discharges. Water rights can also present problems.

As we have said, most surface water results from surface runoff. The amount and flow rate of this surface water are highly variable, which comes into play for two main reasons: (1) human interference (influences) and (2) natural conditions. In some cases, surface water dashes off land surfaces. This is generally undesirable from a water resources standpoint because quick runoff does not provide enough time for the water to infiltrate the ground and recharge groundwater aquifers. Surface water that quickly runs off land also causes erosion and flooding problems. The only good thing that can be said about surface water that runs off quickly is that it usually does not have enough contact time to increase mineral content. Slow surface water off land has all the opposite effects.

Drainage basins collect surface water and direct it on its gravitationally influenced path to the ocean. The drainage basin is generally characterized as an area measured in square miles, acres, or sections. If a community is drawing water from a surface water source, the size of its drainage basin is an important consideration. Surface water runoff, like the flow of electricity, flows or follows the path of least resistance. Surface water within the drainage basin normally flows toward one primary watercourse (river, stream, brook, creek, etc.) unless some man-made distribution system (canal or pipeline) diverts the flow. Surface water runoff on land surfaces depends on several factors:

- **Rainfall Duration**: Even a light, gentle rain, if it lasts long enough, can, with time, saturate the soil and allow runoff to take place.
- **Rainfall Intensity**: The soil surface quickly becomes saturated with increased rainfall intensity. This saturated soil can no longer hold water; as more rain falls and water builds up on the surface, it creates surface runoff.
- **Soil Moisture:** The amount of existing moisture in the soil has a definite impact on surface runoff. Soil already wet or saturated from a previous rain causes surface runoff to occur sooner than if the soil were dry. Surface runoff from frozen soil can be up to 100% snow melt or rain runoff because frozen ground is impervious (Figure 5.1).
- **Soil Composition:** The composition of the surface soil directly affects the amount of runoff. For example, hard rock surfaces result in 100% runoff.

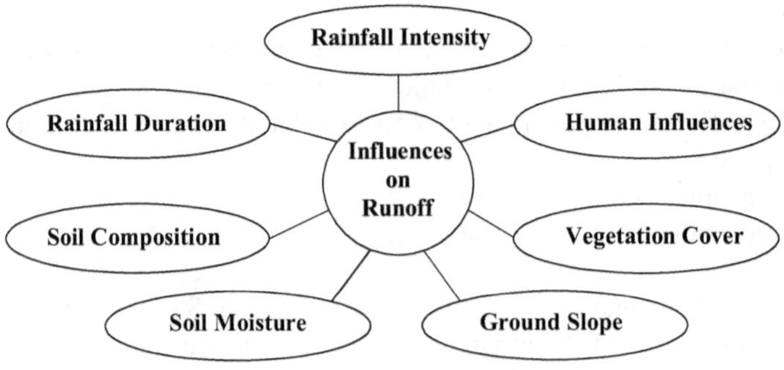

FIGURE 5.1 Influences of runoff.

Clay soils have very small void spaces that swell when wet; the void spaces close and do not allow infiltration. Coarse sand possesses large void spaces that enable the easy flow of water through, which produces the opposite effect, even in a torrential downpour.

- **Vegetation Cover:** Ground cover limits runoff. Roots of vegetation and pine needles, pine cones, leaves, and branches create a porous layer (a sheet of decaying natural organic substances) above the soil. This porous "organic" sheet readily allows water into the soil. Vegetation and organic waste also act as cover to protect the soil from hard, driving rains, which can compact bare soils, close off void spaces, and increase runoff. Vegetation and ground cover work to maintain the soil's infiltration and water-holding capacity and reduce soil moisture evaporation.
- **Ground Slope:** When rain falls on steeply sloping ground, up to 80% or more may become surface runoff. Gravity moves the water down the surface more quickly than it can infiltrate the surface. Water flow off flat land is usually slow enough to allow a higher percentage of the rainwater to penetrate the ground.
- **Human Influences:** Various human activities have a definite impact on surface water runoff. Most human activities tend to increase the rate of water flow. For example, canals and ditches are usually constructed to provide steady flow, and agricultural activities generally remove ground cover that would work to retard the runoff rate. On the opposite extreme, man-made dams are generally built to retard the flow of runoff.

DID YOU KNOW?

Paved streets, tarmac, paved parking lots, and buildings are impervious to water infiltration, greatly increasing the amount of stormwater runoff from precipitation events. These human-made surfaces, which hasten the flow of surface water, often cause flooding, sometimes with devastating consequences. In poorly planned areas, even relatively light precipitation can cause local flooding. Impervious surfaces not only present flooding problems but also do not allow water to percolate into the soil to recharge groundwater supplies—often another devastating blow to a location's water supply.

Groundwater Supply

Unbeknownst to most of us, our Earth possesses an unseen ocean. Unlike the surface oceans that cover most of the globe, this ocean is freshwater: the groundwater in aquifers beneath the Earth's crust. This gigantic water source forms a reservoir that feeds all the natural fountains and springs of the Earth. But how does water travel into the aquifers that lie under the Earth?

Groundwater sources are replenished from a percentage of the average approximately 3 ft of water that falls to Earth each year on every square foot of land. Water falling to Earth as precipitation follows three courses. Some runs off directly to rivers and streams (roughly 3 inches of that 3 ft), eventually working back to the sea. Evaporation and transpiration through vegetation take up about 2 ft. The remaining 6 inches seep into the ground, entering and filling every interstice, hollow and cavity. Although groundwater comprises only 1/6 of the total (1,680,000 miles of water), spreading this water over the land would blanket it to a depth of 1,000 ft.

Almost all groundwater is in constant motion through the pores and crevices of the aquifer in which it occurs. The water table is rarely level; it generally follows the shape of the ground surface. Groundwater flows in the downhill direction of the sloping water table. The water table sometimes intersects low points of the ground, seeping into springs, lakes, or streams.

Usual groundwater sources include wells and springs that are not influenced by surface water or local hydrologic events.

Groundwater has several advantages over surface water as a potable water source. Unlike surface water, groundwater is not easily contaminated. Groundwater sources usually have lower bacteriological contamination than surface waters. Groundwater quality and quantity typically remain stable throughout the year. In the United States, groundwater is available in most locations.

Groundwater presents some disadvantages as a potable water source compared to surface water sources. Operating costs are usually higher because groundwater supplies must be pumped to the surface. Contamination is often hidden from view, and removing contaminants is very difficult. Groundwater frequently possesses high mineral levels, and thus an increased hardness level, because it is in contact longer with minerals. Near coastal areas, groundwater sources may be subject to saltwater intrusion.

Note: The quality of its source influences groundwater quality. Changes in source waters or degraded quality of source supplies may seriously impair the quality of the groundwater supply.

Our freshwater supplies are constantly renewed through the hydrologic cycle, but the balance between freshwater and saltwater's standard ratio is not subject to our ability to change. As our population grows and we move into lands without ready freshwater supplies, we place ecological strain upon those areas and their ability to support life. Communities that build in regions without adequate local water supply are at risk in the event of an emergency. Proper attention to our surface and groundwater sources, including remediation, pollution control, and water reclamation and reuse, can help to ease the strain. Still, technology cannot fully replace adequate local freshwater supplies from surface or groundwater sources.

WATER TREATMENT

We live on a planet with three-fourths of its surface covered with water, so we recognize the irony inherent in the fact that many areas of the world face critical water shortages. Most of Earth's water is seawater—far too saline for human consumption. Of the little "fresh" water that remains, most is trapped in polar ice caps, and harnessing it for use by the world's population is difficult.

Much of the accessible natural supply of potable water faces stress from a growing world population, which increases the primary demand for this natural resource while reducing the supply further through contamination.

Major population centers in developing nations (those without established waste treatment or water treatment infrastructures) often suffer from epidemics of waterborne disease. Raw sewage often directly contaminates these areas' rivers and streams used for drinking, washing, and cooking. In other cases, unchecked industrialization leads to water contamination through improper disposal of chemical and nuclear wastes.

Thus, the drinking water purveyor must ensure the water supplied is safe for human consumption. The primary reason for developing and installing a public water system is to protect public health. Basically, an adequately operated water system serves as a line of defense against disease and the public.

Properly operated water treatment and supply systems are defined as those that:

- remove or inactivate pathogenic microorganisms, including bacteria, viruses, and protozoans
- reduce and/or remove chemicals that can be detrimental to health
- provide quality water, thus discouraging the customer from seeking better-tasting or better-looking water that may be contaminated

This last point is critical but often overlooked in the operation and management of public water systems. When the water produced by a system is objectionable because of odor, taste, or appearance, the customer will often seek another source for their drinking water. Ironically, these sources, while appearing, tasting, and smelling fine (better than "city water"), could contain harmful microorganisms and/or chemicals.

In this chapter, we discuss the drinking water practitioner's most important function: ensuring that water delivered to the public is treated correctly and arrives as the clean, wholesome, safe product that it must be.

A typical water treatment plant (treating stream or river water—turbid surface water with organics) processes raw water using various unit processes, including (1) screening, (2) coagulation, (3) flocculation, (4) sedimentation or settling, (5) filtration, (6) hardness treatment, (7) disinfection, and (8) fluoridation (see Figure 5.2). This briefly overviews these unit processes—a typical drinking water treatment system for surface water supplies.

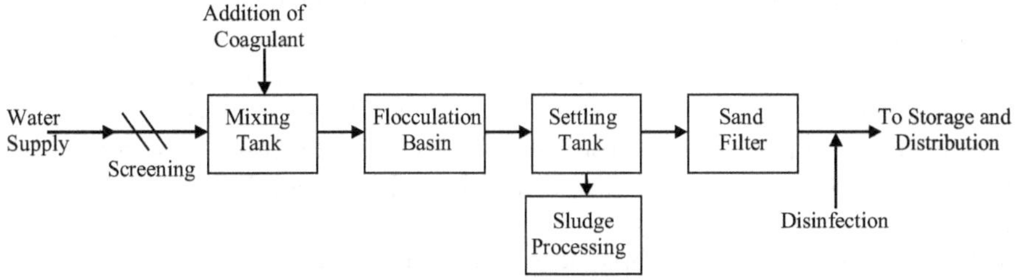

FIGURE 5.2 Typical water treatment processes.

Screening

Earlier, we discussed water intakes and some of the screening devices included. *Screening* (the first important step in treating water containing large solids) is when relatively large and suspended debris is removed from the water before it enters the plant. River water (the water source used in our discussion) frequently contains suspended and floating debris varying in size from small rags to logs. Removing these solids is essential, not only because these items have no place in potable water, but also because this river trash may cause damage to downstream equipment (clog and damage pumps, etc.), may increase chemical requirements, may impede hydraulic flow in open channels or pipes, or may hinder the treatment process (Pankratz, 1995). The most important criteria for selecting a particular water treatment technology screening system are the screen opening size and flow rate. Other important criteria include costs related to operation and equipment, plant hydraulics, debris handling requirements, operator qualifications, and availability.

Significant surface water treatment plants may employ various screening devices, including trash screens (or trash rakes), traveling water screens, drum screens, bar screens, or passive screens. In the following sections, we briefly discuss each of these screening devices.

Trash Screens (Rakes)

Trash screens or *trash rakes* are used to remove rough or large debris retained on a trash rack. For example, they protect pumping equipment and may be a preliminary screening device to protect finer screens—drums or traveling water screens.

A trash screen consists of one or more stationary trash rakes and a screen raking device. Trash rack bar spacings range from 1.5 to 4 inches and are mainly constructed of steel bars. Those constructed of high-density polyethylene polymers are beginning to replace many of the older steel bar models. These synthetic screens are lighter and less prone to microbial growth, corrosion, and ice.

Raking mechanisms are available in various intake configurations, including installation on vertical buildings and dam walls. They are typically mounted on fixed structures designed to clean a single trash rack, wheel-mounted to traverse the entire width of an intake structure, or suspended from an overhead gantry to clean individual sections of a wide trash rack.

Traveling Water Screens

Traveling water screens (band screens) are placed in a flowing water channel to remove floating or suspended debris. These automatically cleaned screening devices protect pumping or other downstream equipment from debris in surface water intakes.

Consisting of a continuous series of wire mesh panels bolted to basket frames or trays and attached to two matched strands of roller chain; the traveling water screen operates vertically over a sprocket assembly through the flow. As raw water passes through the revolving baskets, debris is collected and retained on the upstream face of the wire mesh panels. The debris-laden baskets are lifted out of the flow and above the operating flow, where a high-pressure water spray directed outward removes the impinged debris. This process can be continuous or intermittent. For intermittent operation, the screen activates when a specified head loss or time elapse has occurred.

When located on a river, traveling water screens may be subject to large fluctuations in flow conditions, debris loading, water depths, and salinity. Depending upon the application, the traveling water screen's size is determined by considering such factors as maximum-average flow, maximum-minimum-average water levels, wire mesh size, velocity through mesh, basket or channel width, number of screens, type of service, and/or starting/operating headloss requirements.

Drum Screens

A *drum screen* (or cylinder screen) has very few moving parts and is mounted on a horizontal axis with a series of wire mesh panels attached or mounted on the periphery of its cylinder. The cylinder slowly rotates on its axis. Because of its simplicity of construction, a drum screen installation's maintenance and operating costs are usually less than those of a traveling water screen.

Bar Screens

Though primarily used in wastewater treatment applications, bar screens are employed in some water treatment facilities. A *bar screen* consists of straight steel bars welded at both ends to two horizontal steel members and is automatically cleaned by one or more power-operated rakes. As a rake is operated up the face of the bar rack, it removes accumulated debris (usually large solid objects and rags) and elevates in and out of the flow. At the top of the rake's operating cycle, the debris is swept from the rake into a debris receptacle by a wiper mechanism. When installed in a waterway, the bar screen assembly is typically placed at an angle of 60°–80° from the horizontal.

Passive Screens

Passive intake screens (stationary screening cylinders) have no moving parts and require no debris handling or removal equipment. They are placed in a surface water body to take advantage of natural ambient currents and controlled through-screen velocities to minimize debris buildup.

Usually mounted on a horizontal axis and oriented parallel to the natural current flow within the water body, current flow action keeps the screen clean. A maximum intake velocity of about 0.5 fps is typical and works to minimize debris impingement on the screen surface.

Coagulation

Coagulation, the second step in water purification, is a unit process that has been used for several years in raw water treatment. It works to settle suspended solid material effectively.

***Note*:** Chemicals employed for coagulation are expected to be safe for drinking water when used according to standards.[2]

Coagulation: The Coagulants

After screening, raw water is typically pumped into large *settling basins* (aka clarifiers or sedimentation tanks). Within the confines of the settling basin, the screened raw water is allowed to sit for some predetermined time. Although screened, the raw water still contains impurities that may be dissolved or suspended. The settling basin provides the most convenient way to remove the suspended matter: it lets the force of gravity do the work. Within the basin, when flow and turbulence are minimal (quiescent conditions), particles become dense, and then settle to the bottom of the tank. This process is called *sedimentation*, and the layer of accumulated solids at the bottom of the tank is called *sludge* (or *biosolids* in some wastewater treatment unit processes).

The suspended particles' size and density directly affect the speed at which they settle toward the basin's bottom. Larger and heavier particles will, of course, settle faster than smaller or lighter particles. The forces opposing the downward force of gravity include buoyancy and drag (friction). The particle-settling rate is also affected by the temperature and viscosity of the water.

Note: The nature of the sedimentation process also varies with the concentration of suspended solids and their tendency to interact with one another.

In the described sedimentation process, not all suspended solids or particles can be removed entirely from water, even when given long detention times. For example, tiny particles called *colloids* (such as bacteria, particles of color, and turbidity, which may be colloidal) will not settle out of suspension by gravity without some help. This is where *coagulants* come into play. The colloidal particles will settle if we rapidly mix chemical coagulants into the water and slowly stir the mixture before allowing sedimentation.

Colloids or finer particles must be chemically coagulated to produce a removable larger floc in subsequent settling and filtration.

The coagulation process (along with flocculation) neutralizes or reduces the natural repelling electrical force of particles in water, keeping them apart and in suspension. Particles in water usually carry a negative electrical charge. Since all these particles carry the same negative electrical charge, they repel each other—in the same way as the poles of a magnet do.

The object of coagulation (and subsequently flocculation) is to turn the small particles into larger flocs, either as precipitates or suspended particles. These flocs are then conditioned for ready removal in subsequent processes. In other words, in this text, we define *coagulation* as a method to alter the colloids so that they can approach and adhere to each other to form larger floc particles.

DID YOU KNOW?

Coagulation and flocculation convert nonsettleable turbidity particles into settable forms for their effective removal by gravity.

Coagulants

Coagulants, both coagulants and coagulant aids, are used in the coagulation process. The plant process scheme generally defines the type of coagulants and aids available. Jar tests are normally used to determine the optimum chemical dosages for treatment.

We discuss jar tests in the following section.

DID YOU KNOW?

Selection of the proper coagulant and coagulant aid is critical for a water supply. A jar test for water should be run to determine which coagulants and flocculant aids are economical and most effective.

Jar Test

Jar tests are widely used to simulate a full-scale coagulation (and flocculation) process to determine optimum chemical dosages—the cost-effective coagulant dose for the time and intensity of agitation selected. Used for many years by the water treatment industry, the test conditions are intended to reflect the regular operation of a chemical treatment facility. In this way, the type and quantity of sludge and the physical properties of the floc can be evaluated.

The test can be used to:

a. select the most effective chemical
b. select the optimum dosage
c. determine the value of a flocculant aid and the proper dose.

The testing procedure requires a series of samples to be placed in testing jars and mixed at 100 rpm. Varying amounts of the process chemical or specified amounts of several flocculants are added (one volume/sample container). The mixture is continued for 1 minute. The mixing slows to 30 rpm to provide gentle agitation; then, the floc is allowed to settle. The flocculation period and settling process are observed to determine the floc strength, settleability, and clarity of the supernatant liquor (defined as the water that remains above the settled floc). The supernatant can then be tested to determine the chemical addition's efficiency for removing TSS, BOD_5, and phosphorus.

The equipment required for the jar test includes a 6-position variable-speed paddle mixer, six two-quart wide-mouth jars, an interval timer, assorted glassware, pipettes, graduates, and so forth.

DID YOU KNOW?

Rapid mixing is the fast and thorough mixing—flash mixing—of the various chemicals, such as coagulants and coagulant aids, with water for their proper chemical reactions

Types of Coagulants

Several different chemicals can be used for coagulation. Commonly used metal coagulants are those based on aluminum (aluminum sulfate) and iron (ferric sulfate).

The most common coagulant is aluminum sulfate [aka alum, $Al_2(SO_4)_3$]. We list other common coagulation chemicals in Table 5.3.

TABLE 5.3
Common Coagulant Chemicals

Common Coagulant Chemicals	
Common Name	**Comments**
Aluminum sulfate	Most common coagulant in the U.S; often used with cationic polymers
Ferric chloride	It may be more effective than alum in some applications.
Ferric sulfate	Often used with lime softening
Ferrous sulfate	Less pH-dependent than alum
Aluminum polymers	Synthetic polyelectrolytes: large molecules
Sodium aluminate	Used with alum to improve coagulation
Sodium silicate	Ingredient of activated silica coagulant aids

Source: Adapted from Spellman (2007).

Types of Coagulant Aids

Difficulties with coagulation often occur because of slow-settling precipitates or fragile flocs that are easily fragmented under hydraulic forces in basins and filters (Hammer and Hammer, 1996). The American Water Works Association (AWWA, 1995) describes a coagulant aid as a chemical added during coagulation to improve coagulation to build stronger, more settleable floc; to overcome the effect of temperature drops that slow coagulation; to reduce the amount of coagulant needed; and to reduce the amount of sludge produced. Coagulant aids in flocculation by improving the settling qualities and toughness of flocs. Polymers are the most widely used materials. Synthetic polymers are water-soluble, high-molecular-weight organic compounds with multiple electrical charges along a molecular chain of carbon atoms. In drinking water treatment, polymers are extensively used as coagulant aids to build large floc before sedimentation and filtration. Other coagulant aids are activated silica, adsorbent-weighting agents, and oxidants.

Coagulation: Process Operations

The standard coagulation unit process involves adding coagulant chemicals through rapid mixing—detention time in the rapid mix tank is typically on the order of minutes (Masters, 1991). During this mixing process, polymer (or some other coagulant aid) is added and blended into the destabilized water before flocculation.

Coagulation removes impurities depending on their nature and concentration, using both coagulants and coagulant aids, and other characteristics of the water, including pH, temperature, and ionic strength. Because of the complex nature of coagulation reactions, chemical treatment is based on empirical data derived from jar testing or other laboratory tests and field studies (Viessman and Hammer, 1998).

FLOCCULATION

The destabilized particles and chemical precipitates resulting from coagulation are designed to enhance their settling qualities and, thus, their removal from water. However, even after coagulation, these particles and chemical precipitates may still settle very slowly (too slowly). The next step in the water treatment process, flocculation, is employed to speed up the settling process.

Note: Flocculation is the "clumping" of coagulated particles resulting from coagulation. While the two terms are often used interchangeably, they are distinct concepts.

Flocculation is the most critical factor affecting particle removal efficiency. Flocculation is a slow mixing process in which these particles are brought into contact to collide, stick together, and grow (agglomerate) to a size that will readily settle. Enough mixing must be provided (gentle agitation for approximately one-half hour) to bring the floc particles into contact with each other and to keep the floc from settling in the flocculation basin. (Note: The heavier the floc and the higher the suspended solids concentration, the more mixing is required to keep the floc in suspension.) The most common type of "mixer" or flocculator is paddle-type, which uses redwood slats mounted horizontally on motor-driven shafts. Rotating slowly at about one revolution per minute, the paddles provide gentle agitation that promotes floc growth. The rate of agglomeration or flocculation depends on the number of particles present, the relative volume they occupy, and the velocity gradient in the basin.

Note: When we state that the rate of agglomeration or flocculation depends on the *velocity gradient*, we refer to the fact that too much mixing shears the floc particles, tearing them apart again so that the floc becomes small and finely dispersed (a situation we are trying to avoid). Thus, the velocity gradient must be controlled within a relatively narrow range.

The flocculation theory is complex and beyond the needs of this text. Still, on an elemental level, we can say that flocculation is generally affected by slowly rotating large-diameter mixers. Recent practice incorporates the dispersion of the coagulant (flash mixing), flocculation, and sedimentation in a single unit called a *contact clarifier.*

Note: Flocculation is a principal mechanism in removing turbidity from water.

Sedimentation

Note: In a conventional water treatment plant, coagulation/flocculation precedes the sedimentation process for better results and improved utilization of the settling basins, followed by the filtration process. In the past, but not necessarily confined to the past, filtration may have been preceded only by coagulation; here, filtration is provided only after a few minutes of contact, with consequent additional stress on the filters. Lack of sedimentation means less reliability in the operation of filters when water quality suddenly changes characteristics (De Zuane, 1997).

Sedimentation (*clarification*) is the gravity-induced removal of particulate matter, chemical floc, precipitates from suspension, and other settleable solids. As stated, sedimentation separates the liquid from the solids. The process occurs in a rectangular, square, or round tank called a settling or sedimentation tank or basin. Flow patterns within such basins may be rectilinear flow in rectangular basins, radial flow in center-feed settling tanks, radial flow in peripheral-feed settling tanks, spiral flow in peripheral-feed settling tanks, or radial flow in square settling tanks.

DID YOU KNOW?

An adequate detention time is important for complete sedimentation. The shorter the detention time, the less the settling and the higher the turbidity.

Sedimentation is typically the step between flocculation and filtration in conventional water treatment. Design criteria are based on empirical data from the performance of full-scale sedimentation tanks. The standard criteria for sizing settling basins are detention time (typically from 1 to 10 hours), overflow rate, weir loading, and, with rectangular tanks, horizontal velocity.

In water treatment, most settling basins are essentially up-flow clarifiers where the water rises vertically for discharge through effluent channels. More specifically, in the idealized sedimentation tank, water flows horizontally through the basin, then rises vertically, overflowing the weir of a discharge channel at the tank surface. Floc settles downward, opposite the water flow, and is removed from the bottom by a continuous mechanical sludge removal apparatus. The particles with a settling velocity more significant than the overflow rate are removed (settled), while lighter flocs are carried out in the effluent. The effluent is then filtered.

Note: Sedimentation basins, either circular or rectangular, are designed for slow, uniform water movement with a minimum of short-circuiting.

DID YOU KNOW?

Short-circuiting causes a short detention time, resulting in inefficient sedimentation.

Filtration

Even after chemical coagulation and gravity sedimentation, not all suspended solids or impurities are removed from water. Non-settleable floc particles (about 5% of the suspended solids) may remain in the water, and with only a percentage left, we might ask ourselves: "Isn't this good enough?" But it is not. This remaining floc would cause problems (including noticeable turbidity), and particles shield microorganisms from the subsequent disinfection processes. The water treatment aims to produce potable, crystal-clear water that satisfies the Safe Drinking Water Act requirement of 0.5 NTU (for turbidity). To accomplish this, water needs an additional treatment step that follows coagulation, flocculation, and sedimentation.

This next step is the physical process of *filtration*. Filtration (sometimes called a polishing process) involves the removal of suspended particles from water by passing it through a layer or *bed* of porous granular material—sand, for example. As water flows through the filter bed, the suspended particles become trapped within the pore spaces of the filter material (or *filter media).* In purifying a surface water source (as we are discussing in this case), filtration is a critical process, even though filtration is only one step in the overall treatment process.

Note: Filtration occurs naturally as surface waters migrate (percolate) through the porous soil layers to recharge groundwater. This natural filtration removes most suspended matter and microorganisms, which is why many wells produce water that does not require further treatment.

The Surface Water Treatment Rule (SWTR) specifies certain filtration technologies. The most common treatment filter systems include rapid gravity filters, either built on-site or packaged plants, and pressure filters. Another type is direct filtration. Two other types are slow sand filters and diatomaceous earth (DE) filters. The SWTR allows the use of "alternate" filtration technologies. Cartridge filters, for example, are an alternative technology.

DID YOU KNOW?

The deeper the filtration bed, the better the filtration.

Filtration treatment unit processes most commonly used in water purification systems include slow or rapid sand filtration, diatomaceous earth filtration, and package filtration systems. Slow and rapid filter systems refer to the flow rate per unit of surface area. Filters are also classified by the type of granular material used in them. Sand, anthracite coal, coal-sand, multi-layered, mixed bed, or diatomaceous earth are examples of different filtering media. The water's direction may also classify filtration systems through the medium. Some examples are downflow, upflow, fine-to-coarse, or coarse-to-fine. Lastly, filters are commonly distinguished by whether they are gravity or pressure filters. Gravity filters rely only on the force of gravity to move the water down through the grains and typically use upflow for washing (backwashing) the filter media to remove the collected foreign material. Gravity filters are free surface filters that are much more commonly used for municipal applications. Pressure filters are completely enclosed in a shell so that most of the water pressure in the lines leading to the filter is not lost and is used to push the water through the filter.

Rapid Filter Systems

Slow fine sand filtration has been used in the U.S. since 1872, but although it is still used in many older plants, it is not commonly used today in most modern water treatment plants because of various associated problems. One of the problems is related to the tiny size of the pore spaces in the fine sand, which slows the water's progress through the filter bed. These filter types also have problems with suspended particles clogging the surface, requiring the filter to be manually scraped clean. These units also occupy a considerable land area because slow filtration rates mean a higher filter surface area is needed to produce the required filtered water qualities.

The rapid filter has primarily replaced the slow sand filter in modern water treatment plants. The rapid filter consists of a layer of carefully sieved silica sand ranging from 0.6 to 0.75 m in depth on top of a bed of graded gravel. The pore openings between the grains of sand are often more significant than the size of the floc particles to be removed, so much of the filtration is accomplished by means other than simple straining.

Note: The ideal filter media possesses the following characteristics: it is coarse enough for large pore openings to retain large quantities of floc yet sufficiently delicate to prevent the passage of suspended solids, has adequate depth to allow relatively long filter runs, and is graded to permit adequate cleaning during backwash.

Adsorption, continued flocculation, and sedimentation in the pore spaces are important removal mechanisms. When the filter becomes clogged with particles (which occurs approximately once a day, depending on the turbidity of the water), it is shut down for a short period and cleaned by forcing water backward through the sand for 10–15 minutes. After washing, the sand settles back in place, and operation resumes.

DID YOU KNOW?

In granular filtration, the higher the temperature, the better the filtration.

Other Common Filter Types

Rapid-flow filters are the most common type for treating water supplies, primarily because they are the most reliable. However, other filters, including pressure and diatomaceous earth filters, are sometimes used to clarify water.

A *pressure filter* is similar to a rapid filter in that the water flows through a granular filter bed. However, instead of being open to the atmosphere and using the force of gravity, the pressure filter is enclosed in a cylindrical steel tank, and the water is pumped through the bed under pressure. They are less reliable than rapid filters because pressure may force solids through the bed in the effluent. Because of this problem, they are seldom employed in municipal water treatment works but instead are used for filtering water for industrial use or in swimming pools.

Diatomaceous earth filters contain a thin layer of a natural powder-like material formed from diatom shells. They are also used primarily for industrial or swimming pool applications because they are less reliable than rapid sand filters.

DID YOU KNOW?

Reverse osmosis is suitable when water has a very high concentration of dissolved substances such as chlorides, nitrates, fluorides, and other contaminants (e.g., salty water).

HARDNESS TREATMENT

The lime-soda process and ion exchange are commonly used methods to reduce hardness. The lime-soda process applies to extensive facilities, while ion exchange is usually employed in smaller water works. The lime-soda process will not remove all the hardness and is typically operated to produce a residual hardness of about 100 mg/L as $CaCO_3$. More significant reductions are not economical and may have adverse health consequences (McGhee, 1991). In this text, our discussion focuses on ion exchange.

Ion exchange is accomplished by charging a resin with sodium ions and allowing the resin to exchange the sodium ions for calcium and magnesium ions.

Standard resins include zeolites—natural and man-made minerals that collect certain ions (sodium or $KMnO_4$) from a solution and either exchange these ions (as in water softening) or use the ions to oxidize a substance (as in iron or manganese removal).

The negative side of ion exchange is that even though the process softens water by removing all (or nearly all) of the hardness and adding sodium ions, the water may be more corrosive. Adding sodium ions to the water may also increase the health risk for those with high blood pressure.

Disinfection

At the turn of the century, 35,000 people per 1,000,000 did not reach 20 years of age. Today, according to USA TODAY (1998), while 101 deaths occur every minute, at the same time, 261 births occur. Curbing waterborne diseases through disinfection has significantly contributed to birthrates outpacing death rates worldwide.

The Safe Drinking Water Act requires that public water supplies be disinfected and that the US Environmental Protection Agency (USEPA) sets standards and establishes processes for treating and distributing disinfected water to ensure no significant risks to human health occur. The USEPA Science Advisory Board ranks pollutants in drinking water as one of the highest health risks meriting USEPA's attention because of large-scale population exposure to contaminants, including lead, disinfectant by-products (DBPs), and disease-causing organisms. Disinfectants are used by virtually all surface water systems in the U.S. and by an unknown percentage of systems that rely on groundwater. Chlorine has been the most widely used and cost-effective disinfectant for nearly a century. However, disinfection treatments can produce a wide variety of by-products, many of which have been shown to cause cancer or have other toxic effects. Recently, concern has been raised over water quality deterioration, which can increase dramatically during distribution unless systems are appropriately designed and operated. While disinfection is integral to water treatment, filtration before disinfection is necessary to reduce pathogen levels, making disinfection more reliable by removing turbidity and other interfering constituents.

To solve the disinfection by-products problem, we need innovation to upgrade existing techniques and develop new approaches to address these problems. Areas of interest include:

- Alternatives to chlorine disinfection for removing pathogenic microorganisms, including innovative applications of ultraviolet radiation and processes that improve overall effectiveness while using reduced amounts of disinfectant.
- Development of innovative unit processes, particularly for small systems, for the removal of organic and inorganic contaminants (such as arsenic), particulates, and pathogens [e.g., cyst-like organisms and emerging pathogens like caliciviruses, microsporidia (septa and Enterocytozoon), hepatitis A virus (HAV), Mycobacterium avium intracellulare (MAC), Helicobacter pylori, Legionella pneumophila, adenovirus 40/41/1–39, and Toxoplasma gondii].

- Development of efficient, cost-effective treatment processes for removing disinfection by-product precursors (e.g., trihalomethanes, haloacetic acids, and for ozonation: bromate, aldehydes; for chlorination: chloropicrin, haloacetonitriles; for chloramination: organic chloramines, cyanogen chloride).
- Improved methods for controlling pathogens through coagulation/settling, filtration, or other cost-effective means.
- Drinking water contamination control between the treatment plant and the user, especially considering potential chemical leaching from distribution system materials and surfaces (e.g., lead, copper, iron, and other pipe materials, protective coatings) as a result of instability, interaction with microorganisms, disinfection agents, and water treatment chemicals.

To this point, the unit processes described—screening, coagulation, flocculation, sedimentation, and filtration—comprise a type of treatment called *clarification*. Clarification removes many microorganisms from the water and removes turbidity and suspended solids. However, clarification alone is insufficient to completely remove pathogenic bacteria and viruses.

Earlier, we stated that one of the primary goals of water treatment is to treat raw water to the point that we can deliver to the consumer a water product that is perceptually crystal clear. The consumer does not want to drink a glass full of mud, slime, metal-colored, foul-smelling water, or even water that looks like it was dipped from the creek. Would you?

Once the water is treated to crystal clarity, the treatment process must be taken a step further—to the point that the water is entirely free of disease-causing microorganisms. To accomplish this, the final treatment process in water treatment plants occurs—*disinfection*, which destroys or inactivates pathogens.

Key Disinfection Terms[3]

Before discussing the primary disinfection methods used in treating water for human consumption, we must first define a few pertinent terms related to disinfection. We need to establish the distinction between primary and secondary disinfection.

1. **Primary Disinfection**: The initial killing of *Giardia* cysts, bacteria, and viruses.
2. **Secondary Disinfection**: The maintenance of a disinfectant residual, which prevents the regrowth of microorganisms in the water distribution system between treatment and consumer.

General Disinfection Terms

Disinfect: To inactivate virtually all recognized pathogenic microorganisms, but not necessarily all microbial life (cf. pasteurize or sterilize).

Disinfectant: (1) Any oxidant, including but not limited to chlorine, chlorine dioxide, chloramine, and ozone, added to water in any part of the treatment or distribution process intended to kill or inactivate pathogenic microorganisms. (2) A chemical or physical process that kills pathogenic organisms in water. Chlorine is often used to disinfect sewage treatment effluent, water supplies, wells, and swimming pools.

Disinfectant Time: The time water moves from one disinfectant application (or the previous point of residual disinfectant measurement) to an endpoint before or at the point where the residual disinfectant is measured.

Disinfectant Contact Time (T in CT Calculation): The time in minutes that water takes to move from the point of disinfectant application or the previous point of disinfection residual measurement to a point before or at the point where residual disinfectant concentration ("C") is measured. Where only one "C" is calculated, "T" is the time in minutes that water takes to move from the point of disinfectant application to a point before or at where residual disinfectant concentration ("C") is measured. Where more than one "C" is calculated, "T" is:

1. For the first measurement of "C," the time in minutes that water takes to move from the first or only point of disinfectant application to an endpoint before or at the point where the first "C" is measured
2. For subsequent measurements of "C," the time in minutes that water takes to move from the previous "C" measurement point to the "C" measurement point for which the particular "T" is being calculated.

Disinfection: A process that inactivates pathogenic organisms in water by chemical oxidants or equivalent agents.

Disinfectant By-Product: A compound formed by the reaction of a disinfectant, such as chlorine, with organic material in the water supply.

Waterborne Disease: Disease caused by pathogenic organisms in water.

Note: Disinfection should not be confused with sterilization.

Sterilization is the destruction of **all** microorganisms. Sterilizing potable water requires the application of a much higher dose of chemical disinfectants, which would significantly increase operating costs and create taste problems for the consumer. Excessive application of disinfectants also generates excessive levels of unwanted disinfection by-products. Because of these problems, present treatment practices are used for turbidity removal and subsequent disinfection to the extent necessary to eliminate known disease-causing organisms sufficient to protect public health.

Presence/Absence of Coliforms: Coliform bacteria in water indicate that pathogenic organisms may contaminate the water. The absence of coliform bacteria is considered sufficient evidence that pathogens are absent. A chlorine residual level is maintained if the source is good and the supply has a good history.

DID YOU KNOW?

Organic matter including microorganisms, reacts with chlorine to produce chloroorganic compounds, which show as residual chlorine and can cause taste and odor problems.

Disinfection Methods

Although chlorination is the best-known and most common disinfection method, other methods are available and can be used in various situations. The three general types of disinfection are

- *Heat treatment*—Boiling water was probably one of the first methods of disinfecting water. For small quantities of water, boiling is still a suitable emergency procedure.
- *Radiation treatment—Employs* ultraviolet (UV) radiation to disinfect water.
- *Chemical treatment—Employs* chemicals to disinfect water. Chemical disinfectants include oxidizing agents such as chlorine, ozone, bromine, iodine, and potassium permanganate; metal ions such as silver, copper, and mercury; and acids and alkalis.

Several disinfectants are available for water treatment, and we will discuss several of these in detail in subsequent sections. For now, you should understand that even though several choices are available, whichever disinfectant is chosen must meet specific criteria. More specifically, to be effective for disinfecting water (and wastewater), the disinfectant must possess specific desirable characteristics.

Desirable Characteristics of a Disinfectant:

1. Must act in a reasonable time.
2. Must act as temperature or pH changes.
3. Must be non-toxic.
4. Must not add unpleasant taste or odor.
5. Must be readily available.
6. Must be safe and easy to handle and apply.
7. Must be easy to determine the concentration of.
8. Must be able to provide residual protection.
9. Pathogenic organisms must be more sensitive to disinfectants than non-pathogens.
10. Must be capable of being applied continually.
11. The amount applied must be sufficient to produce safe water.

In addition to the desirable characteristics of a disinfectant listed above, the disinfectant chosen must be able to kill or deactivate pathogenic microorganisms by one of several possible methods, including (1) damaging the cell wall; (2) altering the ability to pass food and waste through the cell membrane; (3) altering the cell protoplasm; (4) inhibiting the cells conversion of food to energy; or (5) inhibiting reproduction.

Chlorination

For the past several decades, chlorine dispensed as either a solid (calcium hypochlorite), liquid (sodium hypochlorite), or gas (elemental chlorine, Cl_2) has been the disinfectant of choice, particularly in the United States. Chlorine (sometimes referred to as the workhorse of disinfection) has proven its worth both because of its effectiveness and because it is relatively inexpensive. It also provides a chlorine residual in the water distribution system, ensuring that the water remains disease-free.

Gaseous chlorine (Cl_2) is a greenish-yellow toxic gas 2.5 times as heavy as air. One volume of liquid chlorine confined in a container under pressure yields about 450 gas volumes. Significant water treatment works usually use chlorine gas, supplied in liquid form, in high-strength, high-pressure steel cylinders. The liquid immediately vaporizes in the form of gas when released from these pressurized containers. Chlorine gas is lethal at concentrations as low as 0.1% air by volume. It irritates the eyes, nasal membranes, and respiratory tract in non-lethal concentrations.

Sodium hypochlorite is most commonly used in smaller systems because it is simpler to use and has less extensive safety requirements than gaseous chlorine; in the form used, it is less toxic. Recently, many larger water facilities that have used chlorine for disinfection are beginning to shift (substitute) sodium hypochlorite for chlorine because of regulatory pressure.

Note: Two regulations (Occupational Safety and Health Administration's [OSHA] Process Safety Management Standard, 29 CFR 1910.119 and the USEPA's Risk Management Program, Clean Air Act Section 112(r)(7)) have come to be known in the industry as the "chlorine killers" because of their effect on industrial processes. The USEPA is attempting to steer industry away from the use of chlorine. While they cannot outlaw this substance from use, they are following the path of simply regulating it to death. At present, that is what is occurring. To avoid compliance with strict (in some cases unworkable) regulations, substitution to some other chemical product that is not regulated (at least for the moment), such as sodium hypochlorite, is being implemented by many water/wastewater facilities in the U.S.

Sodium hypochlorite provides 5%–15% available chlorine (Common laundry bleach is a 5% sodium hypochlorite solution). It is usually diluted with water before application as a disinfectant. It is very corrosive and should be handled and stored with care and kept away from equipment that can be damaged by corrosion. Sodium hypochlorite solution is more costly per pound of available chlorine and does not provide the level of chlorine gas protection (the "killing" power against pathogens).

Calcium hypochlorite is a solid in granular, powdered, or tablet form containing 65% available chlorine. In packaged form, calcium hypochlorite is more stable than sodium hypochlorite solutions, which deteriorate over time. However, calcium hypochlorite is *hygroscopic*, which means it readily absorbs moisture. It reacts slowly with moisture in the air to form chlorine gas. It is a corrosive material with a strong odor and requires proper handling. Some practical difficulty is involved in dissolving calcium hypochlorite. It must be kept from organic materials such as wood, cloth, and petroleum products. Reactions between it and organic materials can generate enough heat to cause a fire or explosion.

Chlorine Use

Whatever form of chlorine is used for disinfection (elemental chlorine, sodium hypochlorite, or calcium hypochlorite), it may be added to the incoming flow (prechlorination) to assist with the oxidation of inorganics or to arrest biological action that may produce undesirable gases in the bottom of clarifiers. More often, however, chlorine is added just before filtration to keep algae from growing at the medium surface and prevent large populations of bacteria from developing within the filter medium. Safe and practical application of chlorine requires specialized equipment, considerable care, and skill from the plant operator. Various means of feeding chlorine have been developed, but probably one of the widest-used and safest types of chlorine feed devices is called an *all-vacuum chlorinator.* Mounted directly on the chlorine cylinder, the gaseous chlorine is always under a partial vacuum in the line that carries it to the point of application. In a typical vacuum chlorine feed system, the vacuum is formed by water flowing through the ejector unit at high velocity.

Hypochlorites are usually applied to water in liquid form using positive displacement-type pumps, which deliver a specific amount of liquid on each stroke of a piston or flexible diaphragm. When added to water, chlorine reacts with various substances or impurities (e.g., organic materials, sulfides, ferrous iron, and nitrites), creating a demand for *chlorine.* Chlorine demand measures the amount of chlorine that will combine with impurities and is, therefore, available as a disinfectant.

Chlorine combines with ammonia or other nitrogen compounds to form chlorine compounds that have some disinfectant properties. These compounds are called combined *available chlorine residuals.* In the context used here, "available" means available to act as a disinfectant. The uncombined chlorine that remains in the water after the combined residual is formed is called *free available chlorine residual.* Free chlorine is a much more effective disinfectant than combined chlorine.

Factors Affecting Successful Chlorination

The factors necessary for successful chlorination are

- Concentration of free chlorine
- Contact time
- Temperature
- pH
- Turbidity

The effectiveness of chlorination is directly related to the contact time and concentration of free available chlorine. At lower chlorine concentrations, contact times must be increased. Maintaining a lower pH will also increase the effectiveness of disinfection. The higher the temperature, the faster the disinfection rate. Chlorine (or any other disinfectant, for that matter) is effective only if it comes into **contact** with the organisms to be killed. Good contact between chlorine and microorganisms is prevented whenever high turbidity levels exist. For this and aesthetic reasons, turbidity should be reduced where necessary through the coagulation and sedimentation methods previously discussed.

Chlorination By-Products

A severe disadvantage of chlorination is the potential formation of by-products. Chlorine, for example, can mix with the organic compounds in water (such as decaying vegetation) to form trihalomethanes (THMs). One THM (chloroform) is a suspected carcinogen. Other common trihalomethanes are similar to chloroform and may cause cancer.

According to estimates, about 90% of U.S. water utilities use chlorine to disinfect water. Although chlorine has virtually eliminated the risks of waterborne diseases such as typhoid fever, cholera, and dysentery, recent studies have shown risks associated with chlorine's by-products—a reason why water utilities have already looked at alternative methods for disinfecting water.

Several approaches for reducing harmful chlorination by-products have been used. For example, one approach is to remove more of the organics before any chlorination takes place. This can be accomplished (to a degree) by not chlorinating the incoming raw water before coagulation and filtration, thus reducing the formation of THMs. Aeration or adsorption on activated carbon will remove organic materials at higher concentrations or those not removed by other techniques. Another approach is to reevaluate the amount of chlorine used—the same degree of disinfection might be possible with lower chlorine dosages. Changing the point in treatment where chlorine is added is another approach commonly employed (i.e., instead of adding chlorine as chemical feed during coagulation, sedimentation, or filtration, it can instead be added after filtration). Another current approach is using alternative disinfection methods.

***Note*:** Earlier, we pointed out that because of OSHA's Process Safety Management Standard (PSM) and USEPA's Risk Management Program (RMP), many facilities currently using elemental chlorine have or are actively pursuing alternative disinfection methods. We further re-emphasize that the problem of THMs is also helping spur interest in alternatives to chlorination as the preferred method of disinfection. In the following section, we briefly discuss a few alternative disinfection methods.

Alternative Disinfection Methods

Several alternative disinfection methods are available for treating water. However, in this text (for water treatment), we focus on two alternatives: ozonation and ultraviolet (UV) radiation. These commonly used alternatives (especially in small water treatment systems) are also increasingly being substituted for existing chlorination systems at larger plants because of regulatory pressure.

***Note*:** Before beginning a brief discussion of the ozonation and UV disinfection alternatives, we must point out that neither of these alternative disinfectants is an easy solution to problems created by chlorination. While each has the advantage of not creating THMs and not being covered by the requirements under PSM/RMP, it is also true that each has uncertainties and known disadvantages that have restricted their more widespread use. In addition, ozonation and UV irradiation cannot be used as disinfectants by themselves. Both require a secondary disinfectant (usually chlorine) to maintain a residual in the distribution system.

Ozonation

Ozone (O_3), a gas at ordinary temperatures and pressures, is a very powerful disinfectant (it disinfects by molecules breaking in water)—even more effective against some viruses and cysts than chlorine. It has the added advantage of leaving no taste or odor and is unaffected by pH or the ammonia content of the water. When ozone reacts with reduced inorganic compounds and with organic material, an oxygen atom instead of a chloride atom is added to the organics, resulting in an environmentally acceptable compound. However, since ozone is unstable and cannot be stored, it must be produced on-site. Ozonation usually costs more than chlorination as well.

UV

UV (ultraviolet) light is electromagnetic radiation just beyond the blue end of the light spectrum, outside the visible light range. It has a much higher energy level than visible light, and it inactivates bacteria and viruses in large doses. UV energy is absorbed by the genetic material in microorganisms, interfering with their ability to reproduce and survive as long as the radiation contacts the microorganisms without interference from turbidity. The big advantage of UV disinfection over chlorine and ozone is that UV does not involve chemical use. Generally, UV light used for disinfecting water is generated by a series of submerged, low-pressure mercury lamps. Continuing advances in UV germicidal lamp technology make UV disinfection a more reliable and economical option for many plants.

Non-Conventional Water Treatment Technologies

Stage 1 of the USEPA's Disinfectant/Disinfection By-product Rule (D/DBP) and the new Interim Enhanced Surface Water Treatment Rule (designed to significantly lower trihalomethane by-products of chlorine disinfection in water) has driven (along with the regulatory requirements of PSM/RMP) many water (and wastewater) utilities to find and use alternative disinfection methodologies. While ozonation and ultraviolet irradiation might be suitable disinfection alternatives, switching from chlorine to chlorine dioxide (a chemical proven to form fewer THMs) might also be another viable disinfection alternative, and several others are possible. Whichever disinfection alternative is ultimately selected, remember that the selection is driven not only by regulatory requirements but also by site-specific requirements.

The disinfection issues covered to this point are essential—the overall ramifications of regulatory pressure and environmental impact cannot be overstated—but other issues besides disinfection must be considered when deciding which water treatment methodology to employ. For example, clarification by coagulation, flocculation, sedimentation, and filtration removes suspended impurities and turbidity from drinking water. Disinfection (the final step in the process) produces potable water free of harmful pathogens. Simply put, the water treatment processes discussed in the previous sections of this part of the text are sufficient to render most natural surface water (such as the river source we have used for illustration) potable. In some instances, however, the water supply may contain materials not removed by conventional water treatment processes. Other treatment processes may be required, mainly to remove many of the dissolved organic and inorganic substances: for example, groundwater with excessive dissolved solids and surface waters containing organic compounds from domestic or industrial wastewaters or naturally occurring organics such as humic acid or products of algal blooms. Processes are available for removing these contaminants.

Note: These additional water treatment processes involve sophisticated equipment, require highly skilled operators, and are expensive (Peavy et al., 1985).

Depending on the source and quality of the raw water, additional water unit treatment processes may be used in addition to clarification or applied separately. Let us take a closer look at groundwater. Does a typical groundwater source require treatment beyond conventional means?

Answer: Groundwater does not usually require processing by the unit treatment steps listed above, other than disinfection, because groundwater is filtered naturally by the soil layers from which it is withdrawn. Disinfection is only applied (in many cases) as a precautionary step required by law for public water systems. Groundwater is usually free of bacteria or other microorganisms. However, that groundwater comes into contact with soil and rock (and it all does) is a cause for concern. With such contact, groundwater may become contaminated by high dissolved minerals that must be removed.

Fluoridation

Fluoride can be beneficial when added to drinking water supplies in small concentrations (about 1.0 mg/L). In some locations, standard practice is to mix a 4% sodium fluoride solution and feed that into the water system's flow. The amount fed depends on the air temperature and the fluoride levels in the raw water.

Experience has shown that drinking water containing a proper amount of fluoride can reduce tooth decay by 65% in children. Fluoride combines chemically with tooth enamel when permanent teeth are forming. The result, of course, is teeth that are harder, stronger, and more resistant to decay. The USEPA sets the upper limits for fluoride in drinking water supplies based on ambient temperatures because people drink more water in warmer climates. Fluoride concentrations should be lower in these areas (Spellman, 2020).

Water Treatment of Organic & Inorganic Contaminants

Man-made compounds containing carbon (SOCs or Synthetic Organic Chemicals) are sometimes detected in U.S. water supplies. Some of these are volatile organic compounds (VOCs), such as the solvent trichloroethylene. The problem with VOCs in a water supply (i.e., any water supply used by the public) is twofold: they are easily absorbed through the skin and can volatilize into gases, which can be inhaled during showers, baths, or while washing dishes.

How do water supplies become contaminated by organic compounds? That is a good question. Sources of organic contaminants typically include improperly disposed wastes, pesticide use, industrial effluents, and leaking fuel oil tanks (particularly gasoline).

Water supplies may also contain inorganic contaminants, which consist of substances occurring naturally in the ground, such as sulfate, fluoride, arsenic, barium, radium, selenium, and radon. Metallic substances from industrial sources can contaminate surface waters. The inorganic nitrate (from fertilizers and feedlot runoff in agricultural areas) occurs frequently in groundwater supplies. Another source of inorganic chemical contamination in drinking water supplies is corrosion or deterioration of water supply appurtenances such as plumbing systems, which release metal and nonmetal substances into the water, such as lead, cadmium, zinc, copper, iron, and plumbing cement.

Corrosion controls and removal techniques can treat inorganic contaminants. Corrosion controls work to reduce the presence of corrosion by-products (lead, for example) at the consumer's tap. Removal technologies, including coagulation/filtration, reverse osmosis (RO), and ion exchange, treat source water contaminated with metals or radioactive substances.

The following sections discuss processes for removing inorganic and organic dissolved solids from water intended for potable use. Remember that (with some modifications) these same processes may act as tertiary treatment for wastewater.

Aeration

Aeration (air stripping) is a physical treatment process in which air is thoroughly mixed with water, a technique effective for removing dissolved gases and highly volatile odorous compounds. Contact with air and oxygen can improve water quality in several ways. For example, when aeration is the first step in processing well water, it may achieve any or all of the following: removal of hydrogen sulfide, reduction of dissolved carbon dioxide, and addition of dissolved oxygen for oxidation of iron and manganese (the oxygen in the air reacts with the iron and manganese to form an insoluble precipitate—rust). One of the most common uses of aeration is for taste and odor control. Sedimentation and filtration are then necessary to clarify the water.

Note: Aeration is rarely adequate in processing surface waters because the odor-producing substances are generally non-volatile.

Several methods to aerate the water are available. The selected method depends primarily on the type and concentration of material to be removed from the water and the available pressure. Aeration in water treatment can be accomplished using spray nozzles, cascade systems, multiple-tray aerators, diffused-air aerators, and mechanical aerators.

Oxidation

Simply stated, *oxidation* is a reaction in which a substance loses electrons, thus increasing its charge. A substance that oxidizes another is called an *oxidizing agent* or *oxidizer.* In water treatment, oxidation removes or destroys undesirable tastes or odors, removes iron and manganese, and helps improve clarification and color removal in source water.

Note: Atmospheric oxygen through aeration can oxidize the organic substances responsible for undesirable tastes and odors, but the process is usually too slow to be of value. However, if dissolved gases such as hydrogen sulfide cause taste and odor problems, aeration will remove them through oxidation and stripping.

Chlorine dioxide, potassium permanganate, and ozone are strong oxidants capable of destroying many odorous compounds. Because they do not produce THMs, these chemicals are favored over heavy chlorination.

Adsorption

When we speak of *adsorption*, we discuss primarily a surface phenomenon—the adsorption that results when one substance attaches itself to the surface of another.

The two most common adsorptive media used in water treatment are activated carbon and alumina. These adsorptive materials are generally most effective for controlling taste and odor and removing organic pollutants. However, the most essential applications of adsorption in water treatment are the removal of arsenic and organic pollutants.

Adsorption of organic materials using *activated carbon* has been a common practice in water treatment for many years. Activated carbon is manufactured from carbonaceous materials such as wood, coal, and petroleum residues. A char is made by burning the material without air and then oxidizing at higher temperatures to create a porous structure. This "activation" step provides irregular channels and pores in the solid mass, resulting in a very large surface-area-per-mass ratio. This large surface area gives activated carbon its effectiveness as an adsorbing agent. The larger the surface area of an adsorber, the greater its power. Each activated carbon contains many pores and crevices into which organic molecules enter and are adsorbed onto the activated carbon surface.

Activated carbon has a powerful attraction for organic molecules such as aromatic solvents benzene, toluene, nitrobenzene), chlorinated aromatics (PCBs, chlorobenzenes, chloronaphthalene), phenol and chlorophenols, polynuclear aromatics (acenaphthene and benzopyrenes), pesticides and herbicides chlorinated aliphatics (carbon tetrachloride, chloroalkyl ethers), and high molecular weight hydrocarbons such as dyes, gasoline, amines, and humic substances.

Two forms of activated carbon are used in water treatment: powdered and granular. Powdered activated carbon is often used for taste and odor control. Its effectiveness depends on the source of the undesirable flavors and odors. It also removes the organic precursors that react with chlorine to form harmful THM compounds after disinfection.

Powdered activated carbon is a finely ground, insoluble black powder that can be added at any point in the treatment process before the filters. It is fed to water as a dry powder or wet slurry. Although adsorption is nearly instantaneous, a contact time of 15 minutes or more is desirable before sedimentation or filtration.

Activated carbon media must periodically be replaced with new or regenerated activated carbon. Replacement cycles can vary from 1 to 3 years for taste and odor treatment to as little as 4 or 5 weeks for organics removal. The activated carbon regeneration process involves (1) removing the spent carbon as a slurry, (2) dewatering the slurry, (3) feeding the carbon into a special furnace where regeneration occurs (i.e., the organics are driven from the carbon surface by heat), and (4) returning it to use.

Activated alumina (a highly porous and granular form of aluminum oxide) is also an adsorptive medium used in water treatment. Water is percolated through a column of alumina media, and the removal of arsenic and excess fluoride ions is performed by a combination of adsorption and ion exchange.

Like the regeneration process used to restore used activated carbon to full potency, activated alumina also requires periodic regeneration, accomplished by passing a caustic soda solution through the media. Excess caustic soda is neutralized by rinsing the activated alumina with an acid. Disposal of these wash waters, laden with toxic arsenic and fluoride ions, must be done following applicable laws.

Note: Powdered activated carbon is much more challenging to regenerate than granular activated carbon.

Note: Granular activated carbon is sometimes used in the filter bed, combining filtration and adsorption in one treatment unit. The major problem associated with granular-activated-carbon systems is plugging the bed with suspended solids in the water.

Demineralization

Demineralization refers to removing dissolved solids (inorganic mineral substances) from water. Dissolved solids contain both cations and anions and, therefore, require two types of ion exchange resins. Cation exchange resins used for demineralization purposes have hydrogen exchange sites and are divided into strong acid and weak acid classes. The anion exchange resin commonly contains hydroxide ions and is divided into strong and weak base classes.

Demineralization is commonly used in waste treatment to remove arsenic, barium, cadmium, chromium, fluoride, sulfate, and zinc. Some general advantages of using ion exchange to remove these contaminants are the low capital investment required and the mechanical simplicity of the process. Additionally, the ion exchange process can recover valuable chemicals for reuse or harmful ones for disposal. For example, it is often used to recover chromic acid from metal finishing waste for reuse in chrome-plating baths. It also has some application in the removal of radioactivity. The major disadvantages are the high chemical requirements needed to regenerate the resins and to dispose of chemical wastes from the regeneration process. These factors make ion exchange more suitable for small systems than for large ones.

Membrane Processes

Membrane processes used in water treatment are primarily demineralization processes. Water demineralization can be accomplished using thin, microporous membranes. Electrodialysis and reverse osmosis are the most common membrane processes.

Before discussing these two membrane processes, you need a basic understanding of osmosis.

During osmosis, two solutions containing different concentrations of minerals are separated by a semipermeable membrane. Water tends to migrate through the membrane from the more dilute solution to the side of the more concentrated solution. This osmosis continues until the build-up of hydrostatic pressure on the more concentrated solution is sufficient to stop the net flow.

In *reverse osmosis*, water flow through the semipermeable membrane is reversed by applying external pressure to offset the hydrostatic pressure. This results in a concentration of minerals on one side of the membrane and pure water on the other. Reverse osmosis can treat various health and aesthetic contaminants in water. Effectively designed, reverse osmosis equipment can treat aesthetic contaminants that cause unpleasant taste, color, and odor problems like a salty or soda taste caused by chlorides or sulfates. Reverse osmosis can also effectively treat arsenic, asbestos, atrazine, fluoride, lead, mercury, nitrate, and radium. When used with appropriate carbon prefiltering, additional treatment can also be provided for such "volatile" contaminants as benzene, trichloroethylene, trihalomethanes, and radon. Some reverse osmosis equipment is also capable of treating *Cryptosporidium*. Reverse osmosis can be expected to play a significant role in water treatment for years.

Reverse osmosis (also called ultrafiltration) is the most common process for reducing the salinity of brackish groundwater. A semipermeable membrane (the most essential element in the reverse osmosis demineralization method) separates salty water into two concentrations. Concentrations naturally become equalized by a flow of water from the dilute side to the concentrated side (osmosis). However, high pressure applied to the membrane's high-concentration side can reverse this flow direction. Freshwater diffuses through the membrane, leaving a more concentrated salt solution behind. The performance of reverse osmosis units highly depends on several water quality parameters. Suspended solids, dissolved organics, hydrogen sulfide, iron, and strong oxidizing agents (chlorine, ozone, and permanganate) harm membranes.

Electrodialysis is the demineralization of water using the principles of osmosis—but it uses ion-selective membranes and an electric field to separate anions and cations in solution. In the past, electrodialysis was most often used for purifying salty water, but it is now finding a role in industrial waste treatment. For example, metal salts from plating rinses are sometimes removed in this way.

AI APPLICATIONS

AI is currently finding its way into several water supply and treatment uses. In other cases, the application of AI in water supplies and treatment is at the exploratory pilot study stage. Where AI has been installed, it is making a significant contribution. One of the applications involves leak detection. AI algorithms analyze sensor data and monitor fluctuations in water pressure to detect leaks in water distribution networks. The benefit of this application is that it helps utility managers and water operators promptly identify and fix leaks. At the water facilities I visited all over the USA, I have been told that the AI leak detection service saves time, effort, and money.

AI assesses water quality in real time by analyzing data from remote sensors. Water purveyors' primary concern is ensuring that water is the safest. This technology also allows authorities to detect pollution events, safeguard public health, and take timely action to protect water resources.

In agriculture, farmers use AI-powered methods to optimize irrigation practices. These methods analyze data from sensors, weather forecasts, and soil moisture levels to determine the optimal amount of water needed for crops; this practice reduces water wastage and can increase crop yields.

I first became involved with AI at water treatment works, pumping stations, and water intake locations when I conducted a pre-EPA/OSHA inspection at a large waterworks. The EPA is concerned with protecting water supplies, and OSHA is concerned with preventing on-the-job hazards, including drinking unsafe water. These agencies are also concerned about cyber threats that can interfere with (or worse, compromise) water infrastructure and ensure the resilience of water systems.

AI-ASSISTED SCADA

In Queensland, Australia, on April 23, 2000, police stopped a car on the road and found a stolen computer and radio inside. Using commercially available technology, a disgruntled former employee had turned his vehicle into a pirate command center of sewage treatment along Australia's Sunshine Coast.

The former employee's arrest solved a mystery that had troubled the Maroochy Shire wastewater system for two months. Somehow, the system leaked hundreds of thousands of gallons of putrid sewage into parks, rivers, and the manicured grounds of a Hyatt Regency hotel—marine life died, the creek water turned black, and the stench was unbearable for residents. Until the former employee's capture—during his 46th successful intrusion—the utility's managers did not know why.

Specialists study this cyber-terrorism case because it is the only known one in which someone deliberately used a digital control system to cause harm. The former employee's intrusion shows how easy it is to break in—and how restrained he was with his power.

To sabotage the system, the former employee set the software on his laptop to identify itself as a pumping station and then suppressed all alarms. The former employee was the "central control station" during his intrusions, with unlimited command of 300 SCADA nodes governing sewage and drinking water.

–Gellman (2002)

The bottom line: As serious as the former employee's intrusions were, they pale compared to what he could have done to the freshwater system—he could have done anything he liked.

In 2000, the Federal Bureau of Investigation (FBI) identified and listed threats to critical infrastructure. These threats are listed and described in Table 5.4. In the past few years, especially since 9/11, it has been routine for us to pick up a newspaper or magazine or view a television news program where a significant topic of discussion is cybersecurity or the lack thereof. Many cyber intrusion incidents we read or hear about have added new terms or uses for old terms to our vocabulary. For example, old terms such as Trojan Horse, worms, and viruses have brought new connotations regarding cyber security issues. Relatively new terms such as scanners, Windows NT hacking tools, ICQ hacking tools, mail bombs, sniffers, logic bombs, nukers, dots, backdoor Trojans, keyloggers, hackers' Swiss knives, password crackers, and BIOS crackers are now commonly encountered.

Not all relatively new and universally recognizable cyber terms have sinister connotations or meanings. Consider, for example, the following digital terms: backup, binary, bit byte, CD-ROM, CPU, database, e-mail, HTML, icon, memory, cyberspace, modem, monitor, network, RAM, Wi-Fi (wireless fidelity), record, software, World Wide Web—none of these terms typically generate thoughts of terrorism in most of us. There is, however, one digital term, SCADA, that most people have not heard of. This is not the case for those who work with the Nation's critical infrastructure, including water/wastewater. SCADA, or **S**upervisory **C**ontrol **A**nd **D**ata **A**cquisition System (sometimes called Digital Control Systems or Process Control Systems), plays a vital role in computer-based control systems. Many water/wastewater systems use computer-based systems to control sensitive processes and system equipment previously controlled manually and remotely. These systems (SCADA)

TABLE 5.4
Threats to Critical Infrastructure Observed by the FBI

Threat	Description
Criminal groups	There is an increased use of cyber intrusions by criminal groups who attack systems for monetary gain
	Foreign intelligence services use cyber tools in their information-gathering and espionage activities
Hackers	Hackers sometimes crack into networks for the thrill of the challenge or bragging rights in the hacker community. While remote cracking once required a fair amount of skill or computer knowledge, hackers can now download attack scripts and protocols from the Internet and launch them against victim sites. Thus, while attack tools have become more sophisticated, they have also become more accessible.
Hacktivists	Hacktivism refers to politically motivated attacks on publicly accessible web pages or e-mail servers. Groups and individuals overload e-mail servers and hack into websites to send a political message.
Information warfare	Several nations are aggressively working to develop information warfare doctrine, programs, and capabilities. Such capabilities enable a single entity to have a significant and severe impact by disrupting the supply, communications, and economic infrastructures that support military power—impacts that, according to the Director of Central Intelligence, can affect the daily lives of Americans across the country.
Inside threat	The insiders of disgruntled organizations are a principal source of computer crimes. Although they may not need much knowledge about computer intrusions, their knowledge of a victim system often allows them to gain unrestricted access to cause damage to the system or steal system data. The insider threat also includes outsourcing vendors.
Virus writers	This is an area of increasing concern because virus writers are posing an increasingly severe threat. Several destructive computer viruses and "worms" have harmed files and hard drives, including the Melissa Macro Virus, the Explore.Zip worm, the CIH (Chernobyl) Virus, Nimda, and Code Red.

Source: FBI (2000).

allow a water/wastewater utility to collect data from sensors and control equipment at remote sites. Standard water/wastewater system sensors measure fluid level, temperature, pressure, water purity, clarity, and pipeline flow rates. Standard water/wastewater system equipment includes valves, pumps, and mixers for mixing chemicals in the water supply.

What Is SCADA?

Utilizing efficient and precise water monitoring systems is critical to providing communities with clean, affordable water. Older waterworks water monitoring systems required significant staff, time, and resources to keep operations online. It is now expected to see digital networks that monitor each step of the treatment process. SCADA is a computer-based system that remotely controls processes previously controlled manually. SCADA allows an operator using a central computer to supervise (control and monitor) multiple networked computers at remote locations. Each remote computer can control mechanical processes (pumps, valves, etc.) and collect data from sensors at its remote location. Thus, the phrase: Supervisory Control and Data Acquisition of **SCADA**. The central computer is called the Master Terminal Unit, or MTU. The operator interfaces with the MTU using software called Human Machine Interface, or HMI. The remote computer is called the Program Logic Controller (PLC) or Remote Terminal Unit (RTU). The RTU activates a relay (or switch) that turns mechanical equipment "on" and "off." The RTU also collects data from sensors.

Initially, utilities ran wires, also known as hardwires or landlines, from the central computer (MTU) to the remote computers (RTUs). Because remote locations can be located hundreds of miles from the central area, utilities began to use public phone lines and modems, leased telephone company lines, and radio and microwave communication. More recently, they have also started to use satellite links, the Internet, and newly developed wireless technologies.

Because SCADA systems' sensors provided valuable information, many utilities established "connections" between their SCADA systems and their business systems. This allowed utility management and other staff access to useful statistics, such as water usage. When utilities later connected their systems to the Internet, they were able to provide stakeholders with water statistics on the utility web pages.

AI-Assisted SCADA Applications in Water Systems

The future of SCADA lies in integrating AI, enabling intelligent automation and monitoring.

As stated above, SCADA systems can be designed to measure a variety of equipment operating conditions and parameters or volumes and flow rates or water quality parameters and to respond to changes in those parameters either by alerting operators or by modifying system operation through a feedback loop system without having personnel physically visit each process or piece of equipment daily to check it and ensure it is functioning correctly. SCADA systems can also be used to automate certain functions so that they can be performed without the need to be initiated by an operator (e.g., injecting chlorine in response to periodic low chlorine levels in a distribution system or turning on a pump in response to low water levels in a storage tank). As described above, in addition to process equipment, SCADA systems can integrate specific security alarms and equipment, such as cameras, motion sensors, lights, and data from card reading systems, providing a clear picture of what is happening in areas throughout a facility. Finally, SCADA systems provide constant, real-time data on processes, equipment, location access, etc., enabling the necessary response to be made quickly. This can be extremely useful during emergencies, such as when distribution mains break or when potentially disruptive BOD spikes appear in wastewater influent.

Because these systems can monitor multiple processes, equipment, and infrastructure and quickly notify or respond to problems or upsets, SCADA systems typically provide the first detection line for atypical or abnormal conditions. For example, a SCADA system connected to sensors that measure specific water quality parameters is measured outside a particular range. A real-time customized operator interface screen could display and control critical systems monitoring parameters.

The system could transmit warning signals back to the operators by initiating a call to a personal pager. This might allow the operators to initiate actions to prevent water supply contamination and disruption. Further automation of the system could ensure that the system initiates measures to rectify the problem. Preprogrammed control functions (e.g., shutting a valve, controlling flow, increasing chlorination, or adding other chemicals) can be triggered and operated based on SCADA utility.

Benefits of AI-Assisted SCADA

The integration of AI into SCADA systems offers numerous benefits, including:

- Real-time monitoring and control
- Enhanced efficiency and productivity
- Improved decision-making
- Predictive maintenance

SCADA Vulnerabilities

According to the USEPA (2005), SCADA networks were developed with little attention paid to security, making the security of these systems often weak. Studies have found that, while technological advancements introduced vulnerabilities, many water/wastewater utilities need more time to secure their SCADA networks. As a result, many SCADA networks may be susceptible to attacks and misuse. Remote monitoring and supervisory control of processes developed in the early 1960s and adopted many technological advancements. The advent of minicomputers enabled the automation of many once-manually operated switches. Advancements in radio technology reduced the communication costs associated with installing and maintaining buried cables in remote areas. SCADA systems continued to adopt new communication methods, including satellite and cellular. As the price of computers and communications dropped, it became economically feasible to distribute operations and expand SCADA networks to include even smaller facilities.

Advances in information technology and the necessity of improved efficiency have resulted in increasingly automated and interlinked infrastructures, creating new vulnerabilities due to equipment failure, human error, weather and other natural causes, and physical and cyber-attacks. Some areas and examples of possible SCADA vulnerabilities include:

- **Human**: People can be tricked, corrupted, or may commit errors.
- **Communications**: Messages can be fabricated, intercepted, changed, deleted, or blocked.
- **Hardware**: Security features are not easily adapted to small self-contained units with limited power supplies.
- **Physical**: Intruders can break into a facility to steal or damage SCADA equipment.

- **Natural**: Tornadoes, floods, earthquakes, and other natural disasters can damage equipment and connections.
- **Software**: Programs can be poorly written.

A survey found that many water utilities needed more to secure their SCADA network vulnerabilities (Ezell, 1998). For example, many respondents reported having remote access, which can allow an unauthorized person to access the system without being physically present. Over 60% of the respondents believed their systems were unsafe from unauthorized access and use. Twenty percent of the respondents even reported known attempts, successful unauthorized access, or use of their system. Yet 22 of 43 respondents reported spending little time ensuring their network is safe, and 18 of 43 respondents reported spending less than 10% of their time ensuring network safety.

SCADA system computers and their connections are susceptible to different types of information system attacks and misuse, such as system penetration and unauthorized access to information. The Computer Security Institute and Federal Bureau of Investigation conduct an annual Computer Crime and Security Survey (FBI, 2004). The survey reported ten types of attacks or misuse and found that viruses and denial of service had the most significant negative economic impact. The same study also found that 15% of the respondents reported abuse of wireless networks, which can be a SCADA component. On average, respondents from all sectors did not believe their organization invested enough in security awareness. Utilities reported a lower average computer security expenditure/investment per employee than other sectors, such as transportation, telecommunications, and finance.

Sandia National Laboratories' *Common Vulnerabilities in Critical Infrastructure Control Systems* described some of the common problems it identified in the following five categories (Stamp et al., 2003):

1. **System Data**: Important data attributes for security include availability, authenticity, integrity, and confidentiality. Data should be categorized according to sensitivity, and ownership and responsibility must be assigned. However, SCADA data is often not classified, making it challenging to identify where security precautions are appropriate.
2. **Security Administration**: Vulnerabilities emerge because many systems lack a properly structured security policy, equipment and system implementation guides, configuration management, training, and enforcement and compliance auditing.
3. **Architecture**: Many standard practices negatively affect SCADA security. For example, while it is convenient to use SCADA capabilities for other purposes, such as fire and security systems, these practices create single points of failure. Also, connecting SCADA networks to other automation and business networks introduces multiple entry points for potential adversaries.
4. **Network** (including communication links): Legacy systems' hardware and software have minimal security capabilities, and the vulnerabilities of contemporary systems (based on modern information technology) are publicized. Wireless and shared links are susceptible to eavesdropping and data manipulation.

5. **Platforms**: Many platform vulnerabilities exist, including retained default configurations, poor password practices, shared accounts, inadequate protection for hardware, and nonexistent security monitoring controls. In most cases, important security patches are not installed, often due to concern about negatively impacting system operation; in some cases, technicians are contractually forbidden from updating systems by vendor agreements.

   The following incident helps to illustrate some of the risks associated with SCADA vulnerabilities.

   - During a vulnerability assessment, a contractor stated that personnel from his company penetrated the utility's information system within minutes. Contractor personnel drove to a remote substation and noticed a wireless network antenna. Without leaving their vehicle, they plugged in their wireless radios and connected to the network within 5 minutes. Within 20 minutes, they had mapped the network, including SCADA equipment, and accessed the business network and data.

This illustrates what a cybersecurity advisor from Sandia National Laboratories specializing in SCADA stated: utilities are moving to wireless communication without understanding the added risks.

The Increasing Risk

According to GAO (2003), security concerns about control systems (SCADA included) were historically related to protecting against physical attacks and the misusing of refining and processing sites or distribution and holding facilities. However, more recently, there has been a growing recognition that control systems are now vulnerable to cyberattacks from numerous sources, including hostile governments, terrorist groups, disgruntled employees, and other malicious intruders. In addition to control system vulnerabilities mentioned earlier, several factors have contributed to the escalation of risk to control systems, including (1) the adoption of standardized technologies with known vulnerabilities, (2) the connectivity of control systems to other networks, (3) constraints on the implementation of existing security technologies and practices, (4) insecure remote connections, and (5) the widespread availability of technical information about control systems.

Adoption of Technologies with Known Vulnerabilities

When a technology is not well-known, widely used, understood, or publicized, it is difficult to penetrate it and thus disable it. Historically, proprietary hardware, software, and network protocols made understanding how control systems operated and how to hack into them difficult. Today, however, to reduce costs and improve performance, organizations have been transitioning from proprietary systems to less expensive, standardized technologies such as Microsoft's Windows and Unix-like operating systems and the standard networking protocols used by the Internet. These widely used standardized technologies have commonly known vulnerabilities, and sophisticated and effective exploitation tools are widely available and relatively easy to use. Consequently, the number of people with the knowledge to wage attacks and the number of systems subject to attack has increased. Also, standard communication protocols and the emerging use of Extensible Markup Language (commonly referred to as XML) can make it easier for a hacker to interpret the content of communications among the components of a control system.

Control systems are often connected to other networks—enterprises usually integrate their control systems with their networks. This increased connectivity has significant advantages, including providing decision-makers access to real-time information and allowing engineers to monitor and control the process control system from different points on the enterprise network. In addition, enterprise networks are often connected to the networks of strategic partners and the Internet. Further, control systems increasingly use wide area networks and the Internet to transmit data to their remote or local stations and individual devices. This convergence of control networks with public and enterprise networks exposes the control systems to additional security vulnerabilities. Unless appropriate security controls are deployed in the enterprise and control system networks, enterprise security breaches can affect the control system's operation. According to industry experts, existing security technologies, as well as substantial user authentication and patch management practices, are generally not implemented in control systems because control systems operate in real-time, typically are not designed with cybersecurity in mind, and usually have limited processing capabilities.

Existing security technologies such as authorization, authentication, encryption, intrusion detection, and filtering of network traffic and communications require more bandwidth, processing power, and memory than control system components typically have. Because controller stations are generally designed to do specific tasks, they use low-cost, resource-constrained microprocessors. Some devices in the electrical industry still use the Intel 8088 processor, introduced in 1978. Consequently, installing existing security technologies without seriously degrading the control system's performance is difficult.

Further, complex passwords and other strong password practices are not always used to prevent unauthorized access to control systems, in part because this could hinder a rapid response to safety procedures during an emergency. As a result, according to experts, weak passwords that are easy to guess, shared, and infrequently changed are reportedly common in control systems, including the use of default passwords or even no password at all.

In addition, although modern control systems are based on standard operating systems, they are typically customized to support control system applications. Consequently, vendor-provided software patches are generally incompatible or cannot be implemented without compromising service, shutting down "always-on" systems, or affecting interdependent operations.

Insecure connections exacerbate potential vulnerabilities in control systems. Organizations often leave access links—such as dial-up modems to equipment and control information—open for remote diagnostics, maintenance, and examination of system status. Such links may not be protected with authentication or encryption, which increases the risk that hackers could use these insecure connections to break into remotely controlled systems. Also, control systems often use wireless communications systems, especially those vulnerable to attack, or leased lines that pass through commercial telecommunications facilities. Without encryption to protect data as it flows through these insecure connections or authentication mechanisms to limit access, there is limited protection for the integrity of the information being transmitted.

Public information about infrastructures and control systems is available to potential hackers and intruders. A university graduate student demonstrated the availability of this infrastructure and vulnerability data, whose dissertation reportedly mapped every business and industrial sector in the American economy to the fiber-optic network that connects them—using material available publicly on the Internet, none of which was classified. Many electric utility officials interviewed for the National Security Telecommunications Advisory Committee's Information Assurance Task Force's Electric Power Risk Assessment expressed concern over the information about their infrastructure readily available to the public.

In the electric power industry, open sources of information—such as product data and educational videotapes from engineering associations—can be used to understand the basics of the electrical grid. Other publicly available information—including filings of the Federal Energy Regulatory Commission (FERC), industry publications, maps, and material available on the Internet—is sufficient to allow someone to identify the most heavily loaded transmission lines and the most critical substations in the power grid.

In addition, significant information on control systems is publicly available—including design and maintenance documents, technical standards for the interconnection of control systems and RTUs, and standards for communication among control devices—all of which could assist hackers in understanding the systems and how to attack them. Moreover, there are numerous former employees, vendors, support contractors, and other end users of the same equipment worldwide with inside knowledge of the operation of control systems.

Cyber Threats to Control Systems

There is a consensus—and increasing concern—among government officials and experts on control systems about potential cyber threats to the control systems that govern our critical infrastructures. As components of control systems increasingly make crucial decisions that humans once made, the possible effect of a cyber threat becomes more devastating. Such cyber threats could come from numerous sources, ranging from hostile governments and terrorist groups to disgruntled employees and malicious intruders. Based on interviews and discussions with representatives throughout the electric power industry, the Information Assurance Task Force of the National Security Telecommunications Advisory Committee concluded that an organization with sufficient resources, such as a foreign intelligence service or a well-supported terrorist group, could conduct a structured attack on the electric power grid electronically, with a high degree of anonymity and without having to set foot in the target nation.

In July 2002, the National Infrastructure Protection Center (NIPC) reported that the potential for compound cyber and physical attacks, referred to as "swarming attacks," is an emerging threat to the U.S. critical infrastructure. As NIPC reports, the effects of a swarming attack include slowing or complicating the response to a physical attack. For instance, a cyberattack that disabled the water supply or the electrical system in conjunction with a physical attack could deny emergency services the necessary resources to manage the consequences—such as controlling fires, coordinating actions, and generating light.

Control systems, such as SCADA, can be vulnerable to cyberattacks. Entities or individuals with malicious intent might take one or more of the following actions to attack control systems successfully:

- Disrupt the operation of control systems by delaying or blocking the flow of information through control networks, thereby denying the networks' availability for control system operations.
- Make unauthorized changes to programmed instructions in PLCs, RTUs, or DCS controllers; change alarm thresholds; or issue unauthorized commands to control equipment, which could potentially result in damage to equipment (if tolerances are exceeded), premature shutdown of processes (such as prematurely shutting down transmission lines), or even disabling control equipment.
- Send false information to control system operators either to disguise unauthorized changes or to initiate inappropriate actions by system operators.
- Modify the control system software, producing unpredictable results.
- Interfere with the operation of safety systems.

In addition, remote sites often need to be staffed and may not be physically monitored in control systems that cover a wide geographic area. If such remote systems are physically breached, the attackers could establish a cyber connection to the control network.

Securing Control Systems

Several challenges must be addressed to secure control systems against cyber threats effectively. These challenges include (1) the limitations of current security technologies in securing control systems, (2) the perception that securing control systems may not be economically justifiable, and (3) the conflicting priorities within organizations regarding the security of control systems. A significant challenge in effectively securing control systems is the need for specialized security technologies. The computing resources needed to perform security functions in control systems tend to be limited, making it very difficult to use security technologies within control system networks without severely hindering performance. Securing control systems may not be perceived as economically justifiable. Experts and industry representatives have indicated that organizations may be reluctant to spend more money to secure control systems. Hardening the security of control systems would require industries to expend more resources, including acquiring more personnel, providing training for personnel, and potentially prematurely replacing current systems that typically have a lifespan of about 20 years. Finally, several experts and industry representatives indicated that the responsibility for securing control systems normally includes two separate groups: IT security personnel and control system engineers and operators. IT security personnel tend to focus on securing enterprise systems, while control system engineers and operators tend to be more concerned with the reliable performance of their control systems. Further, they indicate that, as a result, those two groups do not always fully understand each other's requirements and collaborate to implement secure control systems.

Steps to Improve SCADA Security

The President's Critical Infrastructure Protection Board and the Department of Energy (DOE) have developed the steps outlined below to help organizations improve the security of their SCADA networks. DOE (2001) points out that these steps are not meant to be prescriptive or all-inclusive. However, they do address essential actions to improve the protection of SCADA networks. The steps are divided into two categories: specific actions to improve implementation and actions to establish essential underlying management processes and policies (DOE, 2001).

21 Steps to Increase SCADA Security

The following steps focus on specific actions to be taken to increase the security of SCADA networks:

1. **Identify all connections to SCADA networks.**
   Conduct a thorough risk analysis to assess the risk and necessity of each connection to the SCADA network. Develop a comprehensive understanding of all connections to the SCADA network and how well those connections are protected. Identify and evaluate the following types of connections:
   - Internal local area and wide area networks, including business networks
   - The Internet
   - Wireless network devices, including satellite uplinks
   - Modem or dial-up connections
   - Connections to business partners, vendors, or regulatory agencies
2. **Disconnect unnecessary connections to the SCADA network.**
   To ensure the highest degree of security for SCADA systems, isolate the SCADA network from other network connections as much as possible. Any connection to another network introduces security risks, especially if the connection creates a pathway to or from the Internet. Although direct connections with other networks may allow vital information to be passed efficiently and conveniently, insecure connections are not worth the risk. Isolation of the SCADA network must be a primary goal to provide the needed protection. Strategies like "demilitarized zones" (DMZs) and data warehousing can facilitate secure data transfer from the SCADA network to business networks. However, they must be designed and implemented correctly to avoid introducing additional risk through improper configuration.

3. **Evaluate and strengthen the security of any remaining connections to the SCADA networks.**

   Conduct penetration testing or vulnerability analysis of any remaining connections to the SCADA network to evaluate the protection posture associated with these pathways. Use this information with risk management processes to develop a robust protection strategy for any paths to the SCADA network. Since the SCADA network is only as secure as its weakest connecting point, it is essential to implement firewalls, intrusion detection systems (IDSs), and other appropriate security measures at each entry point. Configure firewall rules to prohibit access from and to the SCADA network and be specific when permitting approved connections. For example, an Independent System Operator (ISO) should not be granted "blanket" network access simply because there is a need for a connection to specific components of the SCADA system. Strategically place IDSs at each entry point to alert security personnel of potential breaches of network security. Organizational management must understand and accept responsibility for risks associated with any connection to the SCADA network.
4. **Harden SCADA networks by removing or disabling unnecessary services.**

   SCADA control servers built on commercial or open-source operating systems can be exposed to attack through default network services. To the greatest extent possible, remove or disable unused services and network daemons to reduce the risk of direct attack. This is particularly important when SCADA networks are interconnected with other networks. Do not permit a service or feature on a SCADA network unless a thorough risk assessment of the consequences of allowing the service/feature shows that the benefits of the service/feature far outweigh the potential for vulnerability exploitation. Examples of services to remove from SCADA networks include automated meter reading/remote billing systems, email services, and Internet access. An example of a feature to disable is remote maintenance—numerous secure configurations such as the National Security Agency's series of security guides. Additionally, work closely with SCADA vendors to identify secure configurations and coordinate any changes to operational systems to ensure that removing or disabling services does not cause downtime, interruption of service, or loss of support.
5. **Do not rely on proprietary protocols to protect your system.**

   Some SCADA systems use unique, proprietary protocols for communication between field devices and servers. The security of SCADA systems is often based solely on the secrecy of these protocols. Unfortunately, obscure protocols provide very little "real" security. Do not rely on proprietary protocols or factor default configuration settings to protect your system. Additionally, demand that vendors disclose any backdoors or vendor interfaces to your SCADA systems and expect them to provide systems that can be secured.
6. **Implement the security features provided by device and system vendors.**

   Older SCADA systems (most systems in use) have no security features; SCADA system owners must insist that their system vendor implement security features in product patches or upgrades. Some newer SCADA devices are shipped with basic security features, usually disabled, to ensure ease of installation.

Analyze each SCADA device to determine whether security features are present. Additionally, factory default security settings (such as computer network firewalls) are often set to provide maximum usability but minimal security. Set all security features to provide maximum security only after a thorough risk assessment of the consequences of reducing the security level.

7. **Establish strong controls over any medium used as a backdoor into the SCADA network.**

   Strong authentication must be implemented to ensure secure communications where backdoors or vendor connections exist in SCADA systems. Modems, wireless, and wired networks used for communications and maintenance represent significant vulnerabilities to the SCADA network and remote sites. Successful "war dialing" or "war driving" attacks could allow attackers to bypass all other controls and directly access the SCADA network or resources. Disable inbound access and replace it with a callback system to minimize the risk of such attacks.

8. **Implement internal and external intrusion detection systems and establish 24-hour-a-day incident monitoring.**

   To respond to cyber-attacks effectively, establish an intrusion detection strategy that includes alerting network administrators of malicious network activity from internal or external sources. Intrusion detection system monitoring is essential 24 hours a day; this capability can be easily set up through a pager. Additionally, incident response procedures must be in place to allow an effective response to any attack. To complement network monitoring, enable daily logging on all systems and audit system logs to detect suspicious activity as soon as possible.

9. **Perform technical audits of SCADA devices, networks, and other connected networks to identify security concerns.**

   Technical audits of SCADA devices and networks are critical to ongoing security effectiveness. Many commercial and open-source security tools allow system administrators to conduct audits of their systems/networks to identify active services, patch levels, and common vulnerabilities. These tools will not solve systemic problems but eliminate the "paths of least resistance" that an attacker could exploit. Analyze identified vulnerabilities to determine their significance and take corrective actions as appropriate. Track corrective actions and analyze this information to identify trends. Additionally, retest systems after corrective actions are taken to ensure that vulnerabilities are eliminated. Scan non-production environments actively to identify and address potential problems.

10. **Conduct physical security surveys and assess the security of all remote sites connected to the SCADA network.**

    Any location connected to the SCADA network is a target, especially unmanned or unguarded remote sites. Conduct a physical security survey and inventory access points at each facility connected to the SCADA system. Identify and assess any sources of information, including remote telephone/computer network/fiber optic cables that could be tapped; radio and microwave links that are exploitable; computer terminals that could be accessed; and wireless local area network access points. Identify and eliminate single points of failure. The site's security must be adequate to detect or prevent unauthorized access. Do not allow "live" network access points at remote, unguarded sites for convenience.

11. **Establish SCADA "Red Teams" to identify and evaluate possible attack scenarios.**

    Establish a "Red Team" to identify potential attack scenarios and evaluate system vulnerabilities. Use a variety of people who can provide insight into weaknesses of the overall network, SCADA system, physical systems, and security controls. People who work on the system daily have great insight into the vulnerabilities of your SCADA network and should be consulted when identifying potential attack scenarios and possible consequences. Also, ensure that the risk from a malicious insider is thoroughly evaluated, given that this represents one of the greatest threats to an organization. Feed information from the "Red Team" evaluation into risk management processes to assess the information and establish appropriate protection strategies.

    The following steps focus on management actions to establish an effective cybersecurity program:

12. **Clearly define cybersecurity roles, responsibilities, and authorities for managers, system administrators, and users.**

    Organization personnel must understand the specific expectations of protecting information technology resources by defining clear and logical roles and responsibilities. In addition, key personnel must be given sufficient authority to carry out their duties. Too often, good cybersecurity is left up to the individual's initiative, leading to consistent implementations and adequate protection. Establish a cybersecurity organizational structure that defines roles and responsibilities, identifies how cybersecurity issues are escalated, and specifies who is notified in an emergency.

13. **Document network architecture and identify systems that serve critical functions or contain sensitive information requiring additional protection.**

    Develop and document robust information security architecture as part of a process to establish an effective protection strategy. Organizations must design their network with security in mind and continue to have a strong understanding of their network architecture throughout its lifecycle. Of particular importance is an in-depth knowledge of the functions that the systems perform and the sensitivity of the stored information. Risk cannot be adequately assessed without this understanding, and protection strategies may not be sufficient. Documenting the information security architecture and its components is critical to understanding the overall protection strategy and identifying single points of failure.

14. **Establish a rigorous, ongoing risk management process.**

    A thorough understanding of the risks to network computing resources from denial-of-service attacks and the vulnerability of sensitive information to compromise is essential to an effective cyber security program. Risk assessments from the technical basis of this understanding are critical to formulating effective strategies to mitigate vulnerabilities and preserve the integrity of computing resources. Initially, perform a baseline risk analysis based on the current threat assessment for developing a network protection strategy. Due to rapidly changing technology and the emergence of new threats daily, an ongoing risk assessment process is also needed so that routine changes can be made to the protection strategy to ensure it remains effective. Fundamental to risk management is identifying residual risk with a network protection strategy and the acceptance of that risk by management.

15. **Establish a network protection strategy based on the principle of defense-in-depth.**

    Defense-in-depth is a fundamental principle that must be part of any network protection strategy. Defense-in-depth must be considered early in the design phase of the development process and must be an integral consideration in all technical decision-making associated with the network. Utilize technical and administrative controls to mitigate threats from identified risks to as great a degree as possible at all levels of the network. Single points of failure must be avoided, and cyber security defense must be layered to limit and contain the impact of any security incidents. Additionally, each layer must be protected against other systems at the same level. For example, to protect against the inside threat, restrict users to access only those resources necessary to perform their job functions.

16. **Identify cybersecurity requirements.**

    Organizations and companies need structured security programs with mandated requirements to establish expectations and allow personnel to be held accountable. Formalized policies and procedures are typically used to establish and institutionalize a cybersecurity program. A formal program is essential for establishing a consistent, standards-based approach to cybersecurity throughout an organization and eliminates sole dependence on individual initiative. Policies and procedures also inform employees of their specific cybersecurity responsibilities and the consequences of failing to meet them. They also provide guidance regarding actions to be taken during a cybersecurity incident and promote efficient and effective actions during a crisis. To identify cybersecurity requirements, include user agreements and notification and warning banners. Establish requirements to minimize the threat from malicious insiders, including the need for conducting background checks and limiting network privileges to those necessary.

17. **Establish effective configuration management processes.**

    Configuration management is a fundamental management process needed to maintain a secure network. It needs to cover both hardware and software configurations. Changes to hardware or software can quickly introduce vulnerabilities that undermine network security. Processes must evaluate and control any change to ensure the network remains secure. Configuration management begins with well-tested and documented security baselines for your various systems.

18. **Conduct routine self-assessments.**

    Robust performance evaluation processes are needed to provide organizations with feedback on the effectiveness of cybersecurity policy and technical implementation. A sign of a mature organization is the ability to self-identify issues, conduct root-cause analyses, and implement practical corrective actions that address individual and systemic problems. Self-assessment processes usually part of an effective cybersecurity program, include routine scanning for vulnerabilities, automated auditing of the network, and self-assessments of organizational and individual performance.

19. **Establish system backups and disaster recovery plans**.

    Establish a disaster recovery plan that allows rapid recovery from any emergency (including a cyber-attack). System backups are essential to any plan and allow rapid network reconstruction. Routinely exercise disaster recovery plans to ensure that they work and that personnel are familiar with them. Based on lessons learned from exercises, make appropriate changes to disaster recovery plans.

20. **Senior organizational leadership should establish expectations for cyber security performance and hold individuals accountable for their performance.**

    Effective cybersecurity performance requires commitment and leadership from senior managers. Senior management must establish an expectation for robust cybersecurity and communicate this to subordinate managers in the organization. It is also essential that senior organizational leadership establish a structure for implementing a cybersecurity program. This structure will promote consistent implementation and the ability to sustain a robust cybersecurity program. It is then essential for individuals to be held accountable for their cybersecurity performance. This includes managers, system administrators, technicians, and users/operators.

21. **Establish policies and conduct training to minimize the likelihood that organizational personnel will inadvertently disclose sensitive information regarding SCADA system design, operations, or security controls.**

    Release data related to the SCADA network on a strict, need-to-know basis and only to persons authorized to receive such information. "Social engineering," gathering information about a computer or computer network via questions to naïve users, is often the first step in a malicious attack on computer networks. The more information revealed about a laptop or computer network, the more vulnerable the computer/network is. Never divulge data related to a SCADA network, including the names and contact information about the system operators/administrators, computer operating systems, and physical and logical locations of computers and network systems over the phone or to personnel unless authorized to receive such information. Any requests for information by unknown persons need to be sent to a central network security location for verification and fulfillment. People can be a weak link in an otherwise secure network. Conduct training and information awareness campaigns to ensure personnel remain diligent in guarding sensitive network information, notably their passwords.

THE BOTTOM LINE

These processes, from preliminary screening to filtration, disinfection, and the advanced processes needed for specialized water problems, are in place to serve one primary, essential purpose: to supply the consumer with safe, potable water. Consumers may use this water for various purposes, from drinking to watering the lawn. Still, those uses are in many ways beside the point; safe, potable water is essential for human life.

Again, when it comes to our nation's security and even water/wastewater treatment facilities, few have summed it better than Governor Ridge (Henry, 2002).

> The further removed we get from September 11, the natural tendency is to let down our guard. Unfortunately, we cannot do that. The government will continue to do everything it can to find and stop those who seek to harm us. We owe it to the American people to remind them to be vigilant.

Improved water systems monitoring and automation with AI-assisted SCADA and the ever-evolving landscape of automated and secure water treatment systems are crucial in monitoring and controlling various water treatment processes.

The *ultimate bottom line* is that the future of SCADA lies in integrating AI, enabling intelligent automation and monitoring.

NOTES

1 Adapted from F.R. Spellman's (2020). *The Science of Water,* 4th ed. Boca Raton, FL: CRC Press.
2 AWWA Standards (Coagulation-No. 42402 to 42407) AWWA, Denver, Colorado.
3 Adapted from F.R. Spellman (1999). *Disinfection Alternatives.* Lancaster, PA: Technomic Publishing Company.

REFERENCES

AWWA. (1995). *Water Treatment: Principles and Practices of Water Supply Operations.* Denver, CO: American Water Works Association.

De Zuane, J. (1997). *Handbook of Drinking Water Quality*, 2nd ed. New York: John Wiley & Sons.

DOE. (2001). *21 Steps to Improve the Cyber Security of SCADA Networks.* Washington, DC: Department of Energy.

Ezell, B.C. (1998). *Risks of Cyber Attack to Supervisory Control and Data Acquisition.* Charlottesville, VA: University of Virginia.

FBI. (2000). *Threat to Critical Infrastructure.* Washington, DC: Federal Bureau of Investigation.

FBI. (2004). *Ninth Annual Computer Crime and Security Survey.* Washington, DC: FBI, Computer Crime Institute and Federal Bureau of Investigations.

GAO. (2003). *Critical Infrastructure Protection: Challenges in Securing Control System.* Washington, DC: United States General Accounting Office.

Gellman, B. (2002). Cyber-Attacks by Al Qaeda Feared: Terrorists at Threshold of Using Internet as Tool of Bloodshed, Experts Say—*Washington Post*, June 27; p. A01.

Hammer, M.J. and Hammer, M.J., Jr. (1996). *Water and Wastewater Technology,* 3rd ed. Englewood Cliffs, NJ: Prentice Hall.

Henry, K. (2002). New Face of Security. *Government Security*. April, pp. 30–31.

Masters, G.M. (1991). *Introduction to Environmental Engineering and Science*. Englewood Cliffs, NJ: Prentice Hall.

McGhee, T.J. (1991). *Water Supply and Sewerage*, 6th ed. New York: McGraw-Hill, Inc.

National Academy of Sciences. (2002). *National Research Council Publications 1000-B*. Washington, DC: National Academy of Sciences.

NIPC. (2002). *National Infrastructure Protection Center Report.* Washington, DC: National Infrastructure Protection Center.

Pankratz, T.M. (1995). *Screening Equipment Handbook: For Industrial and Municipal Water and Wastewater Treatment*, 2nd ed. Lancaster, PA: Technomic Publishing Company.

Peavy, H.S. et al. (1985). *Environmental Engineering*. New York: McGraw-Hill, Inc.

Spellman, F.R. (2007). *The Science of Water: Concepts and Applications*, 2nd ed. Boca Raton, FL: CRC Press.

Spellman, F.R. (2020). *The Science of Water*, 4th ed. Boca Raton, FL: CRC Press.

Stamp, J. et al. (2003). *Common Vulnerabilities in Critical Infrastructure Control Systems*, 2nd ed. Sandia National Laboratories.

USA TODAY. (October 1998). *Birth Rate: This Is How Many People Were Norn each Year since 1998.* Accessed 10/5/24 @ https://www.usatoday.com/ Story/money/2020/06/12…

Viessman, W. and Hammer, M.J. (1998). *Water Supply and Pollution Control*, 6th ed. Menlo Park, CA: Addison-Wesley.

6 Water Resource Management

INTRODUCTION

Water resource management is complicated. Why? Water is everywhere. Well, not exactly. Climate change and population growth burden water managers because water supplies and infrastructure are stressed. Managing water resources is challenging.

What is involved in water resource management? Water resource management involves conserving water resources, harvesting the water, planning the available net water resources, and distributing it appropriately to the consumers. It controls and moves water resources for efficient and beneficial use while also minimizing damage to life and property. Parameters of water resource management include rainwater harvesting, irrigation, reuse, sustainability, and wastewater management. USEPA's Water Management Plans and Best Practices provide excellent guidelines for water resource management, helping to reduce water use.

The water resource management techniques put forward by USEPA (2020) include:

- Meter/Measure/Manage
- Optimize Cooling Towers
- Replacc Rcstroom Fixtures
- Eliminate Single-Pass Cooling
- Use Water-Smart Landscaping
- Reduce Steam Sterilizer Tempering Water Use
- Reuse Laboratory Culture Water
- Control Reverse Osmosis System Operation
- Recover Rainwater
- Recover Air Handler Condensate

Let's look at each of these USEPA water management techniques. *Metering and measuring* facility water use helps analyze saving opportunities. This also ensures that equipment is run and maintained correctly to help prevent waste from leaks or malfunctioning mechanical equipment.

Cooling towers provide air conditioning for laboratories and are large consumers of water. To ensure maximum efficiency, cooling tower operations can be optimized by carefully controlling the ratio of water discharged (blowdown) to water evaporated—this is known as the cycle of concentration. Cooling towers should be operated at six or more cycles of concentration. Metering the water input and discharge from the cooling tower ensures the cooling tower is operating and can help identify leaks or other malfunctions.

DOI: 10.1201/9781003601562-8

The U.S. Department of Energy established federal water-efficiency standards in the 1990s, *replacing restroom fixtures.* These standards for plumbing fixtures and appliances, such as showerheads, washing machines, and toilets, have saved taxpayers billions of dollars while protecting water resources. For example, toilets used 3.5 gallons per flush (gpf). New toilets have been installed in EPA laboratories with 1.28 or 1.6 gpf flush rates.

Single-pass cooling circulates a continuous flow of water just once through the system for cooling purposes before it goes down the drain. EPA strives to eliminate single-pass cooling in its laboratories; instead, facilities have air-cooled or recirculating chilled water systems.

Using *water-smart landscaping and irrigation* involves planting native and drought-resistant plant species, which minimizes the need for supplemental irrigation. An irrigation water audit can also reduce landscape water use by 10%–20%. WaterSense-labeled weather-based irrigation controllers or soil moisture sensors water only when plants need it.

Control of steam sterilizer water is essential. Steam sterilizers use cooling water to temper steam condensate discharge from the sterilizer to the laboratory drain. The problem has been that older sterilizers discharge a continuous flow of tempering water to the drain, even when it is only tempering water when needed.

Laboratory culture water is reused instead of being pumped into laboratory specimen tanks from local bodies of water, such as rivers and lakes. It is then discharged into the wastewater system or treated and returned to the body of water.

Note that up to 10% of a laboratory's water consumption can be related to the multi-step process of generating deionized (DI) purified water through *reverse osmosis (RO).* Water savings can be achieved by carefully regulating purified water generation rates to meet laboratory demand and ensuring systems are sized accordingly.

Rainwater recovery systems capture rainwater from the roof and redirect it to a storage tank. This water is used for flushing toilets, supplying cooling towers, and irrigating landscapes.

When an air conditioning system's evaporator coil warms, air passes over it, absorbing heat and moisture from the air and forming condensation. Many EPA laboratories capture *air handler condensate* for use in cooling tower makeup water.

SIDEBAR 6.1 NOT WANDERING LIKE A CLOUD

Ignore mushroom clouds; there are only three types of clouds: real clouds in the sky, the clouds Wordsworth spoke about in his famous "Daffodil" poem, and, finally, digital cloud platforms. In my college-level environmental health/engineering/nursing and training sessions for state and federal regulators (Environmental Protection Agency [EPA], Occupational Safety and Health Administration [OSHA], National Institute of Health [NIH], Centers for Disease Control [CDC], State Water Control personnel), when my lectures turned toward including the digital world as it relates to the subjects I was teaching, I always asked my students: What is "the" cloud?

Almost all the answers I received over the years were murky. So, I would wait until they all ran out of replies and answers, then give them my definition of the cloud as I see it and use it in my lectures.

"It's simply a network of servers, each with an individual, unique function."

Then I would tell them, "There is nothing lonely about the Cloud, but it certainly does much wandering … hopefully at the behest of the human prompter."

I thought I had the final word on this explanation until, in one of my classes, a military officer in one of my graduate courses held up her hand and said, "Professor, there are three types of Clouds … first, there is the Public Cloud, then the Private Cloud, and thirdly, there is the Hybrid Cloud."

Wow, you could have heard a feather hit the cement floor.

But I quickly responded to the student's statement: "You know, she is exactly right … those are the three types of Clouds."

Well, hells, and I was not shipping out … passing the helm to the student but not shipping out—not admitting that I could not recall the three types and their differences. However, in a classroom with more than 100 brilliant students, I have learned over time to allow the student who brought up the subject to have the helm, steer the course, and proceed with an explanation, and she did.

So, she stood up and said, "One thing about cloud computing is that it is extensive, vast, and probably infinite. Anyway, the three types of cloud are public—where a third party offers resources via the internet; private cloud—computing dedicated to a single organization; and, finally, the hybrid cloud—a combo of the two."

After she finished, I clapped, and the entire class joined in with a few whistles added to the thunderous applause.

After everything quieted down again, I said, "Great … right on … and you are exactly correct. Thanks for your contribution."Sometimes, it is the professor who learns from the students.

Then, I quickly changed the subject, moved back to the helm, and steered all the discussion toward subject material that I knew.

AI-ASSISTED WATER RESOURCE MANAGEMENT

The water resource management of a waterworks operation requires water management professionals and licensed operators to perform duties during a typical work shift. From monitoring, operating, and overseeing the operation of pumps, valves, and other equipment to ensure the proper operation of the treatment unit processes in the system to conducting routine tests for water quality and ensuring the treated water meets all health and safety requirements, to adjusting chemical feeders when necessary, to performing maintenance and repairs, to record-keeping, to ensuring compliance with applicable permits, and to emergency response, a waterworks operator (as with wastewater operators) is a Jack and Jill of all trades.

Why do I say this?

I say this because these duties are critical for providing the community with a safe and reliable water supply.

Based on my experience, I have used the Jack and Jill reference title for water and wastewater operators and practitioners because it more closely characterizes the waterworks and wastewater specialist: A Jack or Jill of all trades. This characterization seems only fitting when you consider the knowledge and skills required of operators to perform their assigned duties properly. Moreover, operating the plant or distribution/collection system is one thing: taking samples, operating equipment, monitoring conditions, determining settings for chemical feed systems and high-pressure pumps, performing laboratory tests, and recording the results in the plant's daily operating log.

However, the non-typical, diverse, and off-the-wall functions cause us to describe operators as Jacks or Jills of all trades. For example, in addition to their regular, routine, daily operating duties, operators may be called upon to make emergency repairs to systems (e.g., making a welding repair to a vital piece of machinery to keep the plant or unit process online), perform material handling operations, make chemical additions to process flow, respond to hazardous materials emergencies, make confined space entries, perform site landscaping duties, and carry out several other assorted functions. Remember, the plant operator's job is to keep the plant running and to make a permit. Keeping the plant running, the flow flowing, and making permits—no matter what—requires talent and performing a wide range of functions—many of which are not called for in written job descriptions.

DID YOU KNOW?

Water and wastewater treatment plant and system operators held about 124,800 jobs in 2023. Almost 4 in 5 operators worked for local governments. Others worked primarily for private water, sewage, and other systems utilities, as well as private waste treatment and disposal and waste management services companies. Private firms are increasingly providing operation and management services to local governments on a contract basis. Along with steady increases in new positions, there is also a loss of about 8,000 positions each year due to retirement, other employment opportunities, and automation (BLS, 2023).

AI-ASSISTED OPERATIONS

USEPA has developed various AI-assisted tools and systems to enhance water resource management. These systems are designed to improve the resilience and efficiency of water infrastructure using artificial intelligence technologies. These tools help waterworks assess risks, plan for emergencies, and enhance their resilience. Moreover, AI is used to optimize water distribution operations, improve service delivery, reduce costs, and address issues like unaccounted-for-water. AI, deep learning (DL), and machine learning (ML) techniques, combined with the Internet of

Things (IoT), are facilitating engineers in designing smart water management systems for efficient and sustainable water usage.

DID YOU KNOW?

Water pollution is increasing worldwide, endangering water quality. Current applications make measuring these pollutants with ML possible. Also, ML predicts which rivers, streams, and wetlands the Clean Water Act regulates.

AI is revolutionizing water resource management by optimizing water utilization and creating a more resilient water infrastructure. AI technologies detect potential water supply failures before they occur. AI algorithms detect trends within vast amounts of data, which helps to predict water demand better and optimize daily water supply. AI predictive analytics can also forecast water demand and quality changes, using proactive measures to address issues. AI has the advantage of real-time monitoring that continuously tracks water quality, instantly detecting pollutants, contaminants, and anomalies. AI optimizes water distribution and usage, reducing waste and ensuring efficient allocation. AI aids decision-making by providing data-driven insights. These benefits work to ensure better, more effective water management.

Speaking of making water management a more compelling experience, this has been demonstrated by the IoT, which has significantly changed and improved how we manage water resources. IoT provides real-time information, helps us use resources efficiently, and protects the environment. With IoT, ML, and DL, AI can ensure that water is used wisely. With IoT, operators can better track water using intelligent systems. In real-time, these innovative systems use sensors, connections, and data to tell them how water is used and where it goes.

So, how can IoT do this? IoT uses a water quality monitoring system via a smart device network that tracks and manages water use, quality, and distribution. Sensors measure flow rate, pressure, and water quality. The collected data is sent wirelessly to a central hub, where it is processed and stored in the cloud. Operators can view real-time information using dashboards and apps, making water management more manageable.

The IoT system comprises various components used in a water monitoring system (see Figure 6.1). These components work together to ensure the efficient management and monitoring of water resources.

IoT technologies connect, in one way or another, with the components shown in Figure 6.1. Additionally, geographic information systems (GIS), digital meters, and data processing and analysis tools that monitor and manage water resources effectively contribute to operating water systems. These technologies allow IoT to profile real-time monitoring and data collection, predictive maintenance, smart irrigation, resource allocation, flood management, innovative distribution, and leak detection. They play a crucial role in optimizing water use by detecting anomalies.

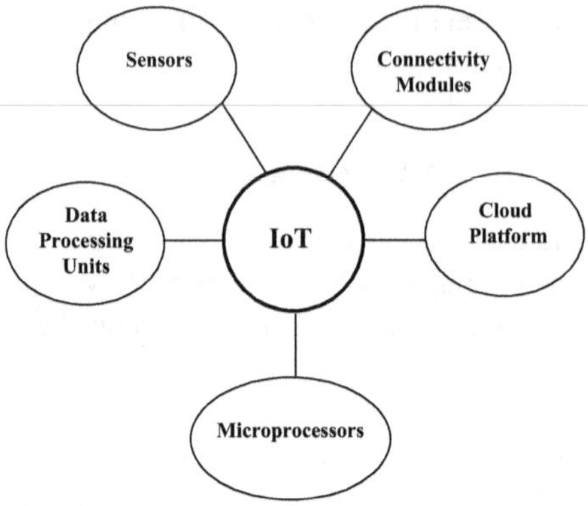

FIGURE 6.1 Components of IoT water monitoring system.

DID YOU KNOW?

In detecting water leaks, IoT uses sensors that monitor pipelines for irregularities and pinpoint leaks. These sensors' early detection minimizes water loss and reduces repair costs by providing real-time data to alert operators quickly. Regarding sensors, IoT water management systems use various devices, including the main types used in IoT applications:

- **Passive sensors** detect changes in their environment with any dedicated power supply. A good example is thermometers.
- **Active sensors** require some form of power source to function. They usually require electricity from the grid or batteries.

These sensors measure the physical environment to collect data about things like temperature or air quality and transmit the information via a network to the cloud. Recall that the cloud is a dedicated wanderer capable of picking up all kinds of data from various sources.

As stated, there is a wide variety of IoT sensors. Some of the sensor types used in IoT include temperature, humidity, fire detection, light, proximity, gas detection, predictive maintenance, IR, and chemical sensors.

THE BOTTOM LINE

Again, water, water, water is everywhere! Indeed, you have noticed this truism. However, even though water is essential for life, it is often ignored, as the old saying goes, "until the well runs dry." Water is one of the most threatened resources, with around 40% of the global population experiencing water shortages. The worst part is that less than 3% of the world's freshwater supply is fit for human consumption or use. Meanwhile, the demand for clean, safe, accessible water continues to rise. Digital technologies such as AI, ML, DL, and IoT can be and are being leveraged as innovative technologies in water management to reduce waste, optimize efficiency, support sustainability, and, more importantly, ensure equitable water distribution.

The problem is that the widespread global adoption of these technologies remains challenging and limited.

Why is adopting IoT technologies for water management challenging?

One of the challenges is finding the money to fund the high initial costs. IoT infrastructure, sensors, networks, and data management systems are not cheap. Funds must be efficiently allocated or available in developing countries and many governments. Finding more technical experts to install, maintain, and operate IoT systems is just as challenging, even if funding can be found. Then, there are the privacy issues. Governments and others hesitate to adopt IoT technologies due to the potential risks of cyberattacks, a current worry. Regulatory standards and requirements may not support integrating new technologies like IoT. Finally, and most importantly [as is being discovered by those who purchase electric vehicles (EVs)], supporting infrastructure may need to be revised or is paltry at best.

REFERENCES

BLS. (2023). Occupational Outlook Handbook: *Water and Wastewater Treatment Plant System Operators.* Accessed 8/30/24 @ https://www.bls.gov.ooh/productions/water-and-wastewater-treatment/plant-and-system-operations.htm.

USEPA. (2020). *Water Management Plans and Best Practices at EPA.* Accessed 9/13/24 @ https//19January2021snapshot.epa.gov/greeningepa/water-management-plans-and-best-practices-epa.html.

7 Wastewater Treatment

INTRODUCTION

This chapter begins by discussing and providing fundamental information about unit processes and their operation in wastewater treatment. After this short but concise introduction, it outlines where artificial intelligence (AI) applications are applied. The AI applications (i.e., the hardware) employed are similar to or the same as those used in water treatment— Supervisory Control and Data Acquisition (SCADA) is engaged in both. The difference lies in the unit processes, except for disinfection and filtration.

WASTEWATER TREATMENT: THE PROCESS[1]

According to the Code of Federal Regulations (CFR) 40 CFR Part 403, regulations were established in the late 1970s and early 1980s to help Publicly Owned Treatment Works (POTW) control industrial discharges to sewers. These regulations were designed to prevent pass-through, interference at the treatment plants, and interference in the collection and transmission systems.

Pass-through occurs when pollutants literally "pass-through" a POTW without being adequately treated, causing the POTW to have an effluent violation or increasing the magnitude or duration of a breach. Interference occurs when a pollutant discharge causes a POTW to violate its permit by inhibiting or disrupting treatment processes, treatment operations, or processes related to sludge use or disposal.

Unit operations (unit processes), which are the components that are linked together to form a process train (as shown in Figure 7.1; keep in mind the caboose attached to this train is treated and cleaned wastewater, which, when outflowed, is usually cleaner than the water in the receiving body), are commonly divided based on the fundamental mechanisms acting on them (i.e., physical, chemical, and biochemical). Physical operations, such as sedimentation, are governed by the laws of physics (gravity). Chemical operations are those in which chemical reactions, such as precipitation, strictly occur. Biochemical operations use living microorganisms to destroy or transform pollutants through enzymatically catalyzed chemical reactions (Grady et al., 2011).

THE WASTEWATER TREATMENT MODEL

Figure 7.1 shows a basic schematic of an example wastewater treatment process providing primary and secondary treatment using the *Activated Sludge process*. This is the model, the prototype, and the paradigm used in this handbook. Though it is true that in secondary treatment (which provides biochemical oxygen demand (BOD) removal beyond what is achievable by simple sedimentation), there are

DOI: 10.1201/9781003601562-9

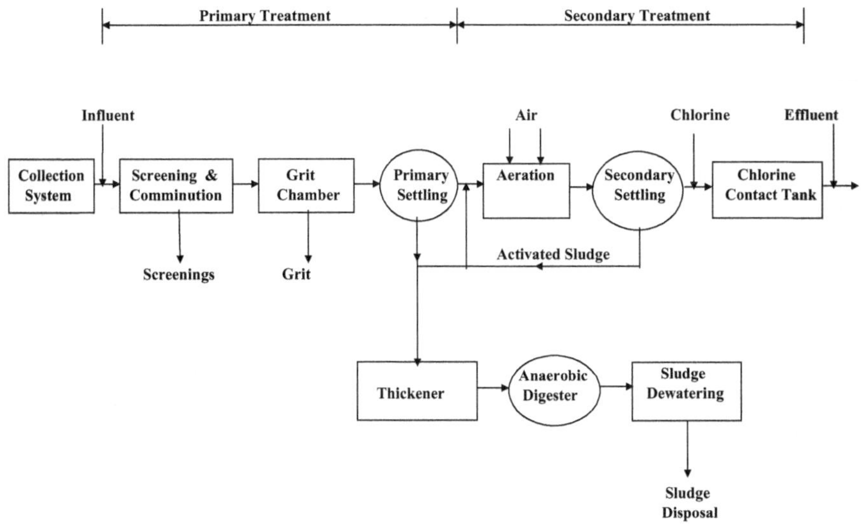

FIGURE 7.1 Schematic of an example wastewater treatment process providing primary and secondary treatment using the activated sludge process.

three commonly used approaches—trickling filter, activated sludge, and oxidation ponds—we focus, for instructive and illustrative purposes, on the activated sludge process throughout this handbook. The purpose of Figure 7.1 is to allow the reader to follow the treatment process step-by-step as it is presented (and configured in the real world) and to assist in understanding how all the various unit processes sequentially follow and tie into each other. Therefore, we begin certain sections (which discuss unit processes) with frequent reference to Figure 7.1. It is crucial to start these sections in this manner because wastewater treatment is a series of individual steps (unit processes) that treat the waste stream as it makes its way through the entire process. Thus, it logically follows that a pictorial presentation and pertinent written information enhance the learning process. However, it should also be pointed out that even though the model shown in Figure 7.1 does not include all unit processes currently used in wastewater treatment, we do not ignore the other significant processes: trickling filters, rotating biological contactors (RBCs), and oxidation ponds.

Wastewater Terminology and Definitions

Like many other technical fields, wastewater treatment technology has unique terms and meanings. Though some terms are unique, many are expected from different professions. Remember that the science of wastewater treatment is a combination of engineering, biology, mathematics, hydrology, chemistry, physics, and other disciplines. Therefore, many terms used in engineering, biology, mathematics, hydrology, chemistry, physics, and others are also used in wastewater treatment. Those terms not listed or defined in the following section will be defined as they appear in the text.

Activated Sludge: The solids formed when microorganisms are used to treat wastewater using the activated sludge treatment process. It includes organisms, accumulated food materials, and waste products from the aerobic decomposition process.

Advanced Waste Treatment: Treatment technology to produce an extremely high-quality discharge.

Aerobic: Conditions in which free, elemental oxygen is present. It is also used to describe organisms, biological activity, or treatment processes that require free oxygen.

Anaerobic: Conditions in which no oxygen (free or combined) is available. It also describes organisms, biological activity, or treatment processes without oxygen.

Anoxic: Conditions in which no free, elemental oxygen is present; the only source of oxygen is combined oxygen, such as that found in nitrate compounds. It also describes biological activity or treatment processes, which function only in combined oxygen.

Average Monthly Discharge Limitation: The highest allowable discharge over a calendar month.

Average Weekly Discharge Limitation: The highest allowable discharge over a calendar week.

Biochemical Oxygen Demand, BOD_5: The amount of organic matter that can be biologically oxidized under controlled conditions (5 days @ 20°C in the dark).

Biosolids: From *Merriam-Webster's Collegiate Dictionary, Tenth Edition* (1998): biosolid *n* (1977)—solid organic matter recovered from sewage treatment and used primarily as fertilizer—usually used in the plural.

Note: In this text, biosolids are used in many places (activated sludge being the exception) to replace the standard term sludge. The author (along with others in the field) views the term sludge as an ugly four-letter word that is inappropriate to describe biosolids. Biosolids are products that can be reused; they have some value. Because biosolids have some value, they should not be classified as a "waste" product, and when biosolids for beneficial reuse are addressed, they are not.

Buffer: A substance or solution that resists changes in pH.

Carbonaceous Biochemical Oxygen Demand, $CBOD_5$: The amount of biochemical oxygen demand attributed to carbonaceous material.

Chemical Oxygen Demand (COD): The amount of chemically oxidizable materials present in the wastewater.

Clarifier: A device that permits solids to settle or rise and be separated from the flow. It is also known as a settling tank or sedimentation basin.

Coliform: A type of bacteria used to indicate possible human or animal contamination of water.

Combined Sewer: A collection system that carries both wastewater and stormwater flows.

Comminution: A process to shred solids into smaller, less harmful particles.

Composite Sample: A combination of individual samples taken in proportion to flow.

Daily Discharge: The discharge of a pollutant measured during a calendar day or any 24-hour period that reasonably represents a calendar day for sampling. Limitations expressed as weight are the total mass (weight) discharged over the day.

Limitations expressed in other units are average measurements of the day.

Daily Maximum Discharge: The highest allowable value for a daily discharge.

Detention Time: The theoretical time water remains in a tank at a given flow rate.

Dewatering: The removal or separation of a portion of water present in a sludge or slurry.

Discharge Monitoring Report (DMR): The monthly report required by the treatment plant's NPDES discharge permit.

Dissolved Oxygen (DO): Free or elemental oxygen dissolved in water.

Effluent: The flow leaving a tank, channel, or treatment process.

Effluent Limitation: Any restriction the regulatory agency imposes on quantities, discharge rates, or concentrations of pollutants discharged from point sources into state waters.

Facultative: Organisms that can survive and function in the presence or absence of free, elemental oxygen.

Fecal Coliform: A type of bacteria found in the bodily discharges of warm-blooded animals and used as an indicator organism.

Floc: Solids that join together to form larger particles that will settle better.

Flume: A flow rate measurement device.

Food-to-Microorganism Ratio (F/M): An activated sludge process control calculation based upon the amount of food (BOD_5 or COD) available per pound of mixed liquor volatile suspended solids.

Grab Sample: An individual sample collected at a randomly selected time.

Grit: Heavy inorganic solids such as sand, gravel, eggshells, or metal filings.

Industrial Wastewater: Wastes associated with industrial manufacturing processes.

Infiltration/Inflow: Extraneous flows in sewers; defined by Metcalf & Eddy in *Wastewater Engineering: Treatment, Disposal, Reuse*, 3rd. ed., New York: McGraw-Hill, Inc., pp. 29–31, 1991 as follows:

- **Infiltration**: Water entering the collection system through cracks, joints, or breaks.
- **Steady Inflow**: Water discharged from cellar and foundation drains, cooling water discharges, and drains from springs and swampy areas. This type of inflow is steady and is identified and measured along with infiltration.
- **Direct Flow**: Those types of inflow that have a direct stormwater runoff connection to the sanitary sewer and cause an almost immediate increase in wastewater flows. Possible sources are roof leaders, yard and areaway drains, maintenance hole covers, cross connections from storm drains and catch basins, and combined sewers.
- **Total Inflow**: The sum of the direct inflow at any point in the system plus any flow discharged from the system upstream through overflows, pumping station bypasses, and the like.
- **Delayed Inflow**: Stormwater that may require several days or more to drain through the sewer system. This category can include the discharge of sump pumps from cellar drainage and the slowed entry of surface water through maintenance holes in ponded areas.

Influent: The wastewater entering a tank, channel, or treatment process.

Inorganic: Mineral materials such as salt, ferric chloride, iron, sand, gravel, etc.

License: A certificate issued by the State Board of Waterworks/Wastewater Works Operators authorizing the holder to perform the duties of a wastewater treatment plant operator.

Mean Cell Residence Time (MCRT): The average length of time a mixed liquor-suspended solids particle remains in the activated sludge process. It may also be known as sludge retention time.

Mixed Liquor: Combines return activated sludge and wastewater in the aeration tank.

Mixed Liquor Suspended Solids (MLSS): The suspended solids concentration of the mixed liquor.

Mixed Liquor Volatile Suspended Solids (MLVSS): The concentration of organic matter in the mixed liquor suspended solids.

Milligrams/Liter (mg/L): A measure of concentration. It is equivalent to parts per million (ppm).

Nitrogenous Oxygen Demand (NOD): Measures the oxygen required to oxidize nitrogen compounds biologically under specified time and temperature conditions.

NPDES Permit: National Pollutant Discharge Elimination System permit that authorizes the discharge of treated wastes and specifies the conditions that must be met for discharge.

Nutrients: Substances required to support living organisms. They usually refer to nitrogen, phosphorus, iron, and other trace metals.

Organic materials: Comprise carbon, hydrogen, oxygen, sulfur, and nitrogen. Many organics are biologically degradable. When subjected to high temperatures, all organic compounds can be converted to carbon dioxide and water.

Pathogenic: Disease-causing. A pathogenic organism is capable of causing illness.

Point Source: Any discernible, defined, and discrete conveyance from which pollutants may be discharged.

Part per Million: An alternative (but numerically equivalent) unit used in chemistry is milligrams per liter (mg/L). As an analogy, imagine a ppm comparable to a full shot glass in a swimming pool.

Return Activated Sludge Solids (RASS): The concentration of suspended solids in the sludge flow being returned from the settling tank to the head of the aeration tank.

Sanitary wastewater, including sewage and industrial waste, is discharged from residences and commercial, institutional, and similar facilities.

Scum: The mixture of floatable solids and water, which is removed from the surface of the settling tank.

Septic: Wastewater that has no dissolved oxygen (DO) present. It is generally characterized by black color and rotten egg (hydrogen sulfide) odors.

Settleability: A process control test used to evaluate the settling characteristics of the activated sludge. Readings taken at 30–60 minutes are used to calculate the settled sludge volume (SSV) and the sludge volume index (SVI).

Settled Sludge Volume: The volume in percent occupied by an activated sludge sample after 30–60 minutes of settling. Generally written as SSV with a subscript to indicate the time of the reading used for calculation (SSV_{60}) or (SSV_{30}).

Sewage: Wastewater containing human wastes.

Sludge: The mixture of settleable solids and water removed from the settling tank's bottom.

Sludge Retention Time (SRT): See Mean Cell Residence Time.

Sludge Volume Index (SVI): A process control calculation used to evaluate the settling quality of activated sludge. The SSV_{30} and mixed liquor suspended solids test results must be calculated.

Storm Sewer: A collection system designed to carry only stormwater runoff.

Stormwater: Runoff resulting from rainfall and snowmelt.

Supernatant: In a digester, it is the amber-colored liquid above the sludge.

Wastewater: The community's water supply after it has been soiled by use.

Waste Activated Sludge Solids (WASS): The concentration of suspended solids in the sludge, which is being removed from the activated sludge process.

Weir: A device used to measure wastewater flow.

Zoogleal Slime: The biological slime that forms on fixed film treatment devices. It contains a wide variety of organisms essential to the treatment process.

Measuring Plant Performance

Performance efficiency or percent (%) removal is used to evaluate the *operating efficiency* of a *plant or treatment* unit process. The results can be compared with those listed in the plant's operation and maintenance manual (O & M) to determine whether the facility performs as expected.

Plant Performance and Efficiency

The calculation for determining a digester's performance (percent removal) differs from that used for performance (percent removal) for other processes. Care must be taken to select the right formula.

$$\%\,\text{Removal} = \frac{\left[\text{Influent Concentration} - \text{Effluent Concentration}\right] \times 100}{\text{Influent Concentration}} \tag{7.1}$$

Equation 7.1 is again used to determine unit process efficiency. The concentration entering and leaving the unit (i.e., primary, secondary, etc.) determines the unit's performance.

PERCENT VOLATILE MATTER REDUCTION IN SLUDGE

The calculation used to determine the *percentage of volatile matter reduction* is more complicated because of the changes occurring during sludge digestion.

$$\%\ \text{V.M. Reduction} = \frac{\left(\%\text{V.M.}_{\text{in}} - \%\text{V.M.}_{\text{out}}\right) \times 100}{\left[\%\text{V.M.}_{\text{in}} - \left(\%\text{V.M.}_{\text{in}} \times \%\text{V.M.}_{\text{out}}\right)\right]} \tag{7.2}$$

V.M. = Volatile Matter

Hydraulic Detention Time

The term detention time or hydraulic detention time (HDT) refers to the average time (theoretical time) a drop of water, wastewater, or suspended particles remains in a tank or channel. It is calculated by dividing the volume of water/wastewater in the tank by the flow rate through the tank. The flow rate units used in the calculation depend on whether the detention time will be calculated in seconds, minutes, hours, or days. Detention time is used with various treatment processes, including sedimentation and coagulation-flocculation. Generally, in practice, detention time is associated with the amount of time required for a tank to empty. The range of detention time varies with the process. For example, detention time is commonly measured in minutes in a tank used for sedimentation.

WASTEWATER SOURCES AND CHARACTERISTICS

Wastewater treatment is designed to use the natural purification processes (self-purification processes of streams and rivers) to the maximum level possible. It is also intended to complete these processes in a controlled environment rather than over many miles of stream or river. Moreover, the treatment plant is also designed to remove contaminants generally not subjected to natural processes and treat the solids generated through the treatment unit steps. The typical wastewater treatment plant is designed to achieve many different purposes:

- Protect public health
- Protect public water supplies
- Protect aquatic life
- Preserve the best uses of the waters
- Protect adjacent lands

Wastewater treatment is a series of steps. Each step can be accomplished using one or more treatment processes or types of equipment. The significant categories of treatment steps are:

- **Preliminary Treatment**: Removes materials that could damage plant equipment or occupy treatment capacity without being treated.
- **Primary Treatment**: Removes settleable and floatable solids (may not be present in all treatment plants).
- **Secondary Treatment**: -- Removes BOD_5 and dissolved and colloidal suspended organic matter by biological action; organics are converted to stable solids, carbon dioxide, and more organisms.
- **Advanced Waste Treatment**: Uses physical, chemical, and biological processes to remove additional BOD_5, solids, and nutrients (not present in all treatment plants).
- **Disinfection**: Removes microorganisms to eliminate or reduce the possibility of disease when the flow is discharged.
- **Sludge Treatment**: Stabilizes the solids removed from wastewater during treatment, inactivates pathogenic organisms, and reduces the volume of the sludge by removing water.

Note: The various treatment processes described above are discussed in detail later.

The principal sources of domestic wastewater in a community are the residential areas and commercial districts. Other significant sources include institutional and recreational facilities, stormwater (runoff), and groundwater (infiltration). Each source produces wastewater with specific characteristics. In this section, wastewater sources and the particular attributes of wastewater are described.

Five significant sources of wastewater are human and animal wastes, household wastes, industrial wastes, stormwater runoff, and groundwater infiltration.

1. **Human and Animal Wastes**: Contain the solid and liquid discharges of humans and animals and are considered by many to be the most dangerous from a human health viewpoint. The primary health hazard is presented by the millions of bacteria, viruses, and other microorganisms (some of which may be pathogenic) in the waste stream.
2. **Household Wastes**: Other than human and animal wastes discharged from the home. Household wastes usually contain paper, household cleaners, detergents, trash, garbage, and other substances the homeowner discharges into the sewer system.
3. **Industrial Wastes:** Include industry-specific materials that can be discharged into the collection system from industrial processes. They typically contain chemicals, dyes, acids, alkalis, grit, detergents, and highly toxic materials.
4. **Stormwater Runoff:** Many collection systems are designed to carry the community's waste and runoff. In this type of system, when a storm occurs, the waste stream can contain large amounts of sand, gravel, other grit, and excessive water.
5. **Groundwater Infiltration**: Groundwater will enter older, improperly sealed collection systems through cracks or unsealed pipe joints. This can add large amounts of water and additional grit to wastewater flows.

Classification of Wastewater

Wastewater can be classified according to the sources of flows:

1. **Domestic (sewage) Wastewater**: Mainly contains human and animal wastes, household wastes, small amounts of groundwater infiltration, and small amounts of industrial wastes.
2. **Sanitary Wastewater**: Consists of domestic and significant industrial wastes. In many cases, industrial waste can be treated without special precautions. However, in some cases, industrial wastes will require special precautions or a pretreatment program to ensure that they do not cause compliance problems for the wastewater treatment plant.
3. **Industrial Wastewater**: Industrial wastes only. Often, the industry determines that treating its waste independently of domestic waste is safer and more economical.

4. **Combined Wastewater**: The combination of sanitary wastewater and stormwater runoff. All the wastewater and stormwater of the community is transported through one system to the treatment plant.
5. **Stormwater:** A separate collection system (no sanitary waste) that carries stormwater runoff, including street debris, road salt, and grit.

Wastewater Characteristics

Wastewater contains many different substances, which can be used to characterize it. The specific substances and amounts or concentrations of each will vary, depending on the source. Thus, it is difficult to "precisely" characterize wastewater. Instead, wastewater characterization is usually based on and applied to average domestic wastewater. Wastewater is described in its physical, chemical, and biological characteristics.

Note: Keep in mind that other sources and types of wastewater can dramatically change the characteristics.

Physical Characteristics

The *physical characteristics* of wastewater are based on color, odor, temperature, and flow.

- **Color**: Fresh wastewater is usually a light brownish-gray color. However, typical wastewater is gray and cloudy. The wastewater color will change significantly if it is allowed to go septic (if travel time in the collection system increases). Typical septic wastewater will have a black color.
- **Odor**: Odors in domestic wastewater usually are caused by gases produced by the decomposition of organic matter or other substances added to the wastewater. Fresh domestic wastewater has a musty odor. If the wastewater is allowed to go septic, this odor will change significantly to a rotten egg odor associated with hydrogen sulfide (H2S) production.
- **Temperature**: The wastewater temperature is commonly higher than that of the water supply because of the addition of warm water from households and industrial plants. However, significant infiltration or stormwater flow can cause major temperature fluctuations.
- **Flow**: The actual volume of wastewater is commonly used as a physical characterization of sewage and is usually expressed in terms of gallons per person per day. Most treatment plants are designed using an expected flow of 100–200 gal/person/day. This figure may have to be revised to reflect the degree of infiltration or storm flow the plant receives. Flow rates will vary throughout the day. This variation, which can be as much as 50%–200% of the average daily flow, is known as the *diurnal flow variation*.

Note: *Diurnal* means occurring in a day or each day; daily.

Chemical Characteristics

When describing the chemical characteristics of wastewater, the discussion generally includes organic matter, the measurement of organic matter, inorganic matter, and gases. For the sake of simplicity, in this handbook, we specifically describe chemical characteristics in terms of alkalinity, BOD, chemical oxygen demand (COD), dissolved gases, nitrogen compounds, pH, phosphorus, solids (organic, inorganic, suspended and dissolved solids), and water.

- **Alkalinity**: Measures the wastewater's capability to neutralize acids. It is measured in terms of bicarbonate, carbonate, and hydroxide alkalinity. Alkalinity is essential to buffer (hold the neutral pH) of the wastewater during the biological treatment processes.
- **Biochemical Oxygen Demand (BOD):** Measures the amount of biodegradable matter in the wastewater. Usually measured by a 5-day test conducted at 20°C. The BOD_5 of domestic waste is generally in the range of 100–300 mg/L.
- **Chemical Oxygen Demand (COD)**: A measure of the amount of oxidizable matter in the sample. The COD is generally in the range of 200–500 mg/L. The presence of industrial wastes can increase this significantly.
- **Dissolved Gases**: Gases that are dissolved in wastewater. The specific gases and average concentrations are based on the composition of the wastewater. Typical domestic wastewater contains oxygen in relatively low concentrations, carbon dioxide, and hydrogen sulfide (if septic conditions exist).
- **Nitrogen Compounds**: The type and amount of nitrogen present will vary from the raw wastewater to the treated effluent. Nitrogen follows a cycle of oxidation and reduction. Most of the nitrogen in untreated wastewater will be in the forms of organic nitrogen and ammonia nitrogen. Laboratory tests exist to determine both of these forms. The sum of these two forms of nitrogen is also measured and is known as *Total Kjeldahl Nitrogen (TKN)*. Wastewater will typically contain between 20 and 85 mg/L of nitrogen. Organic nitrogen will generally be in the range of 8–35 mg/L, and ammonia nitrogen will be in the range of 12–50 mg/L.
- **pH**: A method of expressing the acid condition of the wastewater. pH is expressed on a scale of 1–14. For proper treatment, wastewater pH should generally be 6.5–9.0 (ideal 6.5 to 8.0).
- **Phosphorus:** Essential for biological activity and must be present in at least minimum quantities, or secondary treatment processes will not be performed. Excessive amounts can cause stream damage and excessive algal growth. Phosphorus will generally range from 6 to 20 mg/L. Removing phosphate compounds from detergents has significantly impacted the amounts of phosphorus in wastewater.
- **Solids**: Most pollutants found in wastewater can be classified as solids. Wastewater treatment is generally designed to remove solids or to convert solids to a form that is more stable or can be removed. Solids can be classified by their chemical composition (organic or inorganic) or physical characteristics (settleable, floatable, and colloidal). The total solids concentration in wastewater usually ranges from 350 to 1,200 mg/L.

Organic Solids: Consist of carbon, hydrogen, oxygen, and nitrogen and can be converted to carbon dioxide and water by ignition at 550°C. They are also known as fixed solids or loss on ignition.

Inorganic Solids: Mineral solids that are unaffected by ignition. They are also known as fixed solids or ash.

Suspended Solids: Will not pass through a glass fiber filter pad. They can be further classified as total suspended solids (TSS), Volatile Suspended Solids, and Fixed Suspended Solids. It can also be separated into three components based on settling characteristics: settleable solids, floatable solids, and colloidal solids. Total suspended solids in wastewater generally range from 100 to 350 mg/L.

Dissolved Solids: Will pass through a glass fiber filter pad. Total Dissolved Solids (TDS), volatile dissolved solids, and fixed dissolved solids can also be classified. Total dissolved solids are usually in the range of 250–850 mg/L.

- **Water**: Always the principal constituent of wastewater. Water usually makes up 99.5%–99.9% of the wastewater. Even in the strongest wastewater, the contamination is less than 0.5% of the total, and in average strength wastes, it is usually less than 0.1%.

Biological Characteristics and Processes

After undergoing physical aspects of treatment (i.e., screening, grit removal, and sedimentation) in preliminary and primary treatment, wastewater still contains some suspended solids and other solids that are dissolved in the water. In a natural stream, such substances are a food source for protozoa, fungi, algae, and several varieties of bacteria. In secondary wastewater treatment, these same microscopic organisms (one of the main reasons for treating wastewater) are allowed to work as fast as they can to biologically convert the dissolved solids to suspended solids, which will physically settle out at the end of secondary treatment.

Raw wastewater influent typically contains millions of organisms. Most of these organisms are non-pathogenic; however, several pathogenic organisms may also be present (these may include the microorganisms responsible for diseases such as typhoid, tetanus, hepatitis, dysentery, gastroenteritis, and others). Many of the organisms in wastewater are microscopic (microorganisms); they include algae, bacteria, protozoans (such as amoeba, flagellates, free-swimming ciliates, and stalked ciliates), rotifers, and viruses. Table 7.1 is a summary of typical domestic wastewater characteristics.

Wastewater Collection Systems

Wastewater collection systems collect and convey wastewater to the treatment plant. The system's complexity depends on the community's size and the type of system selected. Methods of wastewater collection and conveyance include gravity systems, force main systems, vacuum systems, and combinations of all three types of systems.

TABLE 7.1
Typical Domestic Wastewater Characteristics

Characteristic	Typical Characteristic
Color	Gray
Odor	Musty
Dissolved oxygen	>1.0 mg/L
pH	6.5–9.0
TSS	100–350 mg/L
BOD_5	100–300 mg/L
COD	200–500 mg/L
Flow	100–200 gal/person/day
Total nitrogen	20–85 mg/L
Total phosphorus	6–20 mg/L
Fecal coliform	500,000–3,000,000 MPN/100 mL

Gravity Collection System

In a gravity collection system, the collection lines are sloped to permit the flow to move through the system with as little pumping as possible. The slope of the lines must keep the wastewater moving at a velocity (speed) of 2–4 ft/s. Otherwise, solids will settle at lower velocities, causing clogged lines, overflows, and offensive odors. To keep collection system lines at a reasonable depth, wastewater must be lifted (pumped) periodically to continue flowing downhill to the treatment plant. Pump stations are installed at selected points within the system for this purpose.

Force Main Collection System

In a typical force primary collection system, wastewater is collected at central points and pumped under pressure to the treatment plant. The system is typically used to convey wastewater long distances. Using the force central system allows the wastewater to flow to the treatment plant at the desired velocity without using sloped lines. It should be noted that the pump station discharge lines in a gravity system are considered force mains since the contents of the lines are under pressure.

Note: Extra care must be taken when performing maintenance on force main systems since the collection system's contents are under pressure.

Vacuum System

In a vacuum collection system, wastewater is collected at central points and drawn toward the treatment plant under a vacuum. The system consists of a lot of mechanical equipment and requires significant maintenance to perform correctly. Generally, vacuum-type collection systems are not economically feasible.

Pumping Stations

Pumping stations provide the motive force (energy) to keep the wastewater moving at the desired velocity. They are used in both the force central and gravity systems. They are designed in several configurations and may use different energy sources to move the wastewater (i.e., pumps, air pressure, or vacuum). One of the more commonly used types of pumping station designs is the wet-well/dry-well design.

Wet-Well/Dry-Well Pumping Stations

The wet-well/dry-well pumping station consists of two separate spaces or sections separated by a common wall. Wastewater is collected in one section (wet-well section), and the pumping equipment (and, in many cases, the motors and controllers) is located in a second section known as the dry well. There are many designs for this type of system, but the pumps selected are centrifugal in most cases. There are a couple of primary considerations in selecting a centrifugal design: (1) It allows for the separation of mechanical equipment (pumps, motors, controllers, wiring, etc.) from the potentially corrosive atmosphere (sulfides) of the wastewater. This type of design is usually safer for workers because they can monitor, maintain, operate, and repair equipment without entering the pumping station wet well.

Note: Most pumping station wet wells are confined spaces. To ensure safe entry into such spaces, compliance with the Occupational Health and Safety Administration's (OSHA) 29 CFR 1910.146 (Confined Space Entry Standard) is required.

PRELIMINARY WASTEWATER TREATMENT

The initial stage in the wastewater treatment process (following collection and influent pumping) is *preliminary treatment*. Raw influent entering the treatment plant may contain many materials (trash). The initial treatment aims to protect plant equipment by removing these materials, which could cause clogs, jams, or excessive wear to plant machinery. In addition, eliminating various materials at the beginning of the treatment process saves valuable space within the treatment plant.

Preliminary treatment may include many different processes, each designed to remove a specific type of material that could potentially cause problems in the treatment process. Processes include wastewater collection—influent pumping, screening, shredding, grit removal, flow measurement, preparation, chemical addition, and flow equalization. The major processes are shown in Figure 7.1. This section discusses these processes and their importance in the treatment process.

Note: As mentioned, not all treatment plants will include all of the processes shown in Figure 7.1. Specific processes have been included to facilitate discussion of major potential problems with each process and its operation; this information may be necessary for environmental engineers and wastewater operators.

SCREENING

The screening process aims to remove large solids, such as rags, cans, rocks, branches, leaves, roots, etc., from the flow before moving on to downstream processes (see Figure 7.1).

Note: Typically, a treatment plant will remove 0.5–12 ft^3 of screenings for each million gallons of influent received.

A *bar screen* traps debris as wastewater influent passes through. Typically, a bar screen consists of a series of parallel, evenly spaced bars or a perforated screen

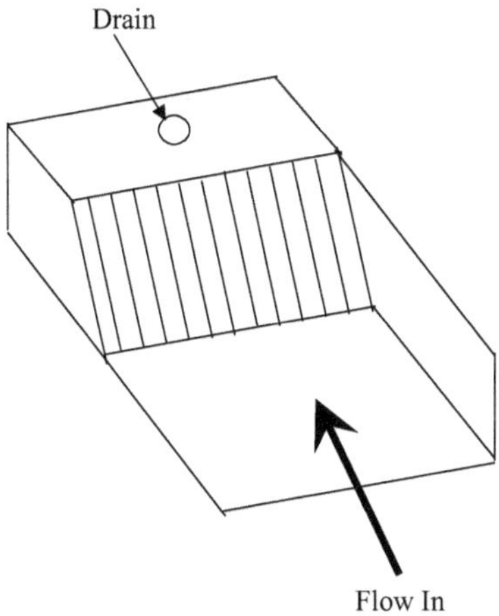

FIGURE 7.2 Basic bar screen.

placed in a channel (see Figure 7.2). The waste stream passes through the screen, and the large solids (*screenings*) are trapped on the bars for removal.

Note: The screenings must be removed frequently enough to prevent accumulation, which will block the screen and cause the water level in front of the screen to build up.

The bar screen may be coarse (2–4-inch openings) or fine (0.75–2.0-inch openings). It may be manually cleaned (bars or screens are placed at 30° for easier solids removal—see Figure 7.2) or mechanically cleaned (bars are placed at 45°–60° angles to improve mechanical cleaner operation).

The screening method depends on the plant's design, the amount of solids expected, and whether the screen is for constant or emergency use only.

Shredding

As an alternative to screening, shredding can reduce solids to a size that can enter the plant without causing mechanical problems or clogging. Shredding processes include comminution (comminution means to cut up) and barminution devices.

Comminution

The *comminutor* is the most common shredding device used in wastewater treatment. In this device, all the wastewater flow passes through the grinder assembly. The grinder consists of a screen or slotted basket, a rotating or oscillating cutter, and a stationary cutter. Solids pass through the screen and are chopped or shredded between the two cutters. The comminutor will not remove solids too large to fit through the slots and will not remove floating objects. These materials must be removed manually. Maintenance requirements for comminutors include aligning, sharpening, and replacing cutters, as well as corrective and preventive maintenance performed according to the plant's O&M manual.

Common operational problems associated with comminutors include output containing coarse solids. When this occurs, it is usually a sign that the cutters are dull or misaligned. If the system does not operate, the unit may be clogged or jammed, a shear pin or coupling may be broken, or electrical power may be shut off. If the unit stalls or jams frequently, this usually indicates cutter misalignment, excessive debris in influent, or dull cutters.

Note: Only qualified maintenance operators should perform shredding equipment maintenance.

Barminution

In barminution, the laminator uses a bar screen to collect solids, which are then shredded and passed through the bar screen for removal at a later process. Cutter alignment and sharpness are critical factors in the effective operation of each device. Cutters must be sharpened or replaced, and alignment must be checked following the manufacturer's recommendations. Solids that are not shredded must be removed daily, stored in closed containers, and disposed of by burial or incineration. Barminutor operational problems are similar to those listed above for comminutors. Preventive and corrective maintenance and lubrication must be performed by qualified personnel following the plant's O&M manual. Because of higher maintenance requirements, the barminutor is less frequently used.

Grit Removal

The purpose of grit removal is to remove heavy inorganic solids, which could cause excessive mechanical wear. Grit is heavier than inorganic solids and includes sand, gravel, clay, eggshells, coffee grounds, metal filings, seeds, and other similar materials. Several processes or devices are used for grit removal. The processes are based on the fact that grit is heavier than organic solids, which should be suspended for treatment. Grit removal may be accomplished in grit chambers (see Figure 7.1) or by the centrifugal separation of sludge.

PRIMARY WASTEWATER TREATMENT (SEDIMENTATION)

The primary treatment (sedimentation or clarification) removes settleable organic and floatable solids. Typically, each primary clarification unit can be expected to remove 90%–95% of settleable solids, 40%–60% of total suspended solids, and 25%–35% BOD_5.

Note: Performance expectations for settling devices used in other areas of plant operation are generally expressed as overall unit performance rather than settling unit performance.

Sedimentation may be used throughout the plant to remove settleable and floatable solids. It is used in primary, secondary, and advanced wastewater treatment processes. This section focuses on primary treatment or clarification, which uses large basins to achieve primary settling under relatively quiescent conditions (see Figure 7.1). Within these basins, mechanical scrapers collect the primary settled solids into a hopper, from which they are pumped to a sludge-processing area. Oil, grease, and other floating materials (scum) are skimmed from the surface. The effluent is discharged over weirs into a collection trough.

In primary sedimentation, wastewater enters a settling tank or basin. Velocity is reduced to approximately 1 ft/min.

Note: The velocity is based on minutes instead of seconds, as was the case in the grit channels. A grit channel velocity of 1 ft/s would be 60 ft/min.

Solids heavier than water, settle to the bottom, while solids lighter than water float to the top. Settled solids are removed as sludge, and floating solids are removed as scum. Wastewater leaves the sedimentation tank over an effluent weir and moves on to the next step in treatment. Detention time, temperature, tank design, and equipment condition control the efficiency of the process.

- Primary treatment reduces the organic loading on downstream treatment processes by removing many settleable, suspended, and floatable materials.
- Primary treatment reduces the velocity of the wastewater through a clarifier to approximately 1–2 ft/min, so settling and flotation can occur. Slowing the flow enhances the removal of suspended solids in wastewater.
- Primary settling tanks remove floated grease and scum, as well as settled sludge solids, and collect them for pumped transfer to disposal or further treatment.
- Clarifiers may be rectangular or circular. In rectangular clarifiers, wastewater flows from one end to the other, and the settled sludge is moved to a hopper at one end, either by flights set on parallel chains or by a single bottom scraper set on a traveling bridge. A surface skimmer collects floating material (mostly grease and oil).
- In circular tanks, the wastewater usually enters in the middle and flows outward. Settled sludge is pushed to a hopper in the middle of the bottom tank, and a surface skimmer removes floating material.
- Factors affecting primary clarifier performance include:
    - Rate of flow through the clarifier
    - Wastewater characteristics (strength, temperature, amount and type of industrial waste, and the density, size, and shapes of particles)
    - Performance of pretreatment processes
- Nature and amount of any wastes recycled to the primary clarifier

Types of Sedimentation Tanks

Sedimentation equipment includes septic, two-story, and plain settling tanks or clarifiers. All three devices may be used for primary treatment, while plain settling tanks are commonly used for secondary or advanced wastewater treatment.

Septic Tanks

Septic tanks are prefabricated tanks that serve as a combined settling and skimming tank and an unheated, unmixed anaerobic digester. Septic tanks provide long settling times (6–8 hours or more) but do not separate decomposing solids from the wastewater flow. When the tank becomes full, solids will be discharged with the flow.

The process is suitable for small facilities (e.g., schools, motels, homes), but it is unsuitable for larger applications due to the extended detention times and lack of control.

Two-Story (Imhoff) Tank

The two-story Imhoff tank, named after German engineer Karl Imhoff (1876–1965), is similar to a septic tank in that it removes settleable solids and allows for the anaerobic digestion of solids. The two-story tank consists of a settling compartment where sedimentation is accomplished, a lower compartment where settled solids and digestion occur, and gas vents. Solids are removed from the wastewater by settling and pass from the settling compartment into the digestion compartment through a slot at the bottom of the settling compartment. The slot's design prevents solids from returning to the settling compartment. Solids decompose anaerobically in the digestion section. Gases produced from the decomposition of the solids are released through the gas vents running along each side of the settling compartment.

Plain Settling Tanks (Clarifiers)

The plain settling tank or clarifier optimizes the settling process. Sludge is removed from the tank for processing in other downstream treatment units. Flow enters the tank, is slowed, and is distributed evenly across the width and depth of the unit. It passes through the unit and leaves over the effluent weir. Detention time within the primary settling tank is 1–3 hours (2-hour average). Sludge removal is accomplished frequently on either a continuous or intermittent basis. Continuous removal requires additional sludge treatment processes to remove the excess water resulting from the removal of sludge that contains less than 2%–3% solids. Intermittent sludge removal requires the sludge to be pumped from the tank on a schedule frequent enough to prevent large clumps of solids from rising to the surface but infrequent enough to obtain 4%–8% solids in the sludge withdrawn.

Scum must be removed from the surface of the settling tank frequently. This is usually a mechanical process but may require manual start-up. The system should be operated frequently enough to prevent excessive buildup and scum carryover but not so often as to cause hydraulic overloading of the scum removal system. Settling tanks require housekeeping and maintenance. Baffles (to prevent floatable solids and scum from leaving the tank), scum troughs, scum collectors, effluent troughs, and effluent weirs require frequent cleaning to avoid heavy biological growths and solids accumulation. Mechanical equipment must be lubricated and maintained as specified in the manufacturer's recommendations or by plant O&M manual procedures.

Process control sampling and testing evaluate the settling process's performance. Settleable solids, DO, pH, temperature, total suspended solids, and BOD_5, as well as sludge solids and volatile matter testing, are routinely carried out.

SLUDGE PUMPING

Determining sludge pumping (the quantity of solids and volatile solids removed from the sedimentation tank) provides accurate information to control the sedimentation process.

EFFLUENT FROM SETTLING TANKS

Upon screening, degritting, and settling in sedimentation basins, large debris, grit, and many settleable materials have been removed from the waste stream. What is left is referred to as *primary effluent*. Usually cloudy and frequently gray, primary effluent still contains large amounts of dissolved food and other chemicals (nutrients). These nutrients are treated in the next step in the treatment process (secondary treatment), which is discussed in the next section.

Note: Two of the most essential nutrients left to remove are phosphorus and ammonia. While we aim to remove these two nutrients from the waste stream, we do not want to remove too much. Carbonaceous microorganisms in secondary treatment (biological treatment) need phosphorus and ammonia.

SECONDARY TREATMENT

The primary purpose of *secondary treatment* (sometimes referred to as biological treatment) is to provide BOD removal beyond what is achievable by primary treatment (see Figure 7.1). There are three commonly used approaches, all of which take advantage of the ability of microorganisms to convert organic wastes (via biological treatment) into stabilized, low-energy compounds. Two of these approaches, the *trickling filter* [and its variation, the *rotating biological contactor (RBC)*] and the *activated sludge* process, sequentially follow regular primary treatment. However, the third approach—ponds (oxidation ponds or lagoons)—can provide equivalent results without preliminary treatment. This section presents a brief overview of the secondary treatment process, followed by a detailed discussion of wastewater treatment ponds (used primarily in smaller treatment plants), trickling filters, and RBCs. We then shift focus to the activated sludge process—the secondary treatment process mainly used in large installations and the primary focus of this handbook.

Secondary treatment refers to treatment processes that use biological processes to convert dissolved, suspended, and colloidal organic wastes into more stable solids that can either be removed by settling or discharged to the environment without causing harm. What is secondary treatment? The Clean Water Act (CWA) defines secondary treatment as producing an effluent with no more than 30 mg/L BOD_5 and 30 mg/L total suspended solids.

Note: The CWA also states that ponds and trickling filters will be included in the definition of secondary treatment even if they do not meet the effluent quality requirements continuously.

Most secondary treatment processes decompose solids aerobically, producing carbon dioxide, stable solids, and more organisms. Since solids are produced, all biological processes must include some form of solids removal (settling tank, filter, etc.). Secondary treatment processes can be separated into two categories: fixed film systems and suspended growth systems.

Fixed film systems use biological growth (biomass or slime) attached to some form of media. Wastewater passes over or around the media and the slime. When the wastewater and slime are in contact, the organisms remove and oxidize the organic solids. The media may be stone, redwood, synthetic materials, or any other durable substance (capable of withstanding weather conditions for many years), providing a large area for slime growth while providing open space for ventilation, and being non-toxic to the organisms in the biomass. Fixed film devices include trickling filters and rotating biological contactors (RBCs). *Suspended growth systems* use biological growth and mix it with wastewater. Typical suspended growth systems consist of various modifications of the activated sludge process.

Treatment Ponds

Wastewater treatment can be accomplished using *ponds* (aka *lagoons*). Ponds are relatively easy to build and manage, accommodate large fluctuations in flow, and can also provide treatment that approaches conventional systems (producing a highly purified effluent) at a much lower cost. The cost (economics) drives many managers to decide on the pond option. The actual degree of treatment provided depends on the type and number of ponds used. Ponds can be used as the sole type of treatment, or they can be used in conjunction with other forms of wastewater treatment—that is, other treatment processes followed by a pond or a pond followed by other treatment processes.

Stabilization ponds (treatment ponds) have been used to treat wastewater for over 3,000 years. The first recorded construction of a pond system in the U.S. was in San Antonio, Texas, in 1901. Today, over 8,000 wastewater treatment ponds are in place, involving more than 50% of the wastewater treatment facilities in the U.S. (CWNS, 2000). Facultative ponds account for 62%, aerated ponds 25%, anaerobic 0.04%, and total containment 12% of the pond treatment systems. They treat a variety of wastewater, from domestic wastewater to complex industrial wastes, and they function under a wide range of weather conditions, from tropical to arctic. Ponds can be used alone or in combination with other wastewater treatment processes. As our understanding of pond operating mechanisms has increased, different types of ponds have been developed for application to specific kinds of wastewater under local environmental conditions. This handbook focuses on municipal wastewater treatment pond systems.

While the tendency in the U.S. has been for smaller communities to build ponds, in other parts of the world, including Australia, New Zealand, Mexico, Latin America, Asia, and Africa, treatment ponds have been built for large cities. As a result, our understanding of the biological, biochemical, physical, and climatic factors that interact to transform the organic compounds, nutrients, and pathogenic organisms found in sewage into less harmful chemicals and unviable microorganisms (i.e., dead or sterile) has grown since 1983. A wealth of experience has been built up as civil, sanitary, or environmental engineers, operators, public works managers, and public health and environmental agencies have gained more experience with these systems. While some of this information makes its way into technical journals and textbooks, there is a need for a less formal presentation of the subject for those working in the field every day (USEPA, 2011).

Ponds are designed to enhance the growth of natural ecosystems that are either anaerobic (providing conditions for bacteria that grow in the absence of oxygen (O_2)), aerobic (promoting the growth of oxygen-producing and oxygen-requiring organisms, such as algae and bacteria), or facultative, which is a combination of the two. Ponds are managed to reduce concentrations of BOD, TSS, and coliform numbers (fecal or total) to meet water quality requirements.

Trickling Filters

Trickling filters have been used to treat wastewater since the 1890s. If settled wastewater was passed over rock surfaces, slime grew on the rocks, and the water became cleaner. We still use this principle today but use plastic media instead of stones in many installations. In most wastewater treatment systems, the *trickling filter* follows primary treatment and includes a secondary settling tank or clarifier, as shown in Figure 7.3. Trickling filters are widely used for the treatment of domestic and industrial wastes. The process is a fixed film biological treatment method to remove BOD_5 and suspended solids.

A trickling filter is a rotating distribution arm that sprays and evenly distributes liquid wastewater over a circular bed of fist-sized rocks, other coarse materials, or synthetic media (see Figure 7.4). The spaces between the media allow air to

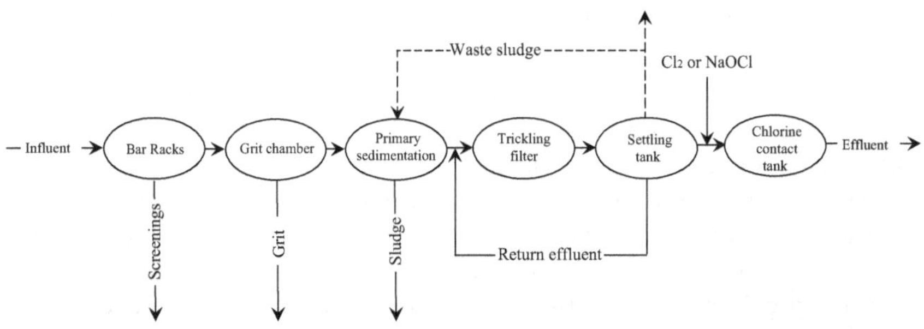

FIGURE 7.3 Simplified flow diagram of trickling filter used for wastewater treatment.

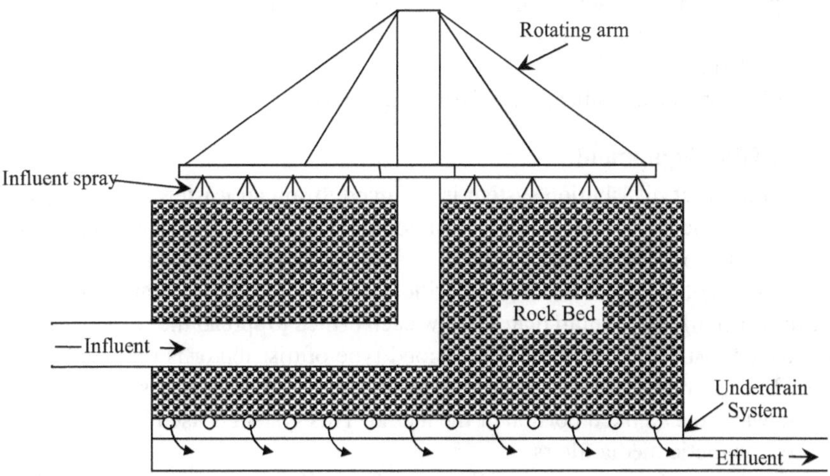

FIGURE 7.4 Schematic of a cross-section of a trickling filter.

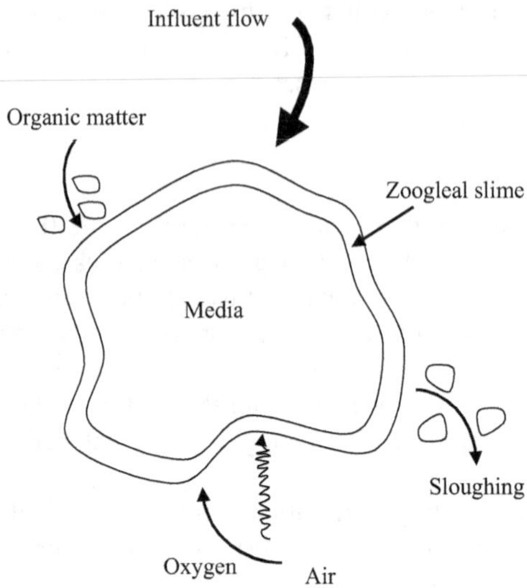

FIGURE 7.5 Filter media showing biological activities that take place on the surface area.

circulate easily to maintain aerobic conditions. The spaces enable wastewater to trickle down through, around, and over the media. A layer of biological slime that absorbs and consumes the wastes trickling through the bed covers the media material. The organisms aerobically decompose the solids, producing more organisms and stable wastes, which either become part of the slime or are discharged back into the wastewater flowing over the media. This slime consists mainly of bacteria but may include algae, protozoa, worms, snails, fungi, and insect larvae. The accumulating slime occasionally sloughs off (*sloughings*) individual media materials (see Figure 7.5). It is collected at the bottom of the filter, along with the treated wastewater, and passed on to the secondary settling tank, where it is removed. The trickling filter's overall performance depends on hydraulic and organic loading, temperature, and recirculation.

Table 7.2 lists the classifications of trickling filters.

Trickling Filter Equipment

The trickling filter distribution system is designed to spread wastewater evenly over the surface of the entire media. The most common system is the rotary distributor, which moves above the surface of the media and sprays the wastewater on the surface. The force of the water leaving the orifices drives the rotary system. The distributor's arms usually have small plates below each orifice to spread the wastewater into a fan-shaped distribution system. The second type of distributor is the fixed nozzle system. In this system, the nozzles are fixed above the media and designed to spray the wastewater over a fixed portion of the media. This system is used frequently with deep-bed synthetic media filters.

Note: Trickling filters that use ordinary rock usually are only about 3 m in depth because of structural problems caused by the weight of stones—which also requires

TABLE 7.2
Trickling Filter Classification

Filter Class	Standard	Intermediate	High-Rate	Super High-Rate	Roughing
Hydraulic loading gpd/ft^2	25–90	90–230	230–900	350–2,100	>900
Organic loading BOD per 1,000 ft^3	5–25	15–30	25–300	Up to 300	>300
Sloughing frequency	Seasonal	Varies	Continuous	Continuous	Continuous
Distribution	Rotary	Rotary fixed	Rotary fixed	Rotary	Rotary fixed
Recirculation	No	Usually	Always	Usually	Not usually
Media depth, ft	6–8	6–8	3–8	Up to 40	3–20
Media type	Rock Plastic Wood	Rock Plastic Wood	Rock Plastic Wood	Plastic Plastic Wood	Rock
Nitrification	Yes	Some	Some	Limited	None
Filter flies	Yes	Variable	Variable	Very few	Not usually
BOD removal	80%–85%	50%–70%	65%–80%	65%–85%	40%–65%
TSS removal	80%–85%	50%–70%	65%–80%	65%–85%	40%–65%

the construction of beds quite broad, in many applications, up to 60 ft in diameter. When synthetic media is used, the bed can be much deeper.

No matter which type of media is selected, the primary consideration is that it must be capable of providing the desired film location to develop the biomass. Depending on the type of media used and the filter classification, the media may be 3–20 ft deep.

The underdrains are designed to support the media, collect the wastewater and sloughings, carry them out of the filter, and provide ventilation to the filter.

Note: To ensure sufficient airflow to the filter, the underdrains should never be allowed to flow more than 50% of wastewater.

The effluent channel is designed to carry the flow from the trickling filter to the secondary settling tank. The secondary settling tank provides 2–4 hours of detention time to separate the sloughing materials from the treated wastewater. Its design, construction, and operation are similar to those of the primary settling tank. Longer detention times are provided because the sloughing materials are lighter and settle more slowly.

Recirculation pumps and piping are designed to recirculate (and thus improve the performance of the trickling filter or settling tank) a portion of the effluent back to be mixed with the filter influent. When recirculation is used, pumps and metering devices must be provided.

Rotating Biological Contactors (RBCs)

The rotating biological contactor (RBC) is a biological treatment system (see Figure 7.6) and is a variation of the attached growth idea provided by the trickling filter. Still relying on microorganisms that grow on the surface of a medium, the

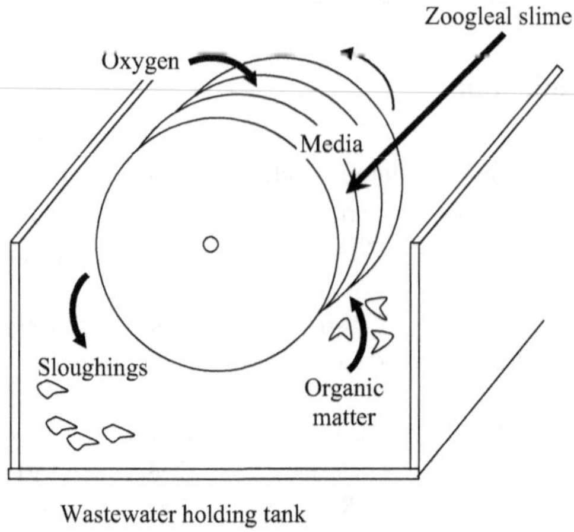

FIGURE 7.6 Rotating biological contactor (RBC) cross-section and treatment system.

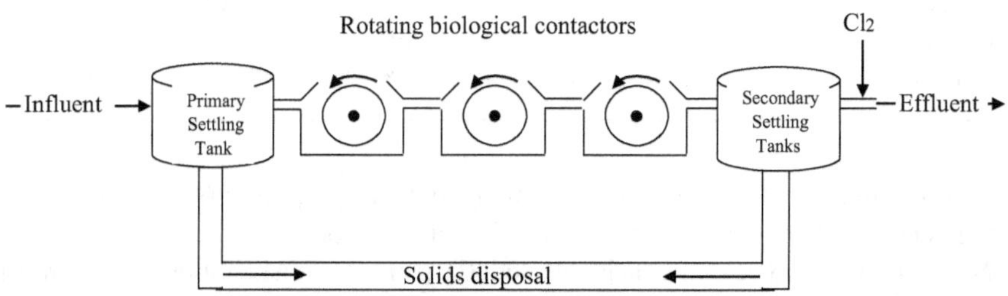

FIGURE 7.7 Rotating biological contactor (RBC) treatment system.

RBC is instead a **fixed-film** biological treatment device—the fundamental biological process, however, is similar to that occurring in the trickling filter. An RBC consists of a series of closely spaced (mounted side by side), circular, plastic (synthetic) disks that are typically about 3.5 m in diameter and attached to a rotating horizontal shaft (see Figure 7.6). Approximately 40% of each disk is submerged in a tank containing the wastewater to be treated. As the RBC rotates, the attached biomass film (zoogleal slime) that grows on the disk's surface moves into and out of the wastewater. While submerged in the wastewater, the microorganisms absorb organics; while rotating out of the wastewater, they are supplied with needed oxygen for aerobic decomposition. As the zoogleal slime reenters the wastewater, excess solids and waste products are stripped off the media as sloughings. These sloughings are transported with the wastewater flow to a settling tank for removal.

Modular RBC units are placed in series (see Figure 7.7). Because a single contactor is insufficient to achieve the desired level of treatment, the resulting treatment exceeds conventional secondary treatment. Each contactor is called a stage, and the group is known as a train. Most RBC systems consist of two or more trains with three or more stages in each. The essential advantage of using RBCs instead of trickling filters is that they are easier to operate under varying load conditions since they are easier to keep the solid medium wet at all times. Moreover, the level of nitrification, which an RBC system can achieve, is significant—especially when multiple stages are employed.

RBC Equipment

The equipment that makes up an RBC includes the rotating biological contactor (the media: either standard or high density), a center shaft, a drive system, a tank, baffles, housing or cover, and a settling tank. The *rotating biological contactor* consists of circular sheets of synthetic material (usually plastic) mounted side by side on a shaft. The sheets (media) contain large amounts of surface area for biomass growth. The *center shaft* supports the media's disks and must be strong enough to support the weight of the media and the biomass. Experience has indicated a significant problem with the collapse of the support shaft. The *drive system* provides the motive force to rotate the disks and shaft. The drive system may be mechanical, air-driven, or a combination. When the drive system does not provide uniform movement of the RBC, major operational problems can arise.

The *tank* holds the wastewater that the RBC rotates in. It should be large enough to permit liquid depth variation and detention time variation. *Baffles* are required to allow proper adjustment of the loading applied to each stage of the RBC process. Adjustments can be made to increase or decrease the submergence of the RBC. RBC stages are usually enclosed in some protective structure (*cover*) to prevent biomass loss due to severe weather changes (snow, rain, temperature, wind, sunlight, etc.). In many instances, this housing dramatically restricts access to the RBC. The *settling tank* is provided to remove the sloughing material created by the biological activity and is similar in design to the primary settling tank. The settling tank provides 2-to-4-hour detention times to permit the settling of lighter biological solids.

During normal operation, operator vigilance is required to observe the RBC movement, slime color, and appearance. However, if the unit is covered, observations may be limited to that portion of the media, which can be viewed through the access door. Slime color and appearance can indicate process conditions. For example:

- Gray, shaggy slime growth—indicates normal operation.
- Reddish brown, golden shaggy growth—indicates nitrification.
- White chalky appearance—indicates high sulfur concentrations.
- No slime—indicates severe temperature or pH changes.

Daily sampling and testing should be conducted for DO content and pH. BOD_5 and suspended solids testing should also be accomplished to assess performance.

The RBC typically produces high-quality effluent with BOD_5 at 85%–95% and Suspended Solids Removal at 85%–95%. The RBC treatment process may also significantly reduce (if designed for this purpose) the organic and ammonia nitrogen levels.

Activated Sludge

The biological treatment systems discussed to this point (ponds, trickling filters, and rotating biological contactors (RBCs)) have been around for years. The trickling filter, for example, has been successfully used since the late 1800s. The problem with ponds, trickling filters, and RBCs is that they are temperature sensitive and remove less BOD. Trickling filters, for example, cost more to build than the activated sludge systems that were later developed.

Note: Although trickling filters and other systems cost more to build than activated sludge systems, activated sludge systems cost more to operate because they require energy to run pumps and blowers.

As shown in Figure 7.1, the activated sludge process follows primary settling. The essential components of an activated sludge sewage treatment system include an aeration tank and a secondary basin, settling basin, or clarifier (see Figure 7.8). Primary effluent is mixed with settled solids recycled from the secondary clarifier and then introduced into the aeration tank. Compressed air is injected continuously into the mixture through porous diffusers at the tank's bottom, usually along one side.

Wastewater is fed continuously into an aerated tank, where the microorganisms metabolize and biologically flocculate the organics. Microorganisms (activated sludge) are settled from the aerated mixed liquor under quiescent conditions in the final clarifier and are returned to the aeration tank. Left uncontrolled, the number of organisms would eventually become too great; therefore, some must periodically be removed (wasted). A portion of the concentrated solids from the bottom of the settling tank must be removed from the process (waste-activated sludge or WAS). The clear supernatant from the final settling tank is the plant effluent.

The equipment requirements for the activated sludge process are more complex than those for the other processes discussed. Equipment includes an *aeration tank, system-settling tank, return sludge*, and *waste sludge system*. These are discussed below.

Aeration Tank: The aeration tank is designed to provide the required detention time (depending on the specific modification) and ensure that the activated sludge and the influent wastewater are thoroughly mixed. Tank design typically attempts to ensure that no dead spots are created.

Aeration: Can be mechanical or diffused. Mechanical aeration systems use agitators or mixers to mix air and mixed liquor. Some systems use a sparge ring to release

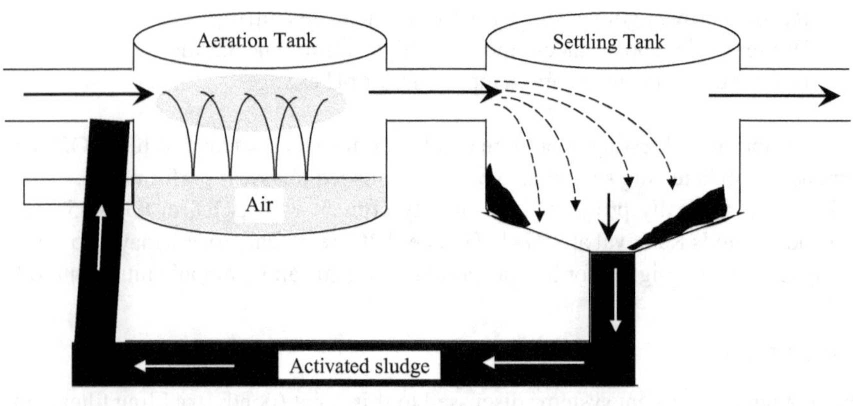

FIGURE 7.8 Activated sludge process.

air directly into the mixer. Diffused aeration systems use pressurized air released through diffusers near the bottom of the tank. Efficiency is directly related to the size of the air bubbles produced. Fine bubble systems have a higher efficiency. The diffused air system has a blower to produce large volumes of low-pressure air (5–10 psi), airlines to carry the air to the aeration tank, and headers to distribute the air to the diffusers, which release the air into the wastewater.

Settling Tank: Activated sludge systems have plain *settling tanks* that provide 2–4 hours of hydraulic detention time.

Return Sludge: The system includes pumps, a timer or variable-speed drive to regulate pump delivery, and a measurement device to determine actual flow rates.

Waste Sludge: In some cases, waste-activated *sludge* withdrawal is accomplished by adjusting valves on the return system. When a separate system is used, it includes pumps, a timer or variable-speed drive, and a flow measurement device.

Activated sludge is a treatment technique in which wastewater and reused biological sludge full of living microorganisms are mixed and aerated. The biological solids are then separated from the treated wastewater in a clarifier and returned to the aeration process or wasted. The microorganisms are mixed thoroughly with the incoming organic material and grow and reproduce using the organic material as food. As they grow and are combined with air, the organisms cling together (flocculate). Once flocculated, they more readily settle in the secondary clarifiers.

The treated wastewater flows continuously into an aeration tank where air is injected to mix the wastewater with the returned activated sludge and supply the oxygen the microbes need to live and feed on the organics. Aeration can be provided by injection through air diffusers at the bottom of the tank or by mechanical aerators at the surface. The mixture of activated sludge and wastewater in the aeration tank is called the "mixed liquor." The mixed liquor flows to a secondary clarifier, where the activated sludge can settle.

The activated sludge is constantly growing, and more is produced than can be returned for use in the aeration basin. Therefore, some of this sludge must be wasted in a sludge handling system for treatment and disposal. The volume of sludge returned to the aeration basins usually is 40%–60% of the wastewater flow. The rest is wasted.

Disinfection of Wastewater

Like drinking water, liquid wastewater effluent is disinfected. Unlike drinking water, wastewater effluent is disinfected not directly (direct end-of-pipe connection) to protect drinking water supplies but instead is treated to protect public health in general. This is particularly important when the secondary effluent is discharged into a body of water used for swimming or a downstream water supply. In treating water for human consumption, treated water is typically chlorinated (although ozonation is also currently being applied in many cases). Chlorination is the preferred disinfection method in potable water supplies because of chlorine's unique ability to provide a residual. This chlorine residual is essential because when treated water leaves the waterworks facility and enters the distribution system, the possibility of contamination is increased—the residual works to disinfect water up to the consumer's tap continuously.

Considering the previously provided in-depth discussion of water disinfection, this section describes primary chlorination and dechlorination as applied to wastewater. In addition, we describe ultraviolet (UV) irradiation, ozonation, bromine chlorine, and no disinfection. Again, remember that much of the chlorination material presented in the following is similar to the information presented earlier in the water disinfection section.

Chlorine Disinfection

As shown in Figure 7.1, chlorination for disinfection follows all other steps in conventional wastewater treatment. The purpose of chlorination is to reduce the population of organisms in the wastewater to levels low enough to ensure that pathogenic microorganisms will not be present in sufficient quantities to cause disease when discharged.

Note: Chlorine gas is heavier than air (vapor density of 2.5). Therefore, exhaust from a chlorinator room should be taken from the floor level.

Note: The safest action to take in case of a significant chlorine container leak is to call the fire department.

Note: You might wonder why chlorinating critical waters such as natural trout streams is not standard practice. This practice is strictly prohibited because chlorine and its by-products (i.e., chloramines) are highly toxic to aquatic organisms.

Though some minor hazards are associated with its use (skin irritation, nose irritation, and burning eyes), hypochlorite is relatively safe to work with. It is usually available in dry form as a white powder, pellet, tablet, or liquid. It can be added directly using a dry chemical feeder or dissolved and fed as a solution.

Note: Disinfection is accomplished in most wastewater treatment systems using combined residual.

Ultraviolet Irradiation

Although ultraviolet (UV) disinfection was recognized as a method for achieving disinfection in the late nineteenth century, its application virtually disappeared with the evolution of chlorination technologies. However, in recent years, there has been a resurgence in its use in the wastewater field, mainly due to concerns about the discharge of toxic chlorine residuals. Even more recently, UV has gained more attention because of the strict new regulations on chlorine use imposed by both OSHA and USEPA. Because of this recently increased regulatory pressure, many facilities are actively substituting chlorine with other disinfection alternatives. Moreover, UV technology has made many improvements, making UV attractive as a disinfection alternative. Ultraviolet light has excellent germicidal qualities and is very effective in destroying microorganisms. It is used in hospitals, biological testing facilities, and other locations. In wastewater treatment, the plant effluent is exposed to ultraviolet light of a specified wavelength and intensity for a specified contact period. The effectiveness of the process depends on:

- UV light intensity
- Contact time
- Wastewater quality (turbidity)
- For any one treatment plant, disinfection success is directly related to the concentration of colloidal and particulate constituents in the wastewater.

The Achilles' heel of UV for disinfecting wastewater is turbidity. If the wastewater quality is poor, the ultraviolet light will be unable to penetrate the solids, and the effectiveness of the process will decrease dramatically. For this reason, many states limit the use of UV disinfection to facilities that can reasonably be expected to produce an effluent containing ≤30 mg/L or less of BOD_5 and total suspended solids. The main components of a UV disinfection system are mercury arc lamps, a reactor, and ballasts. The source of UV radiation is either the low-pressure or medium-pressure mercury arc lamp with low or high intensities. Note that UV lamps must be readily available when replacements are required to operate UV systems. The best lamps are those with a stated operating life of at least 7,500 hours and those that do not produce significant amounts of ozone or hydrogen peroxide. The lamps must also meet technical specifications for intensity, output, and arc length. If the UV light tubes are submerged in the waste stream, they must be protected inside quartz tubes, which not only protect the lights but also make cleaning and replacement easier.

Contact tanks must be used with UV disinfection. They must be designed with the banks of UV lights in a horizontal position, either parallel or perpendicular to the flow, or with banks of lights placed in a vertical position perpendicular to the flow.

Note: The contact tank must provide a minimum of 10 seconds of exposure time.

We stated earlier that turbidity has been a problem with using UV in wastewater treatment—and this is the case. However, if turbidity is its Achilles' heel, then the need for increased maintenance (compared to other disinfection alternatives) is the toe of the same foot. UV maintenance requires that the tubes be cleaned regularly or as needed. In addition, periodic acid washing is also necessary to remove chemical buildup.

Routine monitoring is required to check bulb burnout, the buildup of solids on quartz tubes, and UV light intensity.

Note: UV light is highly hazardous to the eyes. Never enter an area where UV lights operate without proper eye protection, and never look directly into the ultraviolet light.

Advantages and Disadvantages

Advantages:

- UV disinfection effectively inactivates most viruses, spores, and cysts.
- UV disinfection is a physical process rather than a chemical disinfectant, eliminating the need to generate, handle, transport, or store toxic, hazardous or corrosive chemicals.
- There is no residual effect that can harm humans or aquatic life.
- UV disinfection is user-friendly for operators.
- UV disinfection has a shorter contact time when compared with other disinfectants (approximately 20–30 seconds with low-pressure lamps).
- UV disinfection equipment requires less space than other methods.

Disadvantages:

- Low dosages may not effectively inactivate some viruses, spores, and cysts.
- Organisms can sometimes repair and reverse the destructive effects of UV through a "repair mechanism" known as photoreactivation or, in the absence of light, known as "dark repairs."
- A preventive maintenance program is necessary to control the fouling of tubes.
- Turbidity and TSS in the wastewater can render UV disinfection ineffective. UV disinfection with low-pressure lamps is less effective for secondary effluent with TSS levels above 30 mg/L.
- UV disinfection is less cost-competitive when chlorination dechlorination is used, and fire codes are met (USEPA, 1999).

When choosing a UV disinfection system, three critical areas must be considered. The manufacturer primarily determines the first, the second by design and Operation and Maintenance (O&M), and the third, which has to be controlled at the treatment facility. Choosing a UV disinfection system depends on three critical factors listed below:

- **Hydraulic Properties of the Reactor**: Ideally, a UV disinfection system should have a uniform flow with enough axial motion (radial mixing) to maximize exposure to UV radiation. An organism's path in the reactor determines the amount of UV radiation it will be exposed to before inactivation. A reactor must be designed to eliminate short-circuiting and/or dead zones, which can result in inefficient power use and reduced contact time.
- **Intensity of the UV Radiation**: Factors affecting the intensity are the age of the lamps, lamp fouling, and the configuration and placement of lamps in the reactor.
- **Wastewater Characteristics**: These include the flow rate, suspended and colloidal solids, initial bacterial density, and other physical and chemical parameters. The concentration of TSS and particle-associated microorganisms determines how much UV radiation ultimately reaches the target organism. The higher these concentrations, the lower the UV radiation absorbed by the microorganisms. UV disinfection can be used in plants of various sizes that provide secondary or advanced levels of treatment.

Any UV disinfection should be pilot-tested before the full-scale platform to ensure it meets discharge permit requirements for a particular site. The average lamp life ranges from 8,760 to 14,000 working hours, and the lamps are usually replaced after 12,000 hours of use. Operating procedures should be set to reduce the on/off cycles of the lamps since efficacy is reduced with repeated cycles. The ballast must be compatible with the lamps and should be ventilated to protect it from excessive heating, which may shorten its life or even result in fires. Although the life cycle of ballasts is approximately 10–15 years, they are usually replaced every 10 years. Quartz sleeves will last about 5–8 years but are generally replaced every 5 years (USEPA, 1999).

Ozonation

Ozone is a potent oxidizing gas that reacts with most organic and many inorganic molecules. It is produced when oxygen molecules separate, collide with other oxygen atoms, and form a molecule consisting of three oxygen atoms. For high-quality effluents, ozone is a very effective disinfectant. Current regulations for domestic treatment systems limit the use of ozonation to filtered effluents unless the system's effectiveness can be demonstrated before installation.

Note: Effluent quality is the critical performance factor for ozonation.

For wastewater ozonation, the facility must have the capability to generate pure oxygen along with an ozone generator. A contact tank with ≥10-minute contact time at design average daily flow is required. Off-gas monitoring for process control is also required. In addition, safety equipment capable of monitoring ozone in the atmosphere and a ventilation system to prevent ozone levels from exceeding 0.1 ppm are required.

The operation of the ozonation process consists of monitoring and adjusting the ozone generator and monitoring the control system to maintain the required ozone concentration in the off-gas. The process must also be evaluated periodically using biological testing to assess its effectiveness.

Note: Ozone is a highly toxic substance. Air concentrations should not exceed 0.1 ppm. It also has the potential to create an explosive atmosphere. Sufficient ventilation and purging capabilities should be provided.

Note: Ozone has certain advantages over chlorine for the disinfection of wastewater: (1) Ozone increases DO in the effluent; (2) ozone has a briefer contact time; (3) ozone has no undesirable effects on marine organisms; and (4) ozone decreases turbidity and odor.

Advantages and Disadvantages

Advantages:

- Ozone is more effective than chlorine in destroying viruses and bacteria.
- The ozonation process utilizes a short contact time (approximately 10–30 minutes).
- There are no harmful residuals to remove after ozonation because ozone decomposes rapidly.
- After ozonation, microorganisms are not regressed, except those protected by the particulates in the wastewater stream.
- Ozone is generated onsite; thus, there are fewer safety problems associated with shipping and handling.
- Ozonation elevates the DO concentration of the effluent. The increase in DO can eliminate the need for reaeration and raise the DO level in the receiving stream.

Disadvantages:

- Low dosage may not effectively inactivate some viruses, spores, and cysts.
- Ozonation is a more complex technology than chlorine or UV disinfection, requiring complicated equipment and efficient contacting systems.

- Ozone is very reactive and corrosive, thus requiring corrosion-resistant material such as stainless steel.
- Ozonation is not economical for wastewater with high levels of suspended soils (SS), BOD, chemical oxygen demand, or total organic carbon.
- Ozone is highly irritating and possibly toxic, so off-gases from the contactor must be destroyed to prevent worker exposure.
- The cost of treatment can be relatively high in capital and power intensiveness.

Bromine Chloride

Bromine chloride is a mixture of bromine and chlorine. When mixed with water, it forms hydrocarbons and hydrochloric acid. Bromine chloride is an excellent disinfectant that reacts quickly and does not produce long-term residuals.

Note: Bromine chloride is a highly corrosive compound in the presence of low moisture concentrations.

The reactions occurring when bromine chloride is added to the wastewater are similar to those occurring when chlorine is added. The significant difference is the production of bromine compounds rather than chloramines. The bromine compounds are excellent disinfectants but are less stable and dissipate quickly. In most cases, the bromamines rapidly decay into other, less toxic compounds and are undetectable in the plant effluent. The factors that affect performance are similar to those affecting the performance of the chlorine disinfection process. Effluent quality, contact time, etc., directly impact the process's performance.

No Disinfection

In a few cases, treated wastewater discharges without disinfection are permitted. These are approved on a case-by-case basis. Each request must be evaluated based on the discharge point, discharge quality, potential for human contact, and many other factors.

ADVANCED WASTEWATER TREATMENT

Advanced wastewater treatment is defined as the method(s) and/or process(es) that remove more contaminants (suspended and dissolved substances) from wastewater than conventional biological treatment. Put another way, advanced wastewater treatment is the application of a process or system that follows secondary treatment or includes phosphorus removal or nitrification in conventional secondary treatment.

Advanced wastewater treatment is used to augment conventional secondary treatment because secondary treatment typically removes only between 85% and 95% of the BOD and TSS in raw sanitary sewage. Generally, this leaves 30 mg/L or less of BOD and TSS in the secondary effluent. To meet stringent water-quality standards, this level of BOD and TSS in secondary effluent may not prevent violations of water-quality standards—the plant may not make a permit. Thus, advanced wastewater treatment is often used to remove additional pollutants from treated wastewater.

In addition to meeting or exceeding the requirements of water-quality standards, treatment facilities use advanced wastewater treatment for other reasons. Conventional secondary wastewater treatment is sometimes insufficient to protect the aquatic environment. For example, when periodic flow events occur, the stream may not provide the effluent dilution needed to maintain the DO levels for aquatic organism survival.

Secondary treatment has other limitations. It does not significantly reduce sewage's effluent concentration of nitrogen and phosphorus (essential plant nutrients). Overusing these nutrients can overstimulate plant and algae growth, creating water quality problems. For example, if discharged into lakes, these nutrients contribute to algal blooms and accelerated eutrophication (lake aging). Also, the nitrogen in the sewage effluent may be present mainly in the form of ammonia compounds. If in high enough concentrations, ammonia compounds are toxic to aquatic organisms. Yet another problem with these compounds is that they exert a *nitrogenous* oxygen demand in the receiving water as they convert to nitrates. This process is called nitrification.

Note: The term *tertiary treatment* is commonly used as a synonym for advanced wastewater treatment. However, these two terms do not have precisely the same meaning. Tertiary suggests a third step that is applied after primary and secondary treatment.

Advanced wastewater treatment can remove more than 99% of the pollutants from raw sewage and produce an almost potable (drinking) water quality effluent. However, advanced treatment is not cost-free. The cost of advanced treatment, operation and maintenance, and retrofitting present conventional processes is very high (sometimes doubling the cost of secondary treatment). Therefore, a plan to install advanced treatment technology calls for careful study—the benefit-to-cost ratio is not always ample enough to justify the additional expense.

Filtration

The purpose of *filtration* processes used in advanced treatment is to remove suspended solids. The specific operations associated with a filtration system depend on the equipment used. In operation, wastewater flows to a filter (gravity or pressurized). The filter contains single, dual, or multimedia. Wastewater flows through the media, which removes solids. The solids remain in the filter. Backwashing the filter as needed removes trapped solids. Backwash solids are returned to the plant for treatment. Processes typically remove 95%–99% of the suspended matter.

Membrane Bioreactors

The use of microfiltration membrane bioreactors (MBRs), a technology that has become increasingly used in the past 10 years, overcomes many of the limitations of conventional systems. These systems can combine a suspended-growth biological reactor with solids removal via filtration. The membranes can be designed for and operated in small spaces with high removal efficiency of contaminants such as nitrogen, phosphorus, bacteria, biochemical oxygen demand, and total suspended solids. The membrane filtration system, in effect, can replace the secondary clarifier and sand filters in a typical activated sludge treatment system. Membrane filtration maintains a higher biomass concentration, allowing smaller bioreactors to be used (USEPA, 2007a).

For new installations, using MBR systems allows for higher wastewater flow or improved treatment performance in a smaller space than a conventional design, i.e., a facility using secondary clarifiers and sand filters. Historically, membranes have been used for smaller-flow systems due to the high capital cost of the equipment and high operation and maintenance (O&M) costs. Today, however, they are receiving increased use in larger systems. MBR systems are also well-suited for some industrial and commercial applications. The high-quality effluent produced by MBRs makes them particularly applicable to reuse applications for surface water discharge applications requiring extensive nutrient (nitrogen and phosphorus) removal (USEPA, 2007a).

Advantages:

The advantages of MBR systems over conventional biological systems include better effluent quality, smaller space requirements, and ease of automation. Specifically, MBRs operate at higher volumetric loading rates, resulting in lower hydraulic retention times. The low retention times mean less space is required than a conventional system. MBRs have often been operated with more extended solids residence times (SRTs), which results in lower sludge production, but this is not a requirement, and more conventional SRTs have been used (Crawford et al., 2000). The effluent from MBRs contains low concentrations of bacteria, TSS, BOD, and phosphorus. This facilitates high-level disinfection. Effluents are readily discharged to surface streams or can be sold for reuse, such as irrigation (USEPA, 2007a).

Disadvantages:

The primary disadvantage of MBR systems is the typically higher capital and operating costs than conventional systems for the same throughput. O&M costs include membrane cleaning, fouling control, and eventual membrane replacement. Energy costs are also higher because air scouring is needed to control bacterial growth on the membranes. In addition, the waste sludge from such a system might have a low settling rate, resulting in the need for chemicals to produce biosolids acceptable for disposal. Fleischer et al. (2005) demonstrated that waste sludges from MBRs can be processed using standard technologies for activated sludge processes.

Biological Nitrification

Biological *nitrification* is the first essential step of *biological nitrification-denitrification*.

In nitrification, the secondary effluent is introduced into another aeration tank, trickling filter, or biodisc. Because most of the carbonaceous BOD has already been removed, the microorganisms that drive this advanced step are the nitrifying bacteria *Nitrosomonas* and *Nitrobacter.* In nitrification, the ammonia nitrogen is converted to nitrate nitrogen, producing a *nitrified effluent.* At this point, the nitrogen has not been removed; it has only been converted to a form that is not toxic to aquatic life and does not cause an additional oxygen demand. The nitrification process can be limited (performance affected) by alkalinity (requires 7.3 parts alkalinity to 1.0 part ammonia nitrogen), pH, DO availability, toxicity (ammonia or other toxic materials), and process mean cell residence time (sludge retention time). As a general rule, biological nitrification is more effective and achieves higher removal levels during the warmer times of the year.

Biological Denitrification

Biological denitrification removes nitrogen from the wastewater. When bacteria come in contact with a nitrified element without oxygen, they reduce the nitrates to nitrogen gas, which escapes the wastewater. The denitrification process can be done in either an anoxic-activated sludge system (suspended growth) or a column system (fixed growth). The denitrification process can remove up to 85% or more of nitrogen. After effective biological treatment, little oxygen-demanding material is left in the wastewater when it reaches the denitrification process. The denitrification reaction will only occur if an oxygen demand source exists when no DO is present in the wastewater. An oxygen demand source is usually added to reduce the nitrates quickly. The most common demand source added is soluble BOD or methanol. Approximately 3 mg/L of methanol is added for every 1 mg/L of nitrate-nitrogen. Suspended growth denitrification reactors are mixed mechanically, but only enough to keep the biomass from settling without adding unwanted oxygen. Submerged filters of different types of media may also be used to provide denitrification. A fine media downflow filter is sometimes used to give both denitrification and effluent filtration. A fluidized sand bed where wastewater flows upward through a media of sand or activated carbon at a rate to fluidize the bed may also be used. Denitrification bacteria grow on the media.

Carbon Adsorption

The primary purpose of *carbon adsorption* used in advanced treatment processes is the removal of refractory organic compounds (non-BOD_5) and soluble organic material that are difficult to eliminate by biological or physical/chemical treatment. In the carbon adsorption process, wastewater passes through a container filled with carbon powder or carbon slurry. Organics adsorb onto the carbon (i.e., organic molecules are attracted to the activated carbon surface and are held there) with sufficient contact time. A carbon system usually has several columns or basins used as contactors. Most contact chambers are open concrete gravity-type systems or steel pressure containers for upflow or downflow operation. With use, carbon loses its adsorptive capacity. The carbon must then be regenerated or replaced with fresh carbon. As head loss develops in carbon contactors, they are backwashed with clean effluent, just like the effluent filters are backwashed. Carbon used for adsorption may be in a granular form or a powdered form.

Note: Powdered carbon is too fine for use in columns and is usually added to the wastewater, then removed by coagulation and settling.

Land Application

The application of secondary effluent onto a land surface can provide an effective alternative to the expensive and complicated advanced treatment methods discussed previously and the biological nutrient removal (BNR) system discussed later. A high-quality polished effluent (i.e., effluent with high levels of TSS, BOD, phosphorus, and nitrogen compounds, as well as refractory organics, is reduced) can be obtained by the natural processes that occur as the effluent flows over the vegetated ground surface and percolates through the soil. Limitations are involved with the land application of wastewater effluent. For example, the process needs large land areas. Soil type and climate are also critical factors in controlling the design and feasibility of a land treatment process.

Three basic types or modes of land application or treatment are commonly used: irrigation (slow rate), overland flow, and infiltration-percolation (rapid rate). The primary objectives of these types of land applications and the conditions under which they can function vary. In *irrigation* (also called slow rate), wastewater is sprayed or applied (usually by ridge-and-furrow spreading or sprinkler systems) to the land's surface. Wastewater enters the soil. Crops growing in the irrigation area utilize available nutrients. Soil organisms stabilize the organic content of the flow. Water returns to the hydrologic (water) cycle through evaporation or by entering surface water or groundwater.

The land irrigation application method provides the best results (compared with the other two types of land application systems) for advanced treatment levels of pollutant removal. Not only are suspended solids and BOD significantly reduced by wastewater filtration, but biological oxidation of the organics in the top few inches of soil occurs. Nitrogen is removed primarily by crop uptake, and phosphorus is removed by adsorption within the soil. Expected performance levels for irrigation include:

- BOD_5—98%
- Suspended solids—98%
- Nitrogen—85%
- Phosphorus—95%
- Metals—95%

The overland flow application method utilizes physical, chemical, and biological processes as the wastewater flows in a thin film down the relatively impermeable surface. In the process, wastewater sprayed over sloped terraces flows slowly over the surface. Soil and vegetation remove suspended solids, nutrients, and organics. A small portion of the wastewater evaporates. The remainder flows to collection channels. Collected effluent is discharged to surface waters. Expected performance levels for overland flow include:

- BOD_5—92%
- Suspended solids—92%
- Nitrogen—70%–90%
- Phosphorus—40%–80%
- Metals—50%

In the infiltration-percolation (rapid rate) land application process, wastewater is sprayed/pumped to spreading basins (aka recharge basins or large ponds).

Some wastewater evaporates. The remainder percolates/infiltrates into the soil. Solids are removed by filtration. Water recharges the groundwater system. Most of the effluent percolates to the groundwater; very little of it is absorbed by vegetation. The filtering and adsorption action of the soil removes most of the BOD, TSS, and phosphorus from the effluent; however, nitrogen removal is relatively poor. Expected performance levels for infiltration-percolation include:

- BOD_5—85%–99%
- Suspended solids—98%
- Nitrogen—0%–50%
- Phosphorus—60%–95%
- Metals—50%–95%

Biological Nutrient Removal (BNR)

Nitrogen and phosphorus are the primary causes of cultural eutrophication (i.e., nutrient enrichment due to human activities) in surface waters. The most recognizable manifestations of this eutrophication are algal blooms that occur during the summer. Chronic symptoms of over-enrichment include low DO, fish kills, murky water, and depletion of desirable flora and fauna. In addition, the increase in algae and turbidity increases the need to chlorinate drinking water, which, in turn, leads to higher levels of disinfection by-products that have been shown to increase the risk of cancer (USEPA, 2007c). Excessive amounts of nutrients can also stimulate the activity of microbes, such as *Pfisteria*, which may harm human health (USEPA, 2007d).

Approximately 25% of all water impairments are due to nutrient-related causes (e.g., nutrients, oxygen depletion, algal growth, ammonia, harmful algal blooms, biological integrity, and turbidity) (USEPA, 2007d). In efforts to reduce the number of nutrient impairments, many point source discharges have received more stringent effluent limits for nitrogen and phosphorus. To achieve these new, lower effluent limits, facilities have begun to look beyond traditional treatment technologies.

Recent experience has reinforced the concept that biological nutrient removal (BNR) systems are reliable and effective in removing nitrogen and phosphorus. The process is based on the principle that, under specific conditions, microorganisms will remove more phosphorus and nitrogen than is required for biological activity; thus, treatment can be accomplished without using chemicals. Not having to use and, therefore, having to purchase chemicals to remove nitrogen and phosphorus potentially has numerous cost-benefit implications. In addition, because chemicals are not required to be used, chemical waste products are not produced, reducing the need to handle and dispose of waste. Several patented processes are available for this purpose. Performance depends on the biological activity and the method employed.

The importance of BNR process configurations cannot be overstated. Some BNR systems are designed to remove only Total Nitrogen (TN) or Total Phosphorus (TP), while others remove both. If retrofitting an existing facility, the configuration most appropriate for any particular system depends on the target effluent quality, operator experience, influent quality, and existing treatment processes.

BNR configuration varies based on the sequencing of environmental conditions (i.e., aerobic, anaerobic, and anoxic) and timing (USEPA, 2007c). Standard BNR system configurations include:

- **Modified Ludzack-Ettinger (MLE) Process**: A continuous-flow suspended-growth process with an initial anoxic stage followed by an aerobic stage used to remove.
- **A^2/O Process**: MLE process preceded by an initial anaerobic stage; used to remove both TN and TP.
- **Step Feed Process**: Alternating anoxic and aerobic stages; however, influent flow is split to several feed locations, and the recycle sludge stream is sent to the beginning of the process, used to remove TN.
- **Bardenpho Process (Four-Stage)**: A continuous-flow suspended-growth process with alternating anoxic/aerobic/anoxic/aerobic stages; used to remove TN.
- **Modified Bardenpho Process**: Bardenpho process with the addition of an initial anaerobic zone; used to remove both TN and TP.
- **Sequencing Batch Reactor (SBR) Process**: Suspended-growth batch process sequenced to simulate the four-stage process; used to remove TN (TP removal is inconsistent).
- **Modified University of Cape Town (UCT) Process**: A^2/O Process with a second anoxic stage where the internal nitrate recycle is returned; used to remove both TN and TP.
- **Rotating Biological Contactor (RBC) Process**: A continuous-flow process using RBCs with sequential anoxic/aerobic stages to remove TN.
- **Oxidation Ditch**: A continuous-flow process using looped channels to create time-sequenced anoxic, aerobic, and anaerobic zones; used to remove both TN and TP.

Although the exact configurations of each system differ, BNR systems designed to remove TN must have an aerobic zone for nitrification and an anoxic zone for denitrification, and BNR systems designed to remove TP must have an anaerobic zone free of dissolved oxygen and nitrate. Often, sand or other media filtration is used as a polishing step to remove particulate matter when low TN and TP effluent concentrations are required.

Choosing which system is most appropriate for a particular facility depends on the target effluent concentrations and whether the facility will be constructed as new or retrofitted with BNR to achieve more stringent effluent limits. New plants have more flexibility and options when deciding which BNR configuration to implement because they are not constrained by existing treatment units and sludge handling procedures.

Retrofitting an existing plant with BNR capabilities should involve consideration of the following factors (Park, 2012):

- Aeration basin size and configuration
- Clarifier capacity

- Type of aeration system
- Sludge processing units
- Operator skills

The aeration basin size and configuration dictate which BNR configurations are the most economical and feasible. Available excess capacity reduces the need for additional basins and may allow for a more configuration (e.g., 5-stage Bardenpho versus 4-stage Bardenpho configuration). Additional basins could result in more land if the space needed is unavailable. If land is unavailable, another BNR process configuration may have to be considered.

Clarifier capacity influences the return activated sludge (RAS) rate and effluent suspended solids, affecting effluent TN and TP levels. Suppose the existing facility configuration does not allow for a pre-anoxic zone so nitrates can be removed before the anaerobic zone. In that case, the clarifier should be modified to have a sludge blanket just deep enough to prevent the release of phosphorus into the liquid.

The aeration system will most likely need to be modified to accommodate an anaerobic zone and reduce the DO concentration in the return sludge. Such modifications could be as simple as removing aeration equipment from the zone designated for anaerobic conditions or changing the type of pump used for the recycled sludge stream (to avoid the introduction of oxygen).

How sludge is processed at a facility is essential in designing nutrient removal systems. Sludge is recycled to provide the organisms necessary for the TN and TP removal mechanism. The content and volume of sludge recycled directly impacts the system's performance. Thus, sludge handling processes may be modified to achieve optimal TN and TP removal efficiencies. For example, some polymers in sludge dewatering could inhibit nitrification when recycled. Also, aerobic digestion of sludge processes nitrates, denitrification, and phosphorus uptake rates, which may be lowered when the sludge is recycled (USEPA, 2007c).

Operators should be able to adjust the process to compensate for constantly varying conditions. BNR processes are susceptible to influent conditions influenced by weather events, sludge processing, and other treatment processes (e.g., recycling after filter backwashing). Therefore, operator skills and training are essential for achieving target TN and TP effluent concentrations (USEPA, 2007c).

Enhanced Biological Nutrient Removal (EBNR)

Removing phosphorus from wastewater in secondary treatment has evolved into innovative *enhanced biological nutrient removal* (EBNR) technologies. An EBNR treatment process promotes the production of phosphorus-accumulating organisms that utilize more phosphorus in their metabolic processes than a conventional secondary biological treatment process (USEPA, 2007b). The average total phosphorus concentrations in raw domestic wastewater are usually between 6 and 8 mg/L, and the total phosphorus concentration in municipal wastewater after conventional secondary treatment is routinely reduced to 3 or 4 mg/L. Whereas EBNR incorporated into the secondary treatment system can often reduce total phosphorus concentrations to 0.3 mg/L or less. EBNR facilities significantly reduced the amount of phosphorus to be removed through the subsequent chemical addition and tertiary filtration process. This improved the efficiency of the tertiary process and significantly reduced the costs of chemicals used to remove phosphorus. Facilities using EBNR reported that their chemical dosing was cut in half after EBNR was installed to remove phosphorus (USEPA, 2007b).

Treatment provided by these WWTPs also removes other pollutants that commonly affect water quality to deficient levels (USEPA, 2007b). BOD and TSS are routinely less than 2 mg/L, and fecal coliform bacteria are less than 10 fcu/100 mL. The turbidity of the final effluent is very low, which allows for effective disinfection using ultraviolet light rather than chlorination. Recent studies report that wastewater treatment plants using EBNR also significantly reduced the amount of pharmaceuticals and personal healthcare products from municipal wastewater compared to the removal accomplished by conventional secondary treatment.

WASTEWATER SOLIDS (SLUDGE/BIOSOLIDS) HANDLING

The wastewater treatment unit processes described to this point remove solids and BOD from the waste stream before the liquid effluent is discharged to its receiving waters. A mixture of solids and wastes remains to be disposed of, called *process residuals*—more commonly called *sludge* or *biosolids*.

Note: *Sludge* is the commonly accepted name for wastewater solids. However, if wastewater sludge is used for beneficial reuse (e.g., as a soil amendment or fertilizer), it is commonly called *biosolids*.

The most costly and complex aspect of wastewater treatment is the collection, processing, and disposal of sludge. This is because the quantity of sludge produced may be as high as 2% of the original volume of wastewater, depending on the treatment process being used.

Because sludge can have as much as 97% water content, and because the cost of disposal is related to the volume of sludge being processed, one of the primary purposes or goals (along with stabilizing it so it is no longer objectionable or environmentally damaging) of sludge treatment is to separate as much of the water from the solids as possible. Sludge treatment methods may be designed to accomplish both of these purposes.

Note: Sludge treatment methods are divided into three major categories: thickening, stabilization, and dewatering. Many of these processes include complex sludge treatment methods (i.e., heat treatment, vacuum filtration, incineration, and others).

When we speak of *sludge* or *biosolids*, we speak of the same substance or material; each is defined as the suspended solids removed from wastewater during sedimentation and then concentrated for further treatment, disposal, or reuse. The difference between the terms *sludge* and *biosolids* is determined by how they are managed.

Note: Disposing, treating, or reusing wastewater solids is *sludge* or *biosolids management.*)

Sludge is typically seen as wastewater solids that are "disposed" of. Biosolids are the same substance managed for reuse—commonly called beneficial reuse (e.g., for land application as a soil amendment, such as biosolids compost). Even as wastewater treatment standards have become more stringent because of increasing environmental regulations, so has the volume of wastewater sludge. Also, before sludge can be disposed of or reused, some form of treatment is required to reduce its volume, stabilize it, and inactivate pathogenic organisms.

Sludge initially forms as a 3%–7% suspension of solids. With each person typically generating about 4 gal of sludge per week, the total quantity generated each day, week, month, and year is significant. Because of the volume and nature of the material, sludge management is a substantial factor in the design and operation of all water pollution control plants.

Note: Wastewater solids treatment, handling, and disposal account for more than half of the total costs in a typical secondary treatment plant.

Sources of Sludge

Wastewater sludge is generated in primary, secondary, and chemical treatment processes. In primary treatment, the solids that float or settle are removed. The floatable material makes up a portion of the solid waste known as scum. Scum is not customarily considered sludge; however, it should be disposed of in an environmentally sound way. The settleable material that collects on the bottom of the clarifier is known as *primary sludge*. Primary sludge can also be referred to as raw sludge because it has not undergone decomposition. Raw primary sludge from a typical domestic facility is quite objectionable and has a high percentage of water, two characteristics that make handling difficult.

Solids not removed in the primary clarifier are carried out of the primary unit. These solids are known as *colloidal suspended solids*. The secondary treatment system (i.e., trickling filter, activated sludge, etc.) is designed to change those colloidal solids into settleable solids that can be removed. Once in the settleable form, these solids are removed in the secondary clarifier. The sludge at the bottom of the secondary clarifier is called *secondary sludge*. Secondary sludges are light and fluffy and more difficult to process than primary sludges—in short, secondary sludges do not dewater well.

Adding chemicals and various organic and inorganic substances before sedimentation and clarification may increase the capture of solids and reduce the amount of solids lost in the effluent. This *chemical addition* results in the formation of heavier solids, which trap the colloidal solids or convert dissolved solids to settleable solids. The resultant solids are known as *chemical sludges*. As chemical usage increases, so does the quantity of sludge that must be handled and disposed of. Chemical sludges can be very difficult to process; they do not de-water well and contain lower percentages of solids.

Sludge Characteristics

The composition and characteristics of sewage sludge vary widely and can change considerably with time. Notwithstanding these facts, the essential components of wastewater sludge remain the same. The only variations occur in the quantity of the various components as the type of sludge and the process from which it originated changes. The main element of all sludges is *water*. Before treatment, most sludge contains 95% to 99+% water (see Table 7.3). This high water content makes sludge handling and processing highly costly in both money and time. Sludge handling may represent up to 40% of the capital cost and 50% of the operation cost of a treatment plant.

TABLE 7.3
Typical Water Content of Sludges

Water Treatment Process	% Moisture of Sludge	lb Water/lb Sludge Solids Generated
Primary sedimentation	95	19
Trickling filter		
Humus—low rate	93	13.3
Humus—high rate	97	32.3
Activated sludge	99	99

Source: USEPA's *Operational Manual: Sludge Handling and Conditioning.* EPA-430/9-78-002, 1978.

TABLE 7.4
Distribution of Water in an Activated Sludge

Water Type	% Volume
Free water	75
Floc water	20
Capillary water	2
Particle water	2.5
Solids	0.5
Total	**100**

Source: USEPA's *Operations Manual, Sludge Handling and Conditioning. EPA-430/9-78-002, 1978.*

As a result, the importance of optimum design for handling and disposal of sludge cannot be overemphasized. The sludge's water content is present in several different forms. Some forms can be removed by several sludge treatment processes, thus allowing the same flexibility in choosing the optimum sludge treatment and disposal method.

The various forms of water and their approximate percentages for a typical activated sludge are shown in Table 7.4. The forms of water associated with sludges include:

- **Free Water***:* Water that is not attached to sludge solids in any way. This can be removed by simple gravitational settling.
- **Floc Water***:* Water trapped within the floc and travels with it. Its removal is possible by mechanical de-watering.
- **Capillary Water***:* Water that adheres to the individual particles and can be squeezed out of shape and compacted.
- **Particle Water**: Water that is chemically bound to the individual particles and can't be removed without incineration.

From a public health view, solids matter is the second and probably more critical sludge component. Representing 1%–8% of the total mixture, these solids are volatile. Wastewater solids can be classified into two categories based on their origin—organic and inorganic. Organic solids in wastewater are materials that were once alive and will burn or volatilize at 550°C after 15 minutes in a muffle furnace. The percent organic material within sludge will determine how unstable it is.

The inorganic material within sludge will determine its stability. Inorganic solids are those that were never alive and will not burn or volatilize at 550°C after 15 minutes in a muffle furnace. They are generally not subject to breakdown by biological action and are considered stable. Certain inorganic solids, however, can create problems when related to the environment, such as heavy metals like copper, lead, zinc, and mercury. These can be extremely harmful if discharged.

Organic solids may be subject to biological decomposition in either an aerobic or anaerobic environment. The decomposition of organic matter (with its production of objectionable by-products) and the possibility of toxic organic solids within the sludge compound the problems of sludge disposal.

The pathogens in domestic sewage are primarily associated with insoluble solids. Primary wastewater treatment processes concentrate these solids into sewage sludge, so untreated or raw primary sewage sludges have higher quantities of pathogens than the incoming wastewater. Biological wastewater treatment processes such as lagoons, trickling filters, and activated sludge treatment may substantially reduce the number of pathogens in wastewater (USEPA, 1989). These processes may also reduce the number of pathogens in sewage sludge by creating adverse conditions for pathogen survival.

Nevertheless, the resulting biological sewage sludges may still contain sufficient levels of pathogens to pose a public health and environmental concern. Moreover, insects, birds, rodents, and domestic animals may transport sewage sludge and pathogens from sewage sludge to humans and animals. Vectors are attracted to sewage sludge as a food source, and reducing the attraction of vectors to sewage sludge to prevent the spread of pathogens is a focus of current regulations. Sludge-borne pathogens and vector attraction are discussed in the following section.

Sludge Pathogens and Vector Attraction

As discussed earlier, a pathogen is an organism capable of causing disease. Pathogens infect humans through several pathways, including ingestion, inhalation, and dermal contact. The infective dose, or the number of pathogenic organisms to which a human must be exposed to become infected, varies depending on the organism and the health status of the exposed individual. Pathogens that propagate in the enteric or urinary system in humans and are discharged in feces or urine pose the most significant risk to public health concerning the use and disposal of sewage sludge. Pathogens are also found in other animals' urinary and enteric systems and may propagate in non-enteric settings.

The four major types of human pathogenic (disease-causing) organisms (bacteria, viruses, protozoa, and helminths) may be present in domestic sewage. The actual species and quantity of pathogens present in the domestic wastewater from a particular municipality (and the sewage sludge produced when treating the domestic wastewater) depend on the health status of the local community and may vary substantially at different times. The level of pathogens in treated sewage sludge (biosolids) also depends on the reductions achieved by the wastewater and sewage sludge treatment processes.

If improperly treated sewage sludge were illegally applied to land or placed on a surface disposal site, humans and animals could be exposed to pathogens directly by coming into contact with sewage sludge or indirectly by consuming drinking water or food contaminated by sewage sludge pathogens. Insects, birds, rodents, and even farm workers could contribute to these exposure routes by transporting sewage sludge and sewage sludge pathogens away from the site. Potential routes of exposure include the following:

Direct Contact:

- Touching the sewage sludge.
- Walking through an area—such as a field, forest, or reclamation area—shortly after sewage sludge application.
- Handling soil from fields where sewage sludge has been applied.
- Inhaling microbes that become airborne (via aerosols, dust, etc.) during sewage sludge spreading or by strong winds, plowing, or soil cultivation after application.

Indirect Contact:

- Consumption of pathogen-contaminated crops grown on sewage sludge-amended soil or other food products contaminated by contact with these crops or field workers, etc.
- Consumption of pathogen-contaminated milk or other food products from animals contaminated by grazing in pastures or fed crops grown on sewage sludge-amended fields.
- Ingestion of drinking water or recreational waters contaminated by runoff from nearby land application sites or by organisms from sewage sludge migrating into groundwater aquifers.
- Consumption of inadequately cooked or uncooked pathogen-contaminated fish from water contaminated by runoff from a nearby sewage sludge application site.
- Contact with sewage sludge or pathogens transported away from the land application or surface disposal site by rodents, insects, or other vectors, including grazing animals or pets.

One of the lesser impacts on public health can be from inhalation of airborne pathogens. Pathogens may become airborne via the spray of liquid biosolids from a splash plate or high-pressure hose or in fine particulate dissemination as dewatered biosolids are applied or incorporated. While high-pressure spray applications may result in some aerosolization of pathogens, this type of equipment is generally used on large, remote sites such as forests, where the impact on the public is minimal.

Fine particulates created by applying dewatered biosolids or incorporating biosolids into the soil may cause very localized fine particulate/dusty conditions. Still, particles in dewatered biosolids are too large to travel far, and the fine particulates do not spread beyond the immediate area. The activity of applying and incorporating biosolids may create dusty conditions. However, the biosolids are moist and do not add to the dusty condition. The pathogens have been reduced when biosolids have dried sufficiently to create fine particulates (Yeager and Ward, 1981).

Vector attraction reduction can be accomplished in two ways: by treating the sewage sludge to the point at which vectors are no longer attracted to it and by placing a barrier between the sewage sludge and vectors.

Note: Before discussing the fundamentals of sludge treatment methods, it is essential to cover sludge pumping calculations. It is important to note that treating the sludge is difficult (if not impossible) unless pumped into the specific sludge treatment process.

Important Note: Releasing wastewater solids without proper treatment could severely damage the environment. We must have a system to treat the volume of material removed as sludge throughout the system. Release without treatment would defeat the purpose of environmental protection. A design engineer can choose from many processes when developing sludge treatment systems. No matter the system or combination of systems chosen, the ultimate purpose will be the same: the conversion of wastewater sludges into a form that can be handled economically and disposed of without damaging the environment or creating nuisance conditions. Leaving either condition unmet will require further treatment. The degree of treatment will generally depend on the proposed method of disposal. Sludge treatment processes can be classified into several major categories. This handbook discusses the processes of thickening, digestion (or stabilization), de-watering, incineration, and land application. Each category has been further subdivided according to the specific methods used for sludge treatment. As mentioned, adequate, efficient sludge treatment cannot be overlooked when designing wastewater treatment facilities. The inadequacies of a sludge treatment system can severely affect a plant's overall performance capabilities. The inability to remove and process solids as fast as they accumulate can lead to the discharge of large quantities of solids to receiving waters. Even with proper design and capabilities, no system can be effective unless properly operated. Correct operation requires adequate engineering design and operator performance. Proper operator performance begins and ends with appropriate training.

Sludge Thickening

The solids content of primary, activated, trickling filter, or even mixed sludge (i.e., primary plus activated sludge) varies considerably, depending on the characteristics of the sludge. Note that the sludge removal and pumping facilities and the method of operation also affect the solids content. *Sludge thickening* (or *concentration*) is a unit process used to increase the solids content of the sludge by removing a portion of the liquid fraction. By increasing the solids content, more economical treatment of the sludge can be effected. Sludge thickening processes include:

- Gravity Thickeners
- Flotation Thickeners
- Solids Concentrators

Gravity Thickening

In gravity thickening, solids are withdrawn from the primary treatment (and sometimes secondary treatment) and pumped into the thickener. The solid buildup in the thickener forms a solid blanket on the bottom. The weight of the blanket compresses the solids on the bottom and "squeezes" the water out. Adjusting the blanket thickness can increase or decrease the percentage of solids in the underflow (solids withdrawn from the bottom of the thickener). The supernatant (clear water) rises to the surface and is returned to the wastewater flow for treatment. Daily operations of the thickening process include pumping, observation, sampling and testing, process control calculations, maintenance, and housekeeping.

Note: The equipment employed in thickening depends on the specific thickening processes used.

Equipment used for gravity thickening consists of a thickening tank, which is similar in design to the settling tank used in primary treatment. Generally, the tank is circular and provides equipment for continuous solids collection. The collector mechanism uses heavier construction than that in a settling tank because the solids being moved are more concentrated. The gravity thickener pumping facilities (i.e., pump and flow measurement) are used to withdraw thickened solids.

Solids concentrations achieved by gravity thickening are typically 8%–10% solids from primary underflow, 2%–4% solids from waste-activated sludge, 7%–9% solids from trickling filter residuals, and 4%–9% from combined primary and secondary residuals. The performance of gravity thickening processes depends on various factors, including:

- Type of sludge
- Condition of influent sludge
- Temperature
- Blanket depth
- Solids loading
- Hydraulic loading
- Solids retention time
- Hydraulic detention time

Flotation Thickening

Flotation thickening is used most efficiently for waste sludges from the suspended-growth biological treatment process, such as the activated sludge process. In operation, recycled water from the flotation thickener is aerated under pressure. During this time, the water absorbs more air than it normally absorbs under normal pressure. The recycled flow and chemical additives (if used) are mixed with the flow. The excess air is released as fine bubbles when the mixture enters the flotation thickener. These bubbles become attached to the solids and lift them toward the surface. The accumulation of solids on the surface is called the *float cake*. As more solids are added to the bottom of the float cake, it becomes thicker, and water drains from the upper levels of the cake. The solids are then moved up an inclined plane by a scraper and discharged. The supernatant leaves the tank below the surface of the float solids and is recycled or returned to the waste stream for treatment. Typically, flotation thickener performance is 3%–5% solids for waste-activated sludge with polymer addition and 2%–4% without polymer addition.

The flotation thickening process requires pressurized air, a vessel for mixing the air with all or part of the process residual flow, a tank for the flotation process to occur, solids collector mechanisms to remove the float cake (solids) from the top of the tank and accumulated heavy solids from the bottom of the tank. Since the process typically requires adding chemicals to improve separation, chemical mixing equipment, storage tanks, and metering equipment are needed to dispense the chemicals at the desired dose. The performance of the dissolved air-thickening process depends on various factors:

- Bubble size
- Solids loading
- Sludge characteristics
- Chemical selection
- Chemical dose

Solids Concentrators

Solids concentrators (belt thickeners) usually consist of a mixing tank, chemical storage, and metering equipment, and a moving porous belt. In operation, the process residual flow is chemically treated and then spread evenly over the surface of the moving porous belt. As the flow is carried down the belt (like a conveyor belt), the solids are mechanically turned or agitated, and water drains through the belt. This process is primarily used in facilities where space is limited.

Sludge Stabilization

Sludge stabilization aims to reduce volume, stabilize the organic matter, and eliminate pathogenic organisms to permit reuse or disposal. The equipment required for stabilization depends on the specific process used. Sludge stabilization processes include:

- Aerobic Digestion
- Anaerobic Digestion
- Composting
- Lime Stabilization
- Wet Air Oxidation (Heat Treatment)
- Chemical Oxidation (Chlorine Oxidation)
- Incineration

Aerobic Digestion

Equipment used for *aerobic digestion* consists of an aeration tank (digester) similar in design to the aeration tank used for the activated sludge process. Using diffused mechanical aeration equipment is necessary to maintain the aerobic conditions in the tank. Solids and supernatant removal equipment are also required. Process residuals (sludge) are added to the digester and aerated to maintain DO concentration of 1.0 mg/L. Aeration also ensures that the tank contents are well mixed. Generally, aeration continues for approximately 20 days retention time. Periodically, aeration is stopped, and the solids are allowed to settle. Sludge and the clear liquid supernatant are withdrawn to provide more room in the digester. When no additional volume is available, mixing is stopped for 12–24 hours before disposing of solids. Process control testing should include alkalinity, pH, % solids, and % volatile solids for influent sludge, supernatant, digested sludge, and digester contents. Normal operating levels for an aerobic digester are listed in Table 7.5. A typical operational problem associated with an aerobic digester is pH control. When pH drops, for example, this may indicate regular biological activity or low influent alkalinity. This problem is corrected by adding alkalinity (lime, bicarbonate, etc.).

Environmental engineers who design treatment plant unit processes and wastewater operators who operate aerobic digesters must make specific process control calculations to ensure that proper operational parameters are engineered into the system and the parameters are operated correctly.

pH Adjustment

The aerobic digester's pH occasionally falls below the levels required for good biological activity. When this occurs, the operator must perform a laboratory test to determine the alkalinity needed to raise the pH to the desired level. The lab test results must then be converted to the actual quantity of chemical (usually lime) required by the digester.

$$\text{Chem. Required, lb} = \frac{\text{Chemical Used in Lab Test, mg}}{\text{Sample Volume, Liters}} \times \text{Dig. Vol, MG} \times 8.34 \tag{7.3}$$

TABLE 7.5
Aerobic Digester Normal Operating Levels

Parameter	Normal Levels
Detention time, Days	10–20
Volatile solids loading lb/ft^3/day	0.1–0.3
DO mg/L	1.0
pH	5.9–7.7
Volatile solids reduction	40%–50%

Anaerobic Digestion

Anaerobic digestion is the traditional method of sludge stabilization. It involves using bacteria that thrive without oxygen. It is slower than aerobic digestion but has the advantage that only a tiny percentage of the wastes are converted into new bacterial cells. Instead, most organics are converted into carbon dioxide and methane gas.

Note: Air entrance should be prevented in an anaerobic digester because air could mix with the gas produced, creating an explosive mixture.

Equipment used in anaerobic digestion includes a sealed digestion tank with either a fixed or a floating cover, heating and mixing equipment, gas storage tanks, solids and supernatant withdrawal equipment, and safety equipment (e.g., vacuum relief, pressure relief, flame traps, explosion-proof electrical equipment).

Process residual (thickened or unthickened sludge) is pumped into the sealed digester. The organic matter is digested anaerobically through a two-stage process. Sugars, starches, and carbohydrates are converted to volatile acids, carbon dioxide, and hydrogen sulfide. The volatile acids are then converted to methane gas. This operation can occur in a single tank (single stage) or two tanks (two stages). In a single-stage system, supernatant and/or digested solids must be removed whenever flow is added. In a two-stage operation, solids and liquids from the first stage flow into the second stage each time fresh solids are added. The supernatant is withdrawn from the second stage to provide additional treatment space. Periodically, solids are withdrawn for dewatering or disposal. The methane gas produced in the process may be used for many plant activities.

Note: The primary purpose of a secondary digester is to allow for solids separation.

Various performance factors affect the operation of the anaerobic digester. For example, the operator must monitor critical operational parameters, such as % Volatile Matter in raw sludge, digester temperature, mixing, volatile acids/alkalinity ratio, feed rate, % solids in raw sludge, and pH. Along with being able to recognize normal/abnormal anaerobic digester performance parameters, wastewater operators must also know and understand standard operating procedures. Routine operating procedures include sludge additions, supernatant withdrawal, sludge withdrawal, pH control, temperature control, mixing, and safety requirements. Essential performance parameters are listed in Table 7.6.

TABLE 7.6
Anaerobic Digester—Sludge Parameters

Raw Sludge Solids	Impact
<4% solids	Loss of alkalinity
	Decreased sludge retention time
	Increased heating requirements
	Decreased volatile acid: alk ratio
4%–8% solids	Normal operation
>8% solids	Poor mixing
	Organic overloading
	Decreased volatile acid: alk ratio

Sludge Additions

Sludge must be pumped (in small amounts) several times each day to achieve the desired organic loading and optimum performance.

Note: In fixed-cover operations, additions must be balanced by withdrawals. Otherwise, structural damage occurs.

Supernatant Withdrawal

Supernatant withdrawal must be controlled for maximum sludge retention time. When sampling, sample all draw-off points and select the level with the best quality.

Sludge Withdrawal

Digested sludge is withdrawn only when necessary—always leave at least 25% seed.

pH Control

pH should be adjusted to maintain 6.8–7.2 by adjusting feed rate, sludge withdrawal, or alkalinity additions.

Note: The volatile acid/alkalinity relationship indicates the buffer capacity of an anaerobic digester. Decreases in alkalinity cause a corresponding increase in the ratio.

Temperature Control

If the digester is heated, the temperature must be controlled to a normal temperature range of 90°F–95°F. Never adjust the temperature by more than 1°F per day.

Mixing

If the digester is equipped with mixers, mixing should be accomplished to ensure organisms are exposed to food materials.

Safety

Anaerobic digesters are inherently dangerous—several catastrophic failures have been recorded. Safety equipment such as pressure relief and vacuum relief valves, flame traps, condensate traps, and gas collection safety devices are installed to prevent such failures. These critical safety devices must be checked and maintained for proper operation.

Note: Because of the inherent danger of working inside anaerobic digesters, they are automatically classified as permit-required confined spaces. Therefore, all internal entry operations must comply with OSHA's confined space entry standard.

Note: Increases in the ratio typically indicate a potential change in the digester's operating condition, as shown in Table 7.7.

Sludge Retention Time

Sludge retention time (SRT) is the time when the sludge remains in the digester.

$$\text{SRT, Days} = \frac{\text{Digester Vol}(\text{Gal})}{\text{Sludge Vol.Added per Day}(\text{gpd})} \tag{7.4}$$

TABLE 7.7
V.A./Alkalinity Ratio

Operating Condition	V.A./Alkalinity Ratio
Optimum	≤ 0.1
Acceptable range	0.1–0.3
% Carbon dioxide in gas increases	≥ 0.5
pH decreases	≥ 0.8

Estimated Gas Production in Cubic Feet/Day

The gas production rate is usually expressed as the volume of gas (ft^3) produced per pound of volatile matter destroyed. The total cubic feet of gas a digester will produce per day can be calculated by:

$$\text{Gas Prod.}\left(\text{ft}^3\right) = \text{Vol. Matter In, lb/day} \times \%\text{Vol. Matter Red.} \times \text{Prod. Rate ft}^3/\text{lb} \tag{7.5}$$

Composting

The purpose of composting sludge is to stabilize the organic matter, reduce volume, and eliminate pathogenic organisms. In a *composting operation,* dewatered solids are usually mixed with a bulking agent (i.e., hardwood chips) and stored until biological stabilization occurs. The composting mixture is ventilated during storage to provide sufficient oxygen for oxidation and to prevent odors. After the solids are stabilized, they are separated from the bulking agent. The composted solids are then stored for curing and applied to farmlands or other beneficial uses. The expected performance of the composting operation for both percent volatile matter reduction and percent moisture reduction ranges from 40% to 60%.

Three methods of composting wastewater biosolids are standard. Each technique involves mixing dewatered wastewater solids with a bulking agent to provide carbon and increase porosity. The resulting mixture is piled or placed in a vessel where microbial activity causes the temperatures of the mixture to rise during the "active composting" period. The specific temperatures that must be achieved and maintained for successful composting vary based on the method and use of the end product. After active composting, the material is cured and distributed. Again, there are three commonly employed composting methods, but we only describe the aerated static pile (ASP) method because it is commonly used.

The ASP Model Composting Facility uses the homogenized mixture of a bulking agent (coarse hardwood wood chips) and dewatered biosolids piled by front-end loaders onto a large concrete composting pad where it is mechanically aerated via PVC plastic pipe embedded within the concrete slab. This ventilation procedure is part of the 26 days of "active" composting when adequate air and oxygen are necessary to support aerobic biological activity in the compost mass and reduce the compost mixture's heat and moisture content. Remember that a compost pile without an appropriately sized air distribution system can lead to the onset of anaerobic conditions and the appearance of putrefactive odors.

For illustration and discussion purposes, we assume a typical overall composting pad area is approximately 200 ft by 240 ft, consisting of eleven (11) blowers and twenty-four (24) pipe troughs (troughs). Three blowers are 20 hp 2,400 cfm, variable speed drive units capable of operating in either the positive or negative aeration mode. Blowers A, B, & C are each connected to two (2) piping troughs that run the entire pad length. The two troughs are attached to the composting pad's opposite end to create an "aeration pipe loop." The other eight (8) blowers are rated at 3 hp 1,200 cfm and are arranged with 1 blower per 6 troughs at half-length, feeding 200 cfm per trough. These blowers can be operated in the positive or negative aeration mode. Aeration piping within the six (6) pipe troughs is perforated PVC plastic pipe, six inches inside diameter and 1/4-inch wall thickness. Perforation holes/orifices vary in size from 7/32 to 1/2 inch, increasing in diameter as the distance from the blower increases.

The variable speed motor drives installed with blowers A, B, and C are controlled by five thermal probes mounted at various depths in the compost pile, and multiple parameters are fed back to the recorder. In contrast, the other eight blowers are at constant speed, controlled by a timer that cycles them on and off. To ensure optimum composting operations, verifying that these thermal probes are calibrated regularly is essential. In the constant-speed system, thermal probes are installed, but all readings are taken and recorded manually.

For water and leachate drainage purposes, all aeration piping within the troughs slopes downward, with the highest point at the center of the composting pad. Drain caps located at each end of the pipe length are manually removed regularly so that any buildup of debris or moisture will not interfere with the airflow.

The construction process involved in building the compost pile will be covered in detail later, but a few key points should be made for now. For example, before the piling of the mixture on the composting pad, an 18-inch layer of wood chips is used as a base material. The primary purpose of the wood chip base is to keep the composting mixture clear of the aeration pipes, which reduces clogging of the air distribution openings in the pipes and allows free air circulation. A secondary benefit is that the wood chips insulate the composting mixture from the pad. The compost pad is like a heat sink; this insulating barrier improves the uniformity of heat distribution within the composting mix.

Lime or alkaline stabilization can achieve the minimum requirements for Class A (no detectable pathogens) and Class B (a reduced level of pathogens) biosolids concerning pathogens, depending on the amount of alkaline material added and other processes employed. Generally, alkaline stabilization meets the Class B requirements when the pH of the mixture of wastewater solids and alkaline material is at 12 or above after 2 hours of contact.

Class A requirements can be achieved when the pH of the mixture is maintained at or above 12 for at least 72 hours, with a temperature of 52°C maintained for at least 12 hours during this time. In one process, the mixture is air-dried to over 50% solids after 72 hours of elevated pH. Alternatively, the process may be manipulated to maintain temperatures at or above 70°C for 30 or more minutes while maintaining the pH requirement of 12. This higher temperature can be achieved by overdosing with lime (adding more than is needed to reach a pH of 12), using a supplemental heat source, or using a combination of the two. Monitoring for fecal coliform or *Salmonella* sp. is required before release by the generator for use.

Materials used for alkaline stabilization include hydrated lime, quicklime (calcium oxide), fly ash, lime and cement kiln dust, and carbide lime. Quicklime is commonly used because it has a high heat of hydrolysis (491 British thermal units) and can significantly enhance pathogen destruction. Because of their availability and relatively low cost, fly ash, lime kiln dust, or cement kiln dust are often used for alkaline stabilization.

The alkaline stabilized product is suitable for application in many situations, such as landscaping, agriculture, and mine reclamation. The product serves as a lime substitute, a source of organic matter, and a specialty fertilizer. The addition of alkaline stabilized biosolids results in more favorable conditions for vegetative growth by improving soil properties such as pH, texture, and water-holding capacity. Appropriate applications depend on the needs of the soil and crops grown and the pathogen classification. For example, a Class B material would not be suitable for blending in a topsoil mix intended for home landscaping but is ideal for agriculture, mine reclamation, and landfill cover where the potential for contact with the public is lower and access can be restricted. Class A alkaline stabilized biosolids are useful as a topsoil blend ingredient in agriculture. Alkaline stabilized biosolids provide pH adjustment, nutrients, and organic matter, reducing reliance on other fertilizers.

Alkaline-stabilized biosolids are also useful as daily landfill cover. They satisfy the federal requirement that landfills be covered with soil or soil-like material at the end of each day (40 CFR 258). In most cases, lime-stabilized biosolids are blended with other soil to achieve the proper consistency for daily cover.

As mentioned, alkaline stabilized biosolids are excellent for land reclamation in degraded areas, including acid mine spills or mine tailings. Soil conditions at such sites are unfavorable for vegetative growth, often due to acid content, lack of nutrients, elevated levels of heavy metals, and poor soil texture. Alkaline stabilized biosolids help to remedy these problems, making conditions more favorable for plant growth and reducing erosion potential. In addition, once a vegetative cover is established, mine drainage quality improves.

Thermal treatment (or wet air oxidation) subjects sludge to high temperature and pressure in a closed reactor vessel. The high temperature and pressure rupture the cell walls of any microorganisms present in the solids and cause chemical oxidation of the organic matter. This process substantially improves dewatering and reduces the volume of material for disposal. It also produces high-strength waste, which must be returned to the wastewater treatment system for further treatment.

Chlorine oxidation also occurs in a closed vessel. In this process, chlorine (100–1,000 mg/L) is mixed with a recycled solids flow. The recycled flow and process residual flow are combined in the reactor. The solids and water are separated after leaving the reactor vessel. The water is returned to the wastewater treatment system, and the treated solids are dewatered for disposal. The main advantage of chlorine oxidation is that it can be operated intermittently. The main disadvantage is the production of extremely low pH and high chlorine content in the supernatant.

Sludge Dewatering

Digested sludge removed from the digester is still mostly liquid. The primary objective of dewatering biosolids is to reduce moisture and, consequently, volume to a degree that will allow for economical disposal or reuse. Probably one of the best summaries of the various reasons why it is essential to dewater biosolids is given by Metcalf and Eddy (1991) in the following: (1) The costs of transporting biosolids to the ultimate disposal site are significantly reduced when biosolids volume is reduced; (2) dewatered biosolids allow for easier handling; (3) dewatering biosolids (reduction in moisture content) allows for more efficient incineration; (4) if composting is the beneficial reuse choice, dewatered biosolids decrease the amount and therefore the cost of bulking agents; (5) with the USEPA's new 503 rule, dewatering biosolids may be required to render the biosolids less offensive; and (6) when landfilling is the ultimate disposal option, dewatering biosolids is required to reduce leachate production.

Again, the point is that the importance of adequately dewatering biosolids for proper disposal/reuse can't be overstated.

The unit processes that are most often used for dewatering biosolids are (1) vacuum filtration, (2) pressure filtration, (3) centrifugation, and (4) drying beds. Solids content achievable by various dewatering techniques is shown in Table 7.8. The biosolids cake produced by common dewatering processes is consistent with dry, crumbly bread pudding (Spellman, 1996). This non-fluid, dewatered, dry, crumbly cake product is easily handled, non-offensive, and can be land applied manually and by conventional agricultural spreaders (Outwater, 1994).

Dewatering processes are usually divided into natural air drying and mechanical methods. Natural dewatering methods include those in which moisture is removed by evaporation and gravity or induced drainage, such as sand beds, biosolids lagoons, paved beds, Phragmites reed beds, vacuum-assisted beds, Wedgewater beds, and dewatering via freezing. These natural dewatering methods are less controllable than mechanical ones but are typically less expensive. Moreover, these natural dewatering methods require less power because they rely on solar energy, gravity, and biological processes as the source of energy for dewatering. Mechanical dewatering processes include pressure filters, vacuum filters, belt filters, and centrifuges. The air drying and mechanical dewatering processes will be discussed in greater detail later in this text.

TABLE 7.8
Solids Content of Dewatered Biosolids

Dewatering Method	Approximate Solids Content, %
Lagoons/ponds	30
Drying beds	40
Filter press	35–45
Vacuum filtration	25
Standard centrifuge	20–25
High G/high solids centrifuge	25–40

Sand Drying Beds

Sand beds have been used successfully for years to dewater sludge. Composed of a sand bed (consisting of a gravel base, underdrains, and 8–12 inches of filter-grade sand), drying beds include an inlet pipe, splash pad containment walls, and a system to return filtrate (water) for treatment. The sand beds are sometimes covered to protect drying solids from the elements.

In operation, solids are pumped to the sand bed and allowed to dry by draining excess water through the sand and evaporation. This is the simplest and cheapest method for dewatering sludge. Moreover, no special training or expertise is required. However, there is a downside: dry beds require a great deal of manpower to clean; they can create odor and insect problems and cause sludge buildup during inclement weather.

According to Metcalf and Eddy (1991), four types of drying are commonly used in dewatering biosolids: (1) sand, (2) paved, (3) artificial media, and (4) vacuum-assisted. In addition to these commonly used dewatering methods, a few innovative techniques of natural dewatering will also be discussed in this section. The innovative natural dewatering methods to be addressed include experimental work on biosolids dewatering via freezing. Moreover, dewatering biosolids with aquatic plants, which have been tested and installed in several sites throughout the U.S., is also discussed.

Drying beds are generally used for dewatering well-digested biosolids. Attempting to air-dry raw biosolids is usually unsuccessful and may result in odor and vector control problems. Biosolids drying beds consist of a perforated or open-joint drainage system in a support medium, usually gravel, covered with a filter medium, usually sand. Still, they can consist of extruded plastic or wire mesh—wood, concrete, or other materials generally separate drying beds into workable sections. Drying beds may be enclosed or open to the weather. They may rely entirely on natural drainage and evaporation processes or use a vacuum to assist the operation (both types are discussed in the following sections).

Traditional Sand Drying Beds

This is the oldest biosolids dewatering technique and consists of 6–12 inches of coarse sand underlain by layers of graded gravel ranging from 1/8 to 1/4 inches at the top and 3/4 to 1–1/2 inches at the bottom. The total gravel thickness is typically about 1 ft. Graded natural earth (4–6 inches) usually makes up the bottom with a web of drain tile placed on 20- to 30-ft centers. Sidewalls and partitions between bed sections generally have wooden planks or concrete and extend about 14 inches above the sand surface (McGhee, 1991).

Large open areas of land are required for sand-drying biosolids. For example, drying beds up to 125+ ft long and 20–35 ft wide are not unusual. Providing at least two drying beds is normal, even at most miniature wastewater treatment plants.

The actual dewatering occurs due to two physical processes: evaporation and drainage. The liquor that drains off the biosolids goes to a central sump, which pumps it back to the treatment process to undergo further treatment. The operation is very much affected by climate. Covering the beds with a translucent material in wet climates may be necessary, allowing at least 85% of the sun's ultraviolet radiation to pass through.

Typical loading rates for primary biosolids in dry climates range up to 200 kg/(m^2 × year) and from 60 to 125 kg/(m^2 × year) for mixtures of primary and waste-activated biosolids.

When a drying bed is put into operation, it is generally filled with digested biosolids ranging from 8 to 12 inches. The actual drying time is climate-sensitive; that is, drying can take from a few weeks to a few months, depending on the climate and the season. After dewatering, the biosolids' solid content will range from about 20%–35%, and, more importantly, the volume will be reduced to 85%. Upon completion of the drying process, the dried biosolids are generally removed from the bed with handheld forks or front-end loaders. Notably, a small amount of sand is lost in the dried biosolids removal process, and the bed must be refilled and graded periodically. Dried solids removed from a biosolids drying bed can be incinerated or landfilled.

Paved Drying Beds

The main reason for using paved drying beds is that they alleviate the problem of mechanical biosolids removal equipment damaging the underlain piping networks. The beds are paved with concrete or asphalt and are generally sloped toward the center, where a sump-like area with underlain pipes is arranged. Like biosolids lagoons, these dewatering beds depend on evaporation to dewater the applied solids. Paved drying beds are usually rectangular with a center drainage strip. They can be heated via buried pipes in the paved section and are generally covered to prevent rain incursion.

In this type of natural dewatering, 45%–50% solids can be achieved within 35 days in dry climates under normal conditions (McGhee, 1991). The operation of paved drying beds involves applying the biosolids to a depth of about 12 inches. The settled surface area is routinely mixed by a vehicle-mounted machine driven through the bed. Mixing is essential because it breaks up the crust and exposes wet surfaces to the environment. The supernatant is decanted like biosolids lagoons. Biosolids loadings in relatively dry climates range from 120 to 260 kg/(m^2/year). The two major disadvantages of paved drying beds are high capital cost and more extensive land requirements than sand beds.

In attempting to determine the bottom area dimensions of a paved drying bed, Metcalf and Eddy (1991) recommend computation by trial using the following equation:

$$A = \frac{1.04S\left[(1-Sd)/Sd-(1-Se)/Se\right]+(62.4)(P)(A)}{(62.4)(Ke)(Ep)}$$

where

A = Bottom area of paved bed, square ft.
S = Annual biosolids production, dry solids, lb
Sd = Percent dry solids in the biosolids after decanting, as a decimal
Se = Percent dry solids required from final disposal, as a decimal
P = Annual precipitation, ft
Ke = Reduction factor for evaporation from biosolids versus a free water surface.

Use 0.6 for a preliminary estimate; pilot test to determine the factor for the final design

Ep = free water pan evaporation rate for the area, ft/year

Although the construction and operation methodologies for biosolids drying beds are well-known and widely accepted, this is not to say that the wastewater industry has not attempted to incorporate further advances into their construction and operation. For example, in an attempt to reduce the amount of dewatered biosolids that must be manually removed from drying beds, attempts have been made to construct drying beds in a specific manner whereby they can be planted with reeds, namely, the *Phragmites communis* variety (to be covered in greater detail later). The intent of augmenting the biosolids drying bed with reeds is to effect further desiccation. Moreover, tests have shown that the plants extend their root systems into the biosolids mass. This extended root system has the added benefit of helping to establish a rich microflora which eventually feeds upon the organic content of the biosolids. It is interesting to note that normal plant activity keeps the system aerobic.

Artificial Media Drying Beds

The first artificial media drying beds, developed in England in 1970, used a stainless-steel medium called Wedgewire. Later, as the technology advanced, the stainless-steel fine wire screen mesh (Wedgewire) was replaced with a high-density polyurethane medium (Wedgewater). The polyurethane is less expensive than the stainless-steel medium but has a shorter life expectancy.

Wedgewire beds are similar in concept to vacuum-assisted drying beds (to be described later). The medium used in Wedgewire beds is a septum with wedge-shaped slots about 0.01 inch wide. Initially, the bed is filled with water to a level above the wire screen. Chemically conditioned biosolids are added and, after a brief holding period, can drain through the screen. Since excess water cannot return to the biosolids through capillary action, the biosolids dewaters faster with this process (McGhee, 1991).

Vacuum-Assisted Drying Beds

For small plants that process small quantities of biosolids and have limited land area, vacuum-assisted drying beds may be the preferred method of dewatering biosolids. Vacuum-assisted drying beds commonly employ a small vacuum to accelerate the dewatering of biosolids applied to a porous medium plate. As the name implies, this porous medium is set above an aggregate-filled support underdrain, which drains to a sump. A small vacuum is applied to this underdrain, which works to extract free water from the biosolids; with biosolids loadings about less than 10 kg/m^2/cycle, the time required to dewater conditioned biosolids is about 1 day (McGhee, 1991). This dewatering method can achieve a solids content of >30%, although 20% solids is a more normal expectation.

Dewatered biosolids are usually removed with mechanized machinery such as front-end loaders. Once the solids have been removed, washing the bed's surface with high-pressure hoses is essential to remove residuals. The main advantage cited for this dewatering method is the reduced time needed for dewatering, which reduces the effects of inclement climatic conditions on biosolids drying. The main disadvantage of this type of dewatering may be its dependence on adequate chemical conditioning for successful operation (Metcalf & Eddy, 1991).

Natural Methods of Dewatering Biosolids

Two innovative methods of natural biosolids dewatering are discussed in this section: (1) Dewatering via freezing and (2) dewatering using aquatic plants.

Note: Dewatering via freezing is primarily in the pilot study stage. Experimental work in this area has led to pilot plants and has yielded various mathematical models but no full-scale operations (Outwater, 1994).

Freeze-Assisted Drying Beds: Low winter temperatures accelerate the dewatering process in freeze-assisted drying. Freezing biosolids works to separate the water from the solids. The free water drains quickly when the granular mass is thawed. It is not unusual to attain a solids concentration greater than 25% when the mass thaws and drains.

The feasibility of freezing biosolids in a particular area depends on the frost penetration depth. The maximum depth of frost penetration for an area can be found in published sources or local records (Reed, 1987).

In attempting to calculate the depth of frost penetration, it may be helpful to use the following equation for 3-inch (75-mm) layers (McGhee, 1991):

$$Y = 1.76F_p - 101$$

where

Y = total depth of biosolids, cm

F_p = the maximum depth of frost penetration, cm

In this example, the biosolids are applied in 75-mm layers. As soon as the first layer is frozen, another layer is applied. The goal is to fill the bed with biosolids by the end of winter. It must be pointed out that in using this layered method of freeze-dewatering biosolids, it is essential to ensure that each layer is frozen before the next one is applied. Moreover, any snow or debris that falls should be removed from the surface; otherwise, it will insulate the biosolids. Outwater (1994) points out that "to ensure successful performance at all times, the design should be based on the warmest winter in the past 20 years and on a layer thickness which will freeze in a reasonable amount of time if freeze-thaw cycles occur during the winter" (p. 86). Another rule of thumb to use in deciding whether or not biosolids freezing is a viable dewatering option is that biosolids freezing is unlikely to be a practical concept unless more than 100 cm frost penetration is assured (Reed, 1987).

Dewatering Using Aquatic Plants: In the 1960s, Germany developed the use of aquatic plants to dewater biosolids. In this first project, reed beds were constructed in submerged wetlands. The reed bed system modifies a typical sand-drying bed for biosolids. Instead of removing the dewatered biosolids from the beds after each application, *Phragmites communis* (reeds) are planted in the sand. Biosolids are added for the next several years (5–10 years), emptying the beds.

In the reed bed operation, biosolids are spread on the bed's surface via troughs or gravity-fed pipes. When the bed is filled to capacity, about 4 inches of standing liquid will remain on the surface until it evaporates or drains down through the bed, where tile drains return it to the treatment process.

In an aquatic reed bed, the reeds develop (as described earlier) near the root zone, a rich microflora that feeds on the organic material in biosolids. Phragmites reeds are particularly suitable for this application because they resist biosolid contaminants. Although the roots penetrate the finer gravel and sand, they do not penetrate lower areas where the larger stones or fragments are located. This is important because root penetration to the lower bed levels could interfere with free drainage.

Another advantage of using *Phragmites* reeds in the aquatic plant drying bed is their growing pattern. *Phragmites* roots grow and extend themselves through rhizomes (Rhizomes defined: an underground horizontal stem, often thickened and tuber-shaped, and possessing bud, nodes, and scale-like leaves). Several plants branch off from each rhizome and grow vertically; this vertical growth aids in dewatering by providing channels through which the water drains. The reeds absorb some of this water, which is then released into the atmosphere through evapotranspiration.

Phragmites reed beds are operated year-round. The reeds are harvested in the fall, leaving their root systems intact. Depending on contaminant concentrations, the harvested reeds can be incinerated, landfilled, or composted.

It takes about 8 years to fill an average bed. When this occurs, it must be taken out of service and allowed to stand fallow for 6 months to a year. This fallow period provides for the stabilization of the top surface layer. The resulting biosolids product is dry and crumbles in the hands (it is friable). The dewatered biosolids product can be land-applied if contamination levels are within acceptable limits per EPA's 503 Rule.

Operation of a Phragmites reed bed has limitations. For example, to ensure a successful dewatering operation, it is prudent to hire the services of an agronomist who is familiar with plant growth, care, and control of plant pests such as aphids. Moreover, this dewatering system is designed for regions subject to four distinct seasons, i.e., northern exposures. Phragmites require a dormancy, wintering-over period for proper root growth. Additionally, reed beds are unsuitable for large-scale operations for operational reasons and may be cost-prohibitive due to the cost of land.

The jury is still out on how effective reed bed dewatering systems are. This is the case because the technology is relatively new, and none of the beds have been emptied yet; thus, it is difficult to predict the quality of the end product.

Rotary Vacuum Filters

Rotary vacuum filters have also been used for many years to dewater sludge. The vacuum filter includes filter media (belt, cloth, or metal coils), media support (drum), vacuum system, chemical feed equipment, and conveyor belt(s) to transport the dewatered solids. In operation, chemically treated solids are pumped to a vat or tank where a rotating drum is submerged. As the drum rotates, a vacuum is applied to the drum. Solids collect on the media and are held there by the vacuum as the drum rotates out of the tank. The vacuum removes additional water from the captured solids. When solids reach the discharge zone, the vacuum is released, and the dewatered solids are discharged onto a conveyor belt for disposal. The media is then washed before returning to the cycle's start.

Types of Rotary Vacuum Filters

The three principal types of rotary vacuum filters are rotary drum, coil, and belt. The *rotary drum* filter consists of a cylindrical drum rotating partially submerged in a vat or pan of conditioned sludge. The drum is divided lengthwise into several sections connected through internal piping to ports in the hub's valve body (plant). This plate rotates in contact with a fixed valve plate with similar ports, which are connected to a vacuum supply, a compressed air supply, and an atmosphere vent. As the drum rotates, each section is thus connected to the appropriate service.

The coil-type vacuum filter uses two layers of stainless-steel coils arranged in a corduroy fashion around the drum. After a dewatering cycle, the two layers of springs leave the drum bed and are separated so that the cake is lifted off the lower layer and discharged from the upper layer. The coils are then washed and reapplied to the drum. The coil filter is used successfully for all types of sludges; however, sludges with fine particles or ones resistant to flocculation dewater poorly with this system.

The media on a belt filter leaves the drum surface at the end of the drying zone and passes over a small-diameter discharge roll to aid cake discharge. Next, the media is washed. Then, the media is returned to the drum and the vat for another cycle. This type of filter normally has a small-diameter curved bar between the point where the belt leaves the drum and the discharge roll. This bar primarily aids in maintaining belt dimensional stability.

Filter Media

Drum and belt vacuum filters use natural or synthetic fiber materials. On the drum filter, the cloth is stretched and secured to the surface of the drum. The fabric is stretched over the drum and through the pulley system in the belt filter. The installation of a blanket requires several days. The cloth will (with proper care) last several hundred to several thousand hours. The blanket's life depends on the fabric selected, the conditioning chemical, backwash frequency, and cleaning (i.e., acid bath) frequency.

Filter Drum

The filter drum is a maze of pipework running from a metal screen and wooden skeleton and connecting to a rotating valve port at each end of the drum. The drum has a variable speed drive to turn the drum from 1/8 to 1 rpm. Normally, solids pickup is inversely related to the drum speed. The drum is partially submerged in a vat containing the conditioned sludge. Normally, submergence is limited to 1/5 or less of the filter surface at a time.

Chemical Conditioning

Sludge dewatered using vacuum filtration is normally chemically conditioned just before filtration. Sludge conditioning increases the percentage of solids captured by the filter and improves the sludge's dewatering characteristics. However, conditioned sludge must be filtered quickly after chemical addition to obtain these desirable results.

Pressure Filtration

Pressure filtration differs from vacuum filtration in that the liquid is forced through the filter media by a positive pressure instead of a vacuum. Several presses are available, but the most commonly used types are plate, frame, and belt presses. *Filter presses* include the belt or plate and frame types. The belt filter includes two or more porous belts, rollers, related chemical makeup and feed handling systems, and supernatant and solids collection and transport.

The plate-and-frame filter consists of a support frame, filter plates covered with a porous material, a hydraulic or mechanical mechanism for pressing plates together, related handling systems for chemical makeup and feed, and supernatant and solids collection and transport. In the plate-and-frame filter, solids are pumped (sandwiched) between plates. Pressure (200–250 psi) is applied to the plates, and water is "squeezed" from the solids. At the end of the cycle, the pressure is released, and as the plates separate, the solids drop out onto a conveyor belt for transport to storage or disposal. Performance factors for plate-and-frame presses include feed sludge characteristics, type and amount of chemical conditioning, operating pressures, and precoat.

The belt filter uses a coagulant (polymer) mixed with the influent solids. The chemically treated solids are discharged between two moving belts. First, water drains from the solids by gravity. Then, as the two belts move between a series of rollers, pressure "squeezes" additional water out of the solids. The solids are then discharged onto a conveyor belt for transport to storage/disposal. Performance factors for the belt press include sludge feed rate, belt speed, belt tension, belt permeability, chemical dosage, and chemical selection.

Filter presses have lower operation and maintenance costs than vacuum filters or centrifuges. They typically produce a good quality cake and can be batch-operated. However, construction and installation costs are high. Moreover, chemical addition is required, and the presses must be operated by skilled personnel.

Centrifugation

Centrifuges of various types have been used in dewatering operations for at least 30 years and appear to be gaining popularity. Depending on the type of centrifuge used, in addition to centrifuge pumping equipment for solids feed and centrate removal, chemical makeup and feed equipment and support systems for removal of dewatered solids are required. Centrifuge operation is dependent upon various performance factors.

- Bowl design: length/diameter ratio; flow pattern
- Bowl speed
- Pool volume
- Conveyor design
- Relative conveyor speed
- Type/condition of sludge
- Type and amount of chemical conditioning
- Operating pool depth
- Relative conveyor speed (if adjustable)

Centrifuge operators often find that centrifuge operation can be simple, clean, and efficient. In most cases, chemical conditioning is required to achieve optimum concentrations. Operators soon discover that centrifuges are noisemakers; units run at very high speeds and produce high-level noise, which can cause hearing loss with prolonged exposure. Therefore, when working in an area where a centrifuge is in operation, special care must be taken to provide hearing protection.

The actual operation of a centrifugation unit requires the operator to control and adjust chemical feed rates, observe unit operation and performance, control and monitor the centrate returned to the treatment system, and perform required maintenance as outlined in the manufacturer's technical manual.

Sludge Incineration

Not surprisingly, incinerators produce the maximum solids and moisture reductions. The equipment required depends on whether the unit is a multiple hearth or fluid-bed incinerator. Generally, the system will need a heat source to reach ignition temperature, a solids feed system, and ash handling equipment. It is important to note that the system must also include all required equipment (e.g., scrubbers) to achieve compliance with air pollution control requirements. Solids are pumped to the incinerator. The solids are dried and then ignited (burned). As they burn, the organic matter is converted to carbon dioxide and water vapor, and the inorganic matter is left behind as ash or "fixed" solids. The ash is then collected for reuse or disposal.

Multiple Hearth Furnace

The *multiple-hearth furnace* consists of a circular steel shell surrounding several hearths. Scrapers (rabble arms) are connected to a central rotating shaft. Units range from 4.5 to 21.5 ft in diameter and have 4–11 hearths. In operation, dewatered sludge solids are placed on the outer edge of the top hearth. The rotating rabble arms move them slowly to the center of the hearth. At the center of the hearth, the solids fall through ports to the second level. The process is repeated in the opposite direction. Hot gases generated by burning on lower hearths dry the solids. The dry solids pass to the lower hearths. The high temperature on the lower hearths ignites the solids. Burning continues to completion. Ash materials are discharged to lower cooling hearths, where they are discharged for disposal. Airflow inside the center column and rabble arms continuously cools internal equipment.

Fluidized Bed Furnace

The *fluidized bed* incinerator consists of a vertical circular steel shell (reactor) with a grid to support a sand bed and an air system to provide warm air to the bottom of the sand bed. The evaporation and incineration process takes place within the super-heated sand bed layer. In operation, air is pumped to the bottom of the unit. The airflow expands (fluidizes) the sand bed inside. The fluidized bed is heated to its operating temperature (1,200°F–1,500°F). Auxiliary fuel is added when needed to maintain the operating temperature. The sludge solids are injected into the heated sand bed. Moisture immediately evaporates. Organic matter ignites and reduces to ash. Residues are ground to fine ash by the sand movement. Fine ash particles flow up and out of the unit with exhaust gases. Ash particles are removed using common air pollution control processes. Oxygen analyzers in the exhaust gas stack control the airflow rate.

Note: Because these systems retain a high amount of heat in the sand, the system can be operated for as little as 4 hours/day with little or no reheating.

Land Application of Biosolids

The purpose of land application of biosolids is to dispose of the treated biosolids in an environmentally sound manner by recycling nutrients and soil conditioners. To be land-applied, wastewater biosolids must comply with state and federal biosolids management/disposal regulations. Biosolids must not contain materials that are dangerous to human health (i.e., toxicity, pathogenic organisms, etc.) or dangerous to the environment (i.e., toxicity, pesticides, heavy metals, etc.). Treated biosolids are land applied by direct injection or application and plowing (incorporation).

Land application of biosolids requires precise control to avoid problems. The quantity and quality of biosolids applied must be accurately determined. For this reason, the operator's process control activities include biosolids sampling/testing functions. Biosolids sampling and testing determine % solids, heavy metals, organic pesticides and herbicides, alkalinity, total organic carbon (TOC), organic nitrogen, and ammonia nitrogen.

Plant Available Nitrogen (PAN)

One factor considered when land-applying biosolids is the amount of nitrogen in the biosolids available to the plants grown on the site. This includes ammonia nitrogen and organic nitrogen. The organic nitrogen must be mineralized for plant consumption. Only a portion of the organic nitrogen is mineralized per year. The mineralization factor (f_1) is assumed to be 0.20. The amount of ammonia nitrogen available is directly related to the time between applying the biosolids and incorporating (plowing) the sludge into the soil. Volatilization rates are provided based on the example below.

$$\text{Pan, lb/dry ton} = \left[(\text{Org. Nit., mg/kg} \times f_1) + (\text{Amm.Nit., mg/kg} \times V_1)\right] \times 0.002\,\text{lb/dry ton} \quad (7.6)$$

Where

f_1 = Mineral rate for organic nitrogen (assume 0.20)

V_1 = Volatilization rate of ammonia nitrogen

V_1 = 1.00 if biosolids are injected

V_1 = 0.85 if biosolids are plowed in within 24 hours

V_1 = 0.70 if biosolids are plowed in within 7 days

AI APPLICATIONS

Several wastewater operations worldwide are incorporating the emerging technologies of Artificial Intelligence (AI) and the Internet of Things (IoT) that potentially (and in reality) are revolutionizing wastewater treatment. The supervisory control and data acquisition (SCADA) system was described earlier in its water treatment application. In wastewater treatment, SCADA systems consist of hardware and software components that collect, transmit, and distribute data across various devices within a network. SCADA provides information about plant operations, which can be used to make necessary unit process adjustments. AI and IoT continuously monitor water quality parameters such as pH, COD, BOD, conductivity, and total dissolved solids. The continuous monitoring allows adjustments to be made to ensure the quality of the processed water.

AI algorithms can predict equipment failures and maintenance needs, improving efficiency and reducing downtime. By analyzing historical and real-time data, the AI system can optimize chemical dosing, resulting in improved effluent quality and significant cost savings.

AI applications in wastewater treatment operations are designed to monitor critical parameters to ensure timely interventions and meet permit requirements.

One of the recently recognized benefits of AI in wastewater treatment is local monitoring for public health purposes. AI technology is being used to monitor wastewater for COVID-19 to detect the presence of the virus in wastewater influent. This practice is known as wastewater-based epidemiology (WBE). I witnessed the implementation of this monitoring for viruses in wastewater influents at Hampton Roads Sanitation District (HRSD), Hampton Roads, Virginia, when they implemented WBE at their 14 major wastewater treatment plants. This flagship practice (now employed at most major city wastewater treatment systems) does not pinpoint who has a virus or other disease in the service area. Still, it does identify the area or region where diseases, such as viruses like COVID-19, are prevalent. This is an essential tool used to detect and track disease.

The WBE of pathogenic organisms can inform us of the presence of disease outbreaks when or where they are not assumed or expected. Recently, I have become concerned with the COVID-19 pandemic. Wastewater-based epidemiology is being employed to test for the presence of SARS-CoV-2 in wastewater. The point is that wastewater can be tested for signatures of viruses.

In August 2020, the author conducted an informal survey of wastewater treatment plants in the United States and found more than 100 sanitation districts and publicly owned treatment works that were conducting wastewater surveillance of SARS-CoV-2 as a potentially important source of information on the prevalence and chronological trends of COVID-19 in communities. For example, in one of the globe's premier wastewater treatment operations, Hampton Roads Sanitation District (HRSD) in southeastern Virginia is actively pursuing collection, study, research, and testing of wastewater influent in significant treatment plants.

Regarding HRSD and its realization that sewage can suggest a wider spread of COVID-19 or no spread at all, environmental scientists, microbiologists, technical services specialists, and chemists are actively taking samples for COVID-19 testing. The fact is wastewater has proven to be an early indicator of the next outbreak. Interestingly, the 2020 summer increase in the Hampton Roads region of southeastern Virginia was detected not through nose swabs or other medical testing but in the wastewater (sewer) pipes—"And what's in the sewage pipes will tell."

The professionals saw an increase in genetic material from the virus surge in wastewater produced by hundreds of thousands of Hampton Roads residents when samples were analyzed from the District's treatment plants. As the District's data was plotted and monitored with the aid of AI analyzing the health department's data, close similarities were seen. However, there was a lag time of one week from the time signals were apparent in the wastewater analyzed and the time clinical data started to change.

HRSD employs AI-assisted wastewater-based epidemiology for sampling, testing, and analyzing raw sewage samples. The advantage of using WBE is that it detects the presence of viruses soon after people flush their toilets. It is much quicker than waiting for people to be tested and get their results back. While SARS-CoV-2—its official scientific name—is primarily thought of as a respiratory pathogenic virus, scientists have found that it can affect the digestive system, and the virus' genetic makeup can be detected in stool specimens even before symptoms manifest themselves and are observable or felt. Tracking the coronavirus through wastewater can be an early indicator of where the next outbreak could happen. In other words, using AI-assisted WBE-provided information allows decisions to be made before things get worse. And that is the beauty, the goal, the objective, and the purpose of employing WBE.

Along with aiding in detecting disease in raw sewage, AI is used in several wastewater treatment plant operations, including predictive maintenance. AI intelligent sensors can detect equipment failures and maintenance needs, reducing downtime and improving efficiency. AI can also optimize the removal of organic and inorganic contaminants, ensuring that the unit processes are effective. AI helps analyze data for proper decision-making.

THE BOTTOM LINE

Wastewater treatment, AI, and IoT with intelligent sensors and actuators (e.g., valve operators) are the future of wastewater and water operations. Treatment facilities that have already integrated and synergized AI algorithms and IoT devices (i.e., sensors and smart meters) receive a stream of real-time data (e.g., pH levels, dissolved oxygen, flow rates, pressure levels, and turbidity) from various points in the treatment process. This data stream enables operators to optimize process operation, improve energy efficiency (wastewater unit processes use large amounts of electricity), reduce costs, and boost overall system performance.

NOTE

1 Material in this section is from F.R. Spellman (2014). *Handbook of Water and Wastewater Treatment Plant Operations*, 3rd ed. Boca Raton, FL: CRC Press.

REFERENCES

Crawford, G., Thompson, D., Lozier, J., Daigger, G., and Fleischer, E. (2000). Membrane Bioreactors—A Designer's Perspective. In *Proceedings of the Water Environment Federation 73rd Annual Conference & Exposition on Water Quality and Wastewater Treatment,* Anaheim, CA, CD-ROM, October 14–18, 2000.

CWNS (Clean Watersheds Needs Survey) (2000). Report to Congress, EPA-832-R-10-002. Washington, DC.

Fleischer, E.J., Broderick, T.A., Daigger, G.T., Fonseca, A.D., Holbrook, R.D., and Murthy, S.N. (2005). Evaluation of membrane bioreactor process capabilities to meet stringent effluent nutrient discharge requirements. *Water Environ. Res.* 77:162–178.

Grady, B. et al. (2011). *Bioremediation Approaches for Organic Pollution.* Atlanta, GA: National Institute of Health.

McGhee, T.J. (1991). *Water Supply and Sewerage.* New York: McGraw-Hill, Inc.

Metcalf & Eddy, Inc. (1991). *Wastewater Engineering: Treatment, Disposal*, 3rd ed. New York: McGraw-Hill.

Outwater, A.B. (1994). *Reuse of Sludge and Minor Wastewater Residuals.* Boca Raton, FL: Lewis Publishers.

Park, J. (2012). *Biological Nutrient Removal Theories and Design.* Online at https://www.dnr.state.wi,us/org/water/wm/ww/biophos/bnr_remvoal.htm.

Reed, L. (1987). Frost on Biosolids. Baltimore, MD: Johns Hopkins Science.

Spellman, F.R. (1996). *Stream Ecology and Self-Purification.* Boca Raton, FL: CRC Press.

USEPA. (1989). Technical support document for pathogen reducing in sewage sludge. NTIS No. PB89-136618. Springfield, VA: National Technical Information Service.

USEPA. (1999). *Turbidity Requirements: OESWTR Guidance Manual: Turbidity Provisions.* Washington, DC: U.S. Environmental Protection Agency.

USEPA. (2007a). *Wastewater Management Fact Sheet: Membrane Bioreactors.* Washington, DC: United States Environmental Protection Agency.

USEPA. (2007b). *Advanced Wastewater Treatment to Achieve Low Concentration of Phosphorus.* Washington, DC: Environmental Protection Agency.

USEPA. (2007c). *Biological Nutrient Removal Processes and Costs.* Washington, DC: United States Environmental Protection Agency.

USEPA. (2007d). *National* Section 303*(d) List Fact Sheet.* Online at https://iaspub.epa.gov/waters/national_rept.control.

USEPA. (2011). *Technologies and Costs for the Removal of Arsenic from Drinking Water,* EPA-815-R-00-028. Washington, DC: US Environmental Protection Agency.

Yeager, J.G. and Ward, R.I. (1981). Effects of moisture content on long-term survival and regrowth of bacteria in wastewater sludge. *Appl. Environ. Microbiol.* 41(5):1117–1122.

8 Air Pollution Management and Monitoring

DINOSAUR TO DIGITAL

Industry, compliance agencies, and regulators are quickly shifting from the old to the new—from the dinosaur age of air pollution management, forecasting, and monitoring to relatively new digital techniques. Most conventional forecasting techniques (i.e., dinosaur techniques) have shown inferior predictive accrual because they attempt to gauge air pollutants' non-linear, dynamic, and complex nature and fate. Recent experience with digital management and air quality monitoring has employed artificial intelligence (AI) methods, which have become the most potent and forward-looking approaches for air pollution forecasting and tracking because of their specific benefits. These benefits include high precision, organic learning (natural learning), robust fault tolerance (i.e., the ability of a system to maintain proper operation despite faults or failures), superior generalization (i.e., the ability of a model to perform well on new data—having learned and not just memorized), and ease of working with high-dimensional data (i.e., a dataset in which the number of features is larger than the number of observations).

The USEPA is leveraging AI in several innovative ways to monitor and manage air pollution. Before discussing how AI is used or leveraged in air quality issues, it is essential to present fundamental information about air.

AIR FUNDAMENTALS[1]

Engineers always seem to have a definition for just about anything and everything (most of which cannot be understood by many of us—and maybe that is their intent). An engineer might refer to air as a fluid (because it is; like water, air is fluid—you can pour it). Engineers are primarily interested in air as a fluid because they deal with fluid mechanics, which is the study of the behavior of fluids (including air) at rest or in motion. Fluids may be either gases or liquids. You are probably familiar with the physical difference between gases and liquids, as exhibited by air and water. Still, for the study of fluid mechanics (and the purposes of this text), it is convenient to classify fluids according to their compressibility.

- Gases are very readily compressible (you've heard of compressed air).
- Liquids are only slightly compressible (it is unlikely you've heard much about compressed water).

DOI: 10.1201/9781003601562-10

TABLE 8.1
Composition of Air/Earth's Atmosphere

Gas	Chemical Symbol	Volume (%)
Nitrogen	N_2	78.08
Oxygen	O_2	20.94
Carbon dioxide	CO_2	0.03
Argon	Ar	0.093
Neon	Ne	0.0018
Helium	He	0.0005
Krypton	Kr	Trace
Xenon	Xe	Trace
Ozone	O_3	0.00006
Hydrogen	H_2	0.00005

What is air? Air is a mixture of gases that constitutes the Earth's atmosphere. What is the Earth's atmosphere? The atmosphere is that thin shell, veil, or envelope of gases surrounding Earth like the skin of an apple—thin, fragile—but very vital. The approximate composition of dry air is, by volume at sea level: nitrogen 78%, oxygen 21% (necessary for life as we know it), argon 0.93%, and carbon dioxide 0.03%, together with very small amounts of numerous other constituents (see Table 8.1). Because of constant mixing by the winds and other weather factors, the percentages of each gas in the atmosphere are normally constant up to 70,000 ft. However, it is important to note that the water vapor content is highly variable and depends on atmospheric conditions. Air is said to be pure when none of the minor constituents is present in sufficient concentration to be injurious to the health of human beings or animals, to damage vegetation, or to cause loss of amenity (e.g., through the presence of dirt, dust, or odors, or by diminution of sunshine).

Where does air come from? Genesis 1:2 states that God separated the water environment into the atmosphere and surface waters on the second day of creation. Many scientists state that 4.6 billion years ago, a cloud of dust and gases forged the Earth and created a dense molten core enveloped in cosmic gases. This was the *proto-atmosphere,* or *proto-air*, composed mainly of carbon dioxide, hydrogen, ammonia, and carbon monoxide. However, it did not last long before a tremendous outburst of charged particles from the Sun stripped it away. As the Earth's outer crust began to solidify, a new atmosphere began to form from the gases outpouring from gigantic hot springs and volcanoes. This created an atmosphere composed of carbon dioxide, nitrogen oxides, hydrogen, sulfur dioxide, and water vapor. As the Earth cooled, water vapor condensed into highly acidic rainfall, which collected to form oceans and lakes.

Only trace amounts of free oxygen were present for much of Earth's early existence (the first half). But then green plants evolved in the oceans, and they began to add oxygen to the atmosphere as a waste gas. Later, oxygen increased to about 1% of the atmosphere and, with time, to its present 21%.

How do we know for sure about the evolution of air on Earth? Are we just guessing, using "voodoo" science? There is no guessing or voodoo involved with the historical geological record. Consider, for example, geological formations dated 2 billion years ago. In these early sediments, there is a clear and extensive band of red sediment ("red bed" sediments), which are sands colored with oxidized (ferric) iron. Previously, ferrous formations had been laid down, showing no oxidation. But there is more evidence. We can look at 4.5 billion years ago when carbon dioxide in the atmosphere was beginning to be lost in sediments. The vast amount of carbon deposited in limestone, oil, and coal indicates that carbon dioxide concentrations must once have been many times greater than today, which stands at only 0.03%. The first carbonated deposits appeared about 1.7 billion years ago, and the first sulfate deposits about 1 billion years ago. An increase in the nitrogen content of the air balanced the decreasing carbon dioxide. The forms of *respiration* advanced from fermentation 4 billion years ago to anaerobic *photosynthesis* 3 billion years ago to aerobic photosynthesis 1.5 billion years ago. The aerobic respiration that is so familiar today only began to appear about 500 million years ago.

Fast-forward to the present. The atmosphere continues to evolve, but human activities—with their highly polluting effects—have overtaken nature in determining the changes.

If we cannot live without air—if air is so precious and necessary for sustaining life—then two questions arise: (1) Why do we ignore air? (2) Why do we abuse it (pollute it)? We ignore air (like water) because it is so common, so accessible (usually), and so unexceptional.

Question: How can air be so 'unexceptional' when life on Earth is impossible without it?

Is air pollution that big of a deal? Isn't pollution relative? Isn't pollution dependent on your point of view—a judgment call? Well, if you could ask the victims of the incidents listed in Table 8.2 [from U.S. Senate Staff Report, *Air Quality Criteria* (1968)], the answer, simply stated, would be: yes, yes, and yes, it is.

TABLE 8.2
Mortality Occurring During Air Pollution Events

Location	Year	Deaths Reported as a Result of Pollution Event
Belgium	1930	63
Pennsylvania	1948	17
London	1948	700–800
London	1952	4,000
London	1956	1,000
London	1957	700–800
London	1959	200–250
London	1962	700
London	1963	700
New York	1963	200–400

Beyond the fact that air is one of our essential resources, sustaining life as it stimulates and pleases the senses—though invisible to the human eye (hopefully this is the case)—, it makes possible such sights as beautiful and dazzling rainbows, heart-pinching sunsets and sunrises, the Northern Lights, and on occasion, a clear view of that high alpine meadow sprinkled with the colors of spring. But air is more than this. Air is capable of many other wondrous things. For example, have you ever felt the light touch of a calm, soothing breeze against your skin? Air is also capable of much more: it carries thousands of scents—both pungent and subtle: salty ocean breezes, approaching rain, fragrances from blooming flowers, and others.

It is the "others" that concern us here: the sulfurous gases from industrial processes—that typical rotten egg odor; the stink of garbage, refuse, and trash—all part of humanity's throwaways; the toxic poison remnants from pesticides, herbicides, and all the other "-cides." We are surrounded by it, but we seldom think about it until it displeases us (remember, humans can put up with just about anything until it displeases them). Pollution—those discarded, sickening leftovers of the heavy hands of humans—causes the problem.

Definition of Key Terms

The reader must be familiar with the vocabulary used in air pollution control activities to work at the cutting edge of air science, air pollution, and the scientific disciplines closely related to air science.

Definitions

Absolute Pressure: The total pressure in a system, including the pressure of a substance and the pressure of the atmosphere (about 14.7 psi at sea level).

Acid: Any substance that releases hydrogen ions (H^+) when mixed into water.

Acid Precipitation: Rain, snow, or fog that contains higher-than-normal levels of sulfuric or nitric acid, which may damage forests, aquatic ecosystems, and cultural landmarks.

Acid Surge: A period of short, intense acid deposition in lakes and streams due to the release (by rainfall or spring snowmelt) of acids stored in soil or snow.

Acidic Solution: A solution that contains significant numbers of (H^+) ions.

Airborne Toxins: Hazardous chemical pollutants released into the atmosphere and carried by air currents.

Albedo: Reflectivity, or the fraction of incident light reflected by a surface.

Arithmetic Mean: Measures average value, calculated by summing all terms and dividing by the number of terms.

Arithmetic Scale: A series of intervals (marks or lines), usually made along the side or bottom of a graph, that represents the range of values of the data. When the marks or lines are equally spaced, it is called an arithmetic scale.

Atmosphere: A 500-km-thick layer of colorless, odorless gases known as air that surrounds the Earth and is composed of nitrogen, oxygen, argon, carbon dioxide, and other gases in trace amounts.

Atom: The smallest particle of an element that still retains the characteristics of that element.

Atomic Number: The number of protons in the nucleus of an atom.

Atomic Weight: The sum of the number of protons and the number of neutrons in the nucleus of an atom.

Base: Any substance that releases hydroxyl ions (OH^-) when it dissociates in water.

Chemical Bond: The force that holds atoms together within molecules. A chemical bond is formed when a chemical reaction takes place. Two types of chemical bonds are ionic bonds and covalent bonds.

Chemical Reaction: A process that occurs when atoms of certain elements are brought together and combined to form molecules or when molecules are broken down into individual atoms.

Climate: The long-term weather pattern of a particular region.

Covalent Bond: A type of chemical bond in which electrons are shared.

Density: The weight of a substance per unit of its volume, e.g., pounds per cubic foot.

Dew Point: The temperature at which a sample of air becomes saturated, i.e., has a relative humidity of 100%.

Element: Any of more than 100 fundamental substances that consist of atoms of only one kind and constitute all matter.

Emission Standards: The maximum amount of a specific pollutant permitted to be legally discharged from a particular source in a given environment.

Emissivity: The relative power of a surface to reradiate solar radiation back into space as heat or long-wave infrared radiation.

Energy: The ability to do work, to move matter from place to place, or to change matter from one form to another.

First Law of Thermodynamics: A natural law that dictates that energy is neither created nor destroyed during physical or chemical change, but it may be changed in form and moved from place to place.

Global Warming: The increase in global temperature predicted to arise from increased levels of carbon dioxide, methane, and other greenhouse gases in the atmosphere.

Greenhouse Effect: The prevention of the reradiation of heat waves to space by carbon dioxide, methane, and other atmospheric gases. The greenhouse effect makes possible the conditions that enable life to exist on Earth.

Ion: An atom or radical in solution carrying an integral electrical charge, either positive (cation) or negative (anion).

Insolation: The solar radiation received by the Earth and its atmosphere—**in**coming **sol**ar radia**tion**.

Lapse Rate: The rate of temperature change with altitude. The average lapse rate in the troposphere is –3.5°F/1,000 ft.

Matter: Anything that exists in time, occupies space, and has mass.

Mesosphere: A region of the atmosphere based on temperature, approximately 35 and 60 miles in altitude.

Meteorology: The study of atmospheric phenomena.

Mixture: Two or more elements, compounds, or both mixed without a chemical reaction.

Ozone: The compound O_3. It is found naturally in the atmosphere in the ozonosphere and is also a constituent of photochemical smog.

pH: A means of expressing hydrogen ion concentration in terms of powers of 10; a measurement of how acidic or basic a substance is. The pH scale runs from 0 (most acidic) to 14 (most basic). The center of the range (7) indicates the substance is neutral.

Photochemical Smog: An atmospheric haze that occurs above industrial sites and urban areas resulting from reactions, which occur in the presence of sunlight, between pollutants produced at high temperatures and pressurized combustion processes (such as fuel combustion in a motor vehicle). The primary component of smog is ozone.

Photosynthesis: *is* the process by which chlorophyll-containing plants use the sun's light energy to convert carbon dioxide (CO_2) and water (H_2O) into complex chemical bonds, forming simple carbohydrates such as glucose and fructose.

Pollutant: A contaminant at a concentration high enough to endanger the environment.

Pressure: The force pushing on a unit area. It is usually measured in atmospheres, Pascals (Pa), or pounds per square inch (psi) in air applications.

Primary Pollutants: Pollutants emitted directly into the atmosphere where they adversely influence human health or the environment. The six primary pollutants are carbon dioxide, carbon monoxide, sulfur oxides, nitrogen oxides, hydrocarbons, and particulates. All but carbon dioxide are regulated in the United States.

Rayleigh Scattering: The preferential scattering of light by air molecules and particles that accounts for the blueness of the sky. The scattering is proportional to $1/\lambda^4$.

Radon: is a naturally occurring radioactive gas arising from the decay of uranium-238, which may harm human health in high concentrations.

Rain Shadow Effect: The phenomenon that occurs due to the movement of air masses over a mountain range. As an air mass rises to clear a mountain, the air cools, and precipitation forms. The precipitation and the pollutant load carried by the air mass will often be dropped on the windward side of the mountain. The air mass is then devoid of most of its moisture; consequently, the lee side of the mountain receives little or no precipitation and is said to lie in the rain shadow of the mountain range.

Relative Humidity: The concentration of water vapor in the air. It is expressed as the percentage that its moisture content represents of the maximum amount the air could contain at the same temperature and pressure. The higher the temperature, the more water vapor the air can hold.

Secondary Pollutants: Are formed from the interaction of primary pollutants with other primary pollutants or atmospheric compounds such as water vapor.

Second Law of Thermodynamics: A natural law that dictates that with each change in form, some energy is degraded to a less helpful form and given off to the surroundings, usually as low-quality heat.

Solute: The substance dissolved in a solution.

Solution: A liquid containing a dissolved substance.

Specific Gravity: The ratio of the density of a substance to a standard density. For gases, the density is compared with air density (=1).

Stratosphere: An atmospheric layer extending from 6 to 7 miles to 30 miles above the Earth's surface.

Stratospheric Ozone Depletion: The thinning of the ozone layer in the stratosphere; occurs when certain chemicals (such as chlorofluorocarbons) capable of destroying ozone accumulate in the upper atmosphere.

Thermosphere: An atmospheric layer that extends from 56 miles to outer space.

Troposphere: The atmospheric layer that extends from the Earth's surface to 6–7 miles above the surface.

Weather: The day-to-day pattern of precipitation, temperature, wind, barometric pressure, and humidity.

Wind: Horizontal air motion.

AIR COMPONENTS: CHARACTERISTICS AND PROPERTIES

Air combines gases (see Table 8.1) and other matter (suspended minute liquid or particulate matter). In this section, we discuss each of the major components.

Note: Much of the information about atmospheric gases was adapted from the Compressed Gas Association's *Handbook of Compressed Gases* (1990) and *Environmental Science and Technology: Concepts and Applications* (2006).

Atmospheric Nitrogen

Nitrogen (N_2) makes up the central portion of the atmosphere (78.03% by volume, 75.5% by weight). It is a colorless, odorless, tasteless, nontoxic, and almost inert gas. Nitrogen is nonflammable, will not support combustion, and is not life-supporting. Nitrogen is part of Earth's atmosphere primarily because, over time, it has simply accumulated and remained in place and balance. This nitrogen accumulation process has occurred because, chemically, nitrogen is not very reactive. When released by any process, it tends not to recombine with other elements and accumulates in the atmosphere. And this is a good thing because we need nitrogen. No, we don't need it for breathing, but we need it for other life-sustaining processes.

Although nitrogen in its gaseous form is of little use to us, it is the most common element in living tissues after oxygen, carbon, and hydrogen. As a chief constituent of chlorophyll, amino acids, and nucleic acids—the "building blocks" of proteins (which are used as structural components in cells)—nitrogen is essential to life. Nitrogen is dissolved in and carried by the blood. It does not appear to enter any chemical combination as it is transported throughout the body. Each time we breathe, the same amount of nitrogen is exhaled. Animals cannot use nitrogen directly, but only when it is obtained by eating plants or animal tissues; plants get the nitrogen they need through inorganic compounds, principally nitrate and ammonium.

Gaseous nitrogen is converted to a form usable by plants (nitrate ions) chiefly through nitrogen fixation via the nitrogen cycle, shown in simplified form in Figure 8.1.

Via the *nitrogen cycle***,** aerial nitrogen is converted into nitrates mainly by microorganisms, bacteria, and blue-green algae. Lightning also converts some aerial nitrogen gas into forms that return to the Earth as nitrate ions in rainfall and other types of precipitation. Ammonia plays a major role in the nitrogen cycle. Excretion by animals and anaerobic decomposition of dead organic matter by bacteria produce ammonia.

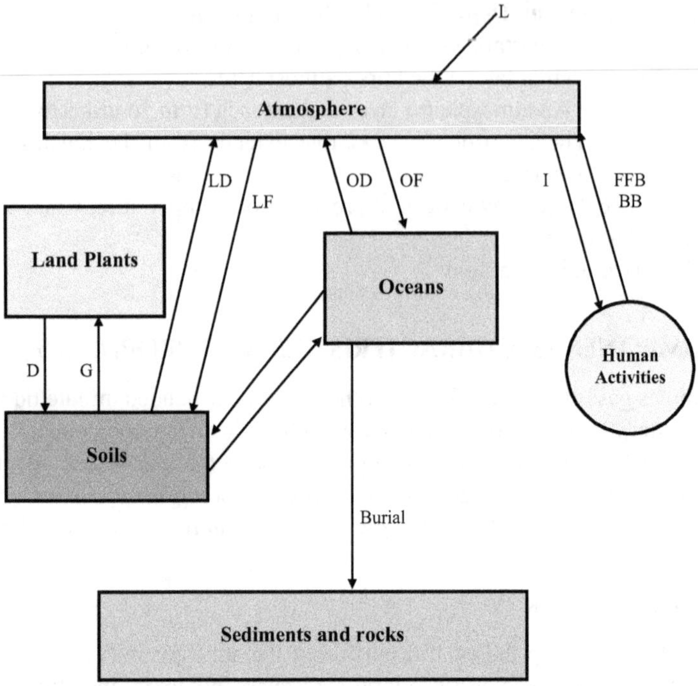

LF = Land Fixation
LD = Land Denitrification
OF = Oceanic Fixation
OD = Oceanic Denitrification
I = Industrial Fixation
FFB = Fossil Fuel Burning
BB = Biomass Burning
L = Lightning
D = Decay
G = Growth

O – L = Ocean-to-Land
L – O = Land-to-Ocean

FIGURE 8.1 Nitrogen cycle.

Ammonia, in turn, is converted by nitrification bacteria into nitrites and then into nitrates. This process is known as nitrification. Nitrification bacteria are aerobic. Bacteria that convert ammonia into nitrites are known as nitrite bacteria (e.g., *Nitrosococcus* and *Nitrosomonas*). Although nitrite is toxic to many plants, it usually does not accumulate in the soil. Instead, other bacteria (such as *Nitrobacter*) oxidize the nitrite to form nitrate (NO_3^-), the most common biologically usable form of nitrogen.

Nitrogen reenters the atmosphere through denitrifying bacteria in nutrient-rich habitats such as marshes and swamps. These bacteria break down nitrates into nitrogen gas and nitrous oxide (N_2O), which then reenter the atmosphere. Nitrogen also reenters the atmosphere from exposed nitrate deposits and emissions from electric power plants, automobiles, and volcanoes.

TABLE 8.3
Nitrogen: Physical Properties

Chemical formula	N_2
Molecular weight	28.01
Density of gas @ 70°F	0.072 lb/ft^3
Specific gravity of the gas @ 70°F & 1 atm (air = 1)	0.967
Specific volume of gas @ 70°F & 1 atm	13.89 ft^3
Boiling point @ 1 atm	−320.4°F
Melting point @ 1 atm	−345.8°F
Critical temperature	−232.4°F
Critical pressure	493 psia
Critical density	19.60 lb/ft^3
Latent heat of vaporization @ boiling point	85.6 Btu/lb
Latent heat of fusion @ melting point	11.1 Btu/lb

Nitrogen: Physical Properties

The physical properties of nitrogen are noted in Table 8.3.

Nitrogen: Uses

In addition to being the preeminent (regarding volume) component of Earth's atmosphere and providing an essential ingredient in sustaining life, nitrogen gas has many commercial and technical applications. As a gas, it is used in heat treating primary metals, the production of semiconductor electronic components, as a blanketing atmosphere, blanketing oxygen-sensitive liquids and volatile liquid chemicals, inhibiting aerobic bacteria growth, and propelling liquids through canisters, cylinders, and pipelines.

Nitrogen Oxides

There are six oxides of nitrogen: nitrous oxide (N_2O), nitric oxide (NO), dinitrogen trioxide (N_2O_3), nitrogen dioxide (NO_2), dinitrogen tetroxide (N_2O_4), and dinitrogen pentoxide (N_2O_5).

Nitric oxide, nitrogen dioxide, and nitrogen tetroxide are fire gases. One or more of them are generated when certain nitrogenous organic compounds (e.g., polyurethane) burn. Nitric oxide is the product of incomplete combustion, whereas a mixture of nitrogen dioxide and nitrogen tetroxide is the product of complete combustion.

The nitrogen oxides are usually collectively symbolized by the formula NO_x. The USEPA, under the Clean Air Act (CAA), regulates the amount of nitrogen oxides that commercial and industrial facilities may emit into the atmosphere. The primary and secondary standards are the same: The annual concentration of nitrogen dioxide may not exceed 100 μg/m^3 (0.05 ppm).

Oxygen Atmosphere

Oxygen (O_2_ Greek *oxys* "acid" *genes* "forming") constitutes approximately a fifth (21% by volume and 23.2% by weight) of the air in Earth's atmosphere. Gaseous oxygen (O_2) is vital to life as we know it. On Earth, oxygen is the most abundant element. Most oxygen on Earth is not found in the free state but in combination with other elements as chemical compounds. Water and carbon dioxide are typical examples of compounds containing oxygen, but countless others exist.

At ordinary temperatures, oxygen is a colorless, odorless, tasteless gas that supports life and combustion. All the elements, except the inert gases, combine directly with oxygen to form oxides. However, oxidation of different elements occurs over a wide range of temperatures.

Oxygen is nonflammable, but it readily supports combustion; that is, all materials that are flammable in air burn much more vigorously in oxygen. Some combustibles, such as oil and grease, burn with nearly explosive violence in oxygen if ignited.

Oxygen: Physical Properties

The physical properties of oxygen are noted in Table 8.4.

Oxygen: Uses

Oxygen's primary uses stem from its life-sustaining and combustion-supporting properties. It also has many industrial applications (when used with other fuel gases such as acetylene), including metal cutting, welding, hardening, and scarfing.

Ozone: Just another Form of Oxygen

Ozone (O_3) is a highly reactive pale-blue gas with a penetrating odor. Ozone is an allotropic modification of oxygen. An allotrope is a variation of an element that possesses a set of physical and chemical properties significantly different from the "normal" form of the component. Only a few elements have allotropic forms; oxygen, phosphorus, and sulfur are some of them. Ozone is just another form of oxygen. It is formed when the molecule of the stable form of oxygen (O_2) is split by ultraviolet

TABLE 8.4
Oxygen: Physical Properties

Chemical formula	O_2
Molecular weight	31.9988
Freezing point	−361.12°F
Boiling point	−297.33°F
Heat of fusion	5.95 Btu/lb
Heat of vaporization	91.70 Btu/lb
The density of gas @ boiling point	0.268 lb/ft^3
The density of gas @ room temperature	0.081 lb/ft^3
Vapor density (air = 1)	1.105
Liquid-to-gas expansion ratio	875

(UV) radiation or electrical discharge; it has three instead of two atoms of oxygen per molecule. Thus, its chemical formula is represented by O_3.

Ozone forms a thin layer in the upper atmosphere, which protects life on Earth from ultraviolet rays, a cause of skin cancer. It is an air pollutant at lower atmospheric levels and contributes to the greenhouse effect. When inhaled at ground level, ozone can cause asthma attacks, stunted plant growth, and corrosion of certain materials. It is produced by the action of sunlight on air pollutants, including car exhaust fumes, and is a major air pollutant in hot summers.

Atmospheric Carbon Dioxide

Carbon dioxide (CO_2) is a colorless, odorless gas (although some people feel it has a slightly pungent odor and biting taste), slightly soluble in water, denser than air (one and a half times heavier than air), and somewhat acidic. Carbon dioxide gas is relatively non-reactive and nontoxic. It will not burn, and it will not support combustion or life.

CO_2 is usually present in atmospheric air at about 0.035% by volume and cycles through the biosphere (Carbon Cycle), as shown in Figure 8.2. Carbon dioxide, along with water vapor, is primarily responsible for the absorption of infrared energy re-emitted by the Earth, and some of this energy is re-radiated back to the Earth's surface. It is also an average end product of human and animal metabolism. The exhaled breath contains up to 5.6% carbon dioxide. In addition, burning carbon-laden fossil fuels releases carbon dioxide into the atmosphere. Much of this carbon dioxide is absorbed by ocean water, some of it is taken up by vegetation through photosynthesis in the carbon cycle (see Figure 8.2), and some remains in the atmosphere. Today, it is estimated that the concentration of carbon dioxide in the atmosphere is approximately 350 parts per million (ppm), rising at approximately 20 ppm every decade. The increasing combustion rate of coal and oil has been primarily responsible for this occurrence, which (as we will see later in this text) may eventually impact global climate.

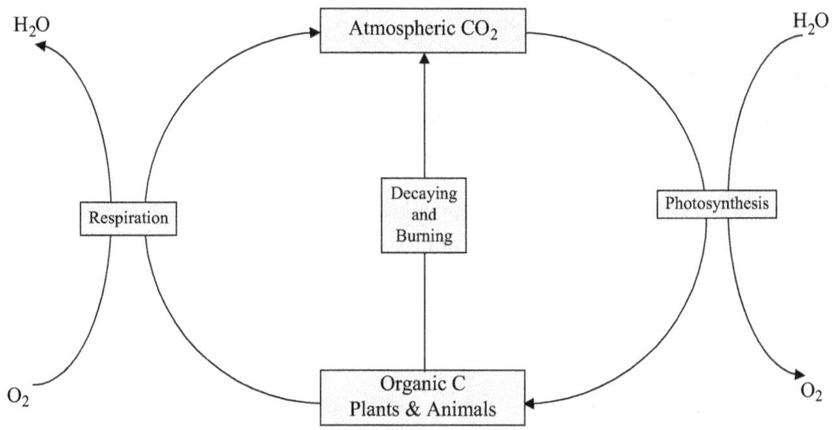

FIGURE 8.2 Carbon cycle.

TABLE 8.5
Carbon Dioxide: Physical Properties

Chemical formula	CO_2
Molecular weight	44.01
Vapor pressure @ 70°F	838 psig
Density of the gas @ 70°F & 1 atm	0.1144 lb/ft^3
The specific gravity of the gas @ 70°F & 1 atm (air = 1)	1.522
The specific volume of the gas @ 70°F & 1 atm	8.741 ft^3/lb
Critical temperature	−109.3°F
Critical pressure	1,070.6 psia
Critical density	29.2 lb/ft^3
Latent heat of vaporization @ 32°F	100.8 Btu/lb
Latent heat of fusion @ −69.9°F	85.6 Btu/lb

Carbon Dioxide: Physical Properties

The physical properties of carbon dioxide are noted in Table 8.5.

Carbon Dioxide: Uses

Solid carbon dioxide is used extensively to refrigerate perishable foods while in transit. It is also used as a cooling agent in many industrial processes, such as grinding, rubber work, cold-treating metals, and vacuum cold traps.

Gaseous carbon dioxide is used to carbonate soft drinks, control pH in water treatment, chemical processing, as a food preservative, and in pneumatic devices.

Atmospheric Argon

Argon (Ar - Greek *argos* 'idle') is a colorless, odorless, tasteless, nontoxic, nonflammable gaseous element (noble gas). It constitutes almost 1% of Earth's atmosphere and is plentiful compared to the other rare atmospheric gases. It is highly inert and forms no known chemical compounds. It is slightly soluble in water.

Argon: Physical Properties

The physical properties of argon are noted in Table 8.6.

Argon: Uses

Argon is used extensively in filling incandescent and fluorescent lamps and electronic tubes, providing a protective shield for growing silicon and germanium crystals, and blanketing the production of titanium, zirconium, and other reactive metals.

Atmospheric Neon

Neon (Ne - Greek *neon* "new") is a colorless, odorless, gaseous, nontoxic, chemically inert element. Air is about two parts per thousand neon by volume.

TABLE 8.6
Argon: Physical Properties

Chemical formula	Ar
Molecular weight	39.95
Density of the gas @ 70°F	0.103 lb/ft^3
The specific gravity of the gas @ 70°F	1.38
The specific volume of the gas @ 70°F	9.71 ft^3/lb
Boiling point at 1 atm	−302.6°F
Melting point at 1 atm	−308.6°F
Critical temperature	−188.1°F
Critical pressure	711.5 psia
Critical density	33.444 lb/ft^3
Latent heat of vaporization @ boiling point and 1 atm	69.8 But/lb
Latent heat of fusion	12.8 Btu/lb

TABLE 8.7
Neon: Physical Properties

Chemical formula	Ne
Molecular weight	20.183
Density of the gas @ 70°F & 1 atm	0.05215 lb/ft^3
The specific gravity of the gas @ 70°F & 1 atm	0.696
The specific volume of the gas @ 70°F & 1 atm	19.18 ft^3/lb
Boiling point at 1 atm	−410.9°F
Melting point at 1 atm	−415.6°F
Critical temperature	−379.8°F
Critical pressure	384.9 psia
Critical density	30.15 lb/ft^3
Latent heat of vaporization @ boiling point	37.08 But/lb
Latent heat of fusion	7.14 Btu/lb

Neon: Physical Properties

The physical properties of neon are noted in Table 8.7.

Neon: Uses

Neon is principally used to fill lamp bulbs and tubes. The electronics industry uses neon singly or in mixtures with other gases in many gas-filled electron tubes.

Atmospheric Helium

Helium (He - Greek *helios* "Sun") is inert (and as a result, does not appear to have any significant effect on, or role in, the atmosphere), non-toxic, odorless, tasteless,

TABLE 8.8
Helium: Physical Properties

Chemical formula	He
Molecular weight	4.00
Density of the gas @ 70°F & 1 atm	0.0103 lb/ft^3
The specific gravity of the gas @ 70°F & 1 atm	0.138
The specific volume of the gas @ 70°F & 1 atm	97.09 ft^3/lb
Boiling point @ 1 atm	−452.1°F
Critical temperature	−450.3°F
Critical pressure	33.0 psia
Critical density	4.347 lb/ft^3
Latent heat of vaporization @ boiling point & 1 atm	8.72 But/lb

non-reactive, forms no compounds, and is colorless. It makes up about 0.00005% (5 ppm) of the volume of air in Earth's atmosphere. Helium, like neon, krypton, hydrogen, and xenon, is a noble gas. Helium is the second lightest element; only hydrogen is lighter. It is one-seventh as heavy as air. Helium is nonflammable and is only slightly soluble in water.

Helium: Physical Properties

The physical properties of helium are noted in Table 8.8.

Atmospheric Krypton

Krypton (Kr - Greek *kryptos* "hidden") is a colorless, odorless, inert gaseous component of Earth's atmosphere. It is present in tiny quantities in the air (about 114 parts per million—ppm).

Krypton: Physical Properties

The physical properties of krypton are listed in Table 8.9.

Krypton: Uses

Krypton is used principally to fill lamp bulbs and tubes. The electronics industry uses it singly or in a mixture in many gas-filled electron tubes.

Atmospheric Xenon

Xenon (Xe - Greek *xenon* "stranger") is a colorless, odorless, non-toxic, inert, heavy gas present in tiny quantities in the air (about 1 part in 20 million).

Xenon: Physical Properties

The physical properties of xenon are listed in Table 8.10.

TABLE 8.9
Krypton: Physical Properties

Chemical formula	Kr
Molecular weight	83.80
Density of the gas @ 70°F & 1 atm	0.2172 lb/ft^3
The specific gravity of the gas @ 70°F & 1 atm	2.899
The specific volume of the gas @ 70°F & 1 atm	4.604 ft^3/lb
Boiling point @ 1 atm	−244.0°F
Melting point @ 1 atm	−251°F
Critical temperature	−82.8°F
Critical pressure	798.0 psia
Critical density	56.7 lb/ft^3
Latent heat of vaporization @ boiling point	46.2 Btu/lb
Latent heat of fusion	8.41 But/lb

TABLE 8.10
Xenon: Physical Properties

Chemical formula	Xe
Molecular weight	131.3
Density of the gas @ 70°F & 1 atm	0.3416 lb/ft^3
The specific gravity of the gas @ 70°F & 1 atm	4.560
The specific volume of the gas @ 70°F & 1 atm	2.927 ft^3/lb
Boiling point at 1 atm	−162.6°F
Melting point at 1 atm	−168°F
Critical temperature	61.9°F
Critical pressure	847.0 psia
Critical density	68.67 lb/ft^3
Latent heat of vaporization at boiling point	41.4 Btu/lb
Latent heat of fusion	7.57 Btu/lb

Xenon: Uses

Xenon is used principally to fill lamp bulbs and tubes. The electronics industry uses it singly or in mixtures in many gas-filled electron tubes.

Atmospheric Hydrogen

Hydrogen (H_2 - Greek *hydros* + *gen* "water generator") is a colorless, odorless, tasteless, non-toxic, flammable gas. It is the lightest of all elements and occurs on Earth chiefly in combination with oxygen as water. Hydrogen is the most abundant element in the universe, accounting for 93% of the total number of atoms and 76% of the total mass. It is the lightest known gas, with a density of approximately 0.07 that of air. Hydrogen is present in the atmosphere, occurring in concentrations of only about 0.5 ppm by volume at lower altitudes.

TABLE 8.11
Hydrogen: Physical Properties

Chemical formula	H_2
Molecular weight	2.016
Density of the gas @ 70°F & 1 atm	0.00521 lb/ft^3
The specific gravity of the gas @ 70° & 1 atm	0.06960
The specific volume of the gas @ 70°F & 1 atm	192.0 ft^3/lb
Boiling point @ 1 atm	−423.0°F
Melting point @ 1 atm	−434.55°F
Critical temperature	−399.93°F
Critical pressure	190.8 psia
Critical density	1.88 lb/ft^3
Latent heat of vaporization @ boiling point	191.7 Btu/lb
Latent heat of fusion	24.97 Btu/lb

Hydrogen: Physical Properties

The physical properties of hydrogen are listed in Table 8.11.

Refineries, petrochemicals, and bulk chemical facilities use hydrogen for hydro-treating, catalytic reforming, and hydro-cracking. Hydrogen is used in the production of a wide variety of chemicals. Metallurgical companies use hydrogen to produce their products. Glass manufacturers use hydrogen as a protective atmosphere in a process whereby molten glass is floated on a surface of molten tin. Food companies hydrogenate fats, oils, and fatty acids to control various physical and chemical properties. Electronic manufacturers use hydrogen at several steps in the complex semiconductor processes.

Atmospheric Water

Leonardo da Vinci understood the importance of water when he said, "Water is the driver of nature." Da Vinci was acknowledging what most scientists and many of us have come to realize: Water, propelled by the varying temperatures and pressures in Earth's atmosphere, allows life as we know it to exist on our planet (Gradel & Crutzen, 1995).

The water vapor content of the lower atmosphere (troposphere) is generally between 1% and 3% by volume, with a global average of about 1%. However, the percentage of water in the atmosphere can vary from as little as 0.1% to as much as 5%, depending upon altitude; water in the atmosphere decreases with increasing altitude. Water circulates in the atmosphere in the hydrologic cycle, as shown in Figure 8.3.

Water vapor contained in Earth's atmosphere plays several vital roles: (1) it absorbs infrared radiation; (2) it acts as a blanket at night, retaining heat from the Earth's surface; and (3) it affects the formation of clouds in the atmosphere.

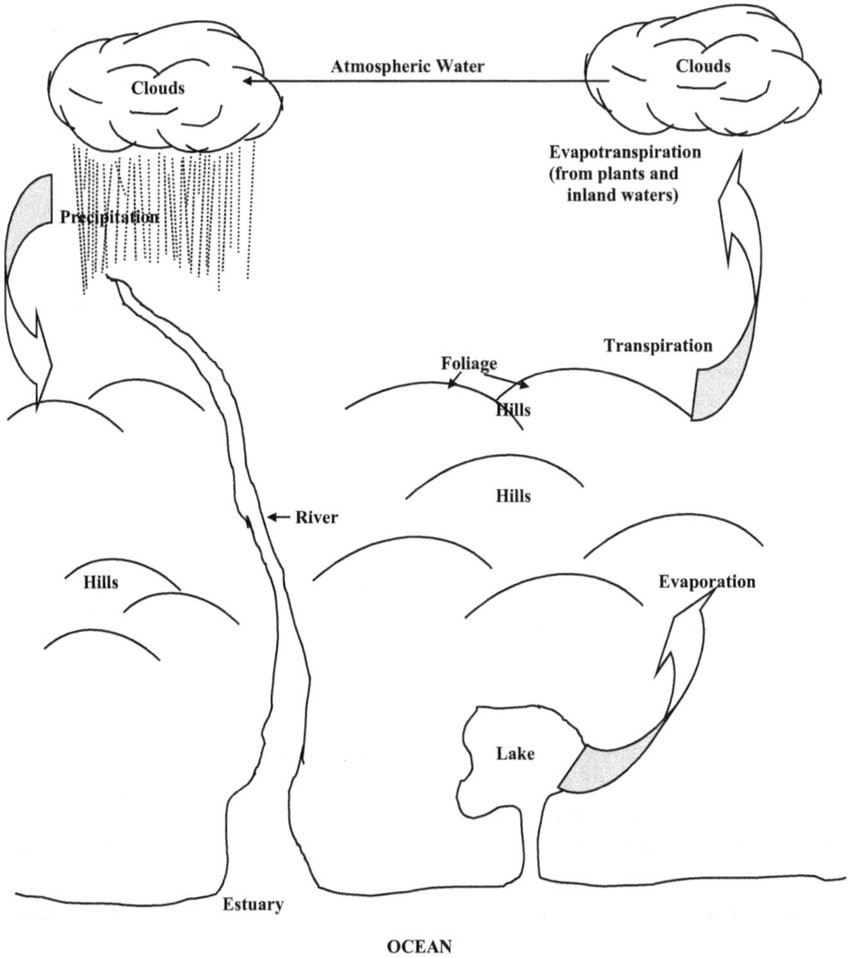

FIGURE 8.3 Water cycle.

Atmospheric Particulate Matter

Significant particles (particulate matter) are suspended in the atmosphere, particularly the troposphere. These particles originate in nature from smoke, sea sprays, dust, and the evaporation of organic materials from vegetation. There is also a wide variety of nature's living or semi-living particles—spores and pollen grains, mites and other tiny insects, spider webs, and diatoms. The atmosphere also contains various anthropogenic (man-made) particles produced by automobiles, refineries, production mills, and many other human activities.

Atmospheric particulate matter varies significantly in size (colloidal-sized particles in the atmosphere are called aerosols—usually less than 0.1 μm in diameter);

the smallest are gaseous clusters and ions and submicroscopic liquids and solids; somewhat larger ones produce the beautiful blue haze in distant vistas; those two to three times larger are highly effective in scattering light; and the largest consist of such things as rock fragments, salt crystals, and ashy residues from volcanoes, forest fires, or incinerators.

The number of particulates in the atmosphere varies greatly, ranging from more than 10,000,000/cubic centimeter to less than 1/L (0.001/cc). Excluding the particles in gases and vegetative material, sizes range from 0.005 to 500 μm, a variation in diameter of 100,000 times.

The most significant number of airborne particulates is always in the invisible range. These numbers vary from less than 1 L to more than half a million per cubic centimeter in heavily polluted air, and up to at least 10 times more when a gas-to-particle reaction occurs (Schaefer & Day, 1981).

The atmosphere has two distinct regions based on particulate level: very clean and dirty. There are so few particulates in the clean parts that they are almost invisible, making them hard to collect or measure. In the dirty parts—the air of a large metropolitan area—the concentration of particles includes an incredible variety of particulates from various sources.

Atmospheric particulate matter performs several functions, undergoes several processes, and is involved in many chemical reactions in the atmosphere. The most essential function of particulate matter in the atmosphere is its action as a nucleus for forming water droplets and ice crystals. Much of the work of Vincent J. Schaefer (inventor of cloud seeding) involved using dry ice in early attempts, but later evolved around the addition of condensing particles to atmospheres supersaturated with water vapor and the use of silver iodide, which forms vast numbers of tiny particles. Another important function of atmospheric particulate matter is that it helps determine the heat balance of the Earth's atmosphere by reflecting light. Particulate matter is also involved in chemical atmospheric reactions such as neutralization, catalytic effects, and oxidation reactions. These chemical reactions will be discussed in greater detail later.

AIR FOR COMBUSTION

It is difficult to imagine where man would be today or how far he would have progressed from his beginning to the present if he had not discovered and developed the use of fire. Today, of course, we are pretty familiar with fire. We use the terms fire, combustion, oxidation, and burning in the same light to mean the same thing. However, there is a subtle difference between combustion and oxidation. During combustion, two or more substances chemically unite. In practice, one is almost always atmospheric oxygen, but combustion reactions are known in which oxygen is not one of the reactants. Thus, it is more correct to describe combustion as a rapid oxidation—or fire.

To state that atmospheric air plays a vital role in combustion is to understate its significance; that is, we are stating the obvious—something readily apparent to most of us. Though air is essential in combustion, most of us give little thought to the chemical reaction of combustion.

Combustion is a chemical reaction in which a fuel combines with air (oxygen) with the evolution of heat: burning. The combustion of fuels containing carbon and hydrogen is said to be complete when these two elements are oxidized to carbon dioxide and water (e.g., the combustion of carbon $C + O_2 = CO_2$). In air pollution control, incomplete combustion is what concerns us. Incomplete combustion may lead to (1) appreciable amounts of carbon remaining in the ash; (2) emission of some of the carbon as carbon monoxide; and (3) reaction of the fuel molecules to give a range of products which are emitted as smoke.

GAS PHYSICS

Material Balance

Material balance is one of the most basic and valuable tools in air pollution engineering. A material balance means "what goes in must come out." Matter is neither created nor destroyed in industrial processes (non-radioactive only).

$$\text{Mass}_{(\text{Out})} = \text{Mass}_{(\text{In})} \tag{8.1}$$

Material balance is used in a wide variety of air pollution control calculations. For example, it is used to evaluate the following:

- Formation of combustion products in boilers
- Rates of air infiltration into air pollution control systems
- Material requirements for process operations
- Rate of ash collection in air pollution control systems
- Humidity of exhaust gas streams
- Exhaust gas flow rates from multiple sources controlled by a single air pollution control system
- Gas flow rates from combustion processes

This principle, called the conservation of matter, can be applied to solve problems involving the quantities of matter moving in various parts of a process.

Compressibility

Air, unlike liquids, is readily compressible, and large quantities can be stored in relatively small containers. The more the air is compressed, the higher its pressure becomes. The higher the pressure in a container, the stronger the container must be. Gases are essential compressible fluids, not only because a gas can be a pollutant but also because gases convey the particles (particulate matter) and gaseous pollutants (Hesketh, 1991).

Gas Laws

Gases can be pollutants as well as conveyors of contaminants. Air (which is mainly nitrogen) is usually the main gas stream. Gas conditions are generally described in two ways: *standard temperature and pressure (STP)* and *Standard Conditions (SC)*. STP represents 0°C (32°F) and 1 atm. SC is more commonly used and represents typical room conditions of 20°C (70°F) and 1 atm; SC is usually measured in cubic meters, Nm^3, or standard cubic feet (scf).

Various physical laws governing the behavior of pressurized gases must be understood to understand the physics of air. One of the more well-known physical laws mentioned earlier is *Pascal's Law*. In addition to one of its previously mentioned parameters, Pascal's Law states that a confined gas (fluid) transmits externally applied pressure uniformly in all directions without change in magnitude. This parameter can be seen in a flexible container; it will assume a spherical (balloon) shape. However, you probably have noticed that most compressed-gas tanks are cylindrical (allowing thinner sheets of steel without sacrificing safety) with spherical ends to contain the pressure more effectively.

Though gases are compressible, note that, for a given mass flow rate, the actual volume of gas passing through the system is not constant within the system due to changes in pressure. This physical property (the fundamental relationship between the pressure of a gas and its volume) is described by *Boyle's Law* (named for its discoverer: Irish physicist and chemist Robert Boyle in 1662), which states: "The absolute pressure of a confined quantity of gas varies inversely with its volume if its temperature does not change." For example, if the pressure of a gas doubles, its volume will be reduced by half, and vice versa. That is, *as pressure goes up, volume goes down*, and vice versa. This means, for example, that if 12 ft^3 of air at 14.7 psia is compressed to 1 ft^3, air pressure will rise to 176.4 psia as long as the air temperature remains the same. This relationship can be calculated as follows:

$$P_1 \times V_1 = P_2 \times V_2 \tag{8.2}$$

where

P_1 = Original pressure (units for pressure must be absolute)
P_2 = New pressure (units for pressure must be absolute)
V_1 = Original gas volume at pressure P_1
V_2 = New gas volume at pressure P_2
This equation can also be written as:

$$\frac{P_2}{P_1} = \frac{V_1}{V_2} \text{ or } \frac{P_1}{P_2} = \frac{V_2}{V_1} \tag{8.3}$$

To allow for the effects of atmospheric pressure, always remember to convert from gage pressure *before* solving the problem, then convert back to gage pressure *after* solving it:

$$\text{Psia} = \text{psig} + 14.7\,\text{psi}$$

and

$$\text{Psig} = \text{psia} - 14.7\,\text{psi}$$

Note that in a pressurized gas system where gas is caused to move through the system by the fact that gases will flow from an area of high pressure to that of low pressure, we will always have a greater actual volume of gas at the end of the system than at the beginning (assuming the temperature remains constant).

Charles's Law

Another physical law dealing with temperature is *Charles's Law* (discovered by French physicist Jacques Charles in 1787). It states, "The volume of a given mass of gas at constant pressure is directly proportional to its absolute temperature." [The temperature in Kelvin (273 + °C) or Rankine (absolute zero = −460°F, or 0°R).]

This is calculated by using the following equation:

$$P_2 = P_1 \times \frac{T_2}{T_1} \tag{8.4}$$

Charles's Law also states, "If the pressure of a confined quantity of gas remains the same, the change in the volume (V) of the gas varies directly with a change in the temperature of the gas," as given in the equation:

$$V_2 = V_1 \times \frac{T_2}{T_1} \tag{8.5}$$

Ideal Gas Law

The *Ideal Gas Law* combines Boyle's and Charles's Laws because air cannot be compressed without temperature changes. The equation expresses the Ideal Gas Law:

$$\frac{P_1 \times V_1}{T_1} = \frac{P_2 \times V_2}{T_2} \tag{8.6}$$

Note that the Ideal Gas Law is still used as a design equation even though the equation shows that the pressure, volume, and temperature of the second state of a gas are equal to the pressure, volume, and temperature of the first state. In actual practice, however, other factors such as humidity, heat of friction, and efficiency losses all affect the gas. Also, this equation uses absolute pressure (psia) and absolute temperatures (°R) in its calculations.

In air science practice, the importance of the Ideal Gas Law cannot be overstated. It is one of the fundamental principles used in calculations involving gas flow in air pollution-related work. This law calculates actual gas flow rates based on the quantity of gas present at standard pressures and temperatures. It is also used to determine the total amount of a contaminant in a gas that can participate in a chemical reaction.

- Number of moles of gas
- Absolute temperature
- Absolute pressure

In practical applications, practitioners generally use the following standard ideal gas law equation:

$$V = \frac{nRT}{P} \text{ or } PV = nRT \tag{8.7}$$

where

V = Volume
n = Number of moles
R = Universal Gas constant
T = Absolute temperature
P = Absolute pressure

Flow Rate

The gas flow rate measures the volume of gas that passes a point in an industrial system during a given time. The ideal gas law states that this flow rate varies depending on the temperature and pressure of the gas stream and the number of moles of gas moving per unit of time.

When gas flow rates are expressed at actual temperature and pressure conditions, the actual gas flow rate is used. As you will learn later, gas flow rates can also be represented at standard temperature and pressure conditions, referred to as the standard gas flow rate.

Gas Conversions

Gases of interest in air pollution control are usually mixtures of several different compounds. For example, air comprises three significant constituents: nitrogen (N_2) at approximately 78.1%, oxygen (O_2) at approximately 20.9%, and argon at 0.9%. Many flue gas streams generated by industrial processes consist of the following significant constituents: (1) nitrogen, (2) oxygen, (3) argon, (4) carbon dioxide (CO_2), and (5) water vapor (H_2O). Both air and industrial gas streams also contain minor constituents, including air pollutants, present at relatively low concentrations compared to these significant constituents.

There is a need for ways to express both the concentrations of the major constituents of the gas stream and the concentrations of the pollutants present as minor constituents at relatively low concentrations. There are a variety of ways to express gas-phase concentrations, which can easily be converted from one type of unit to another. They include the following:

Major Constituents

- **Volume Percent**: This is one of the most common formats used to express the concentrations of significant gas stream constituents such as oxygen, nitrogen, carbon dioxide, and water vapor. The format is widespread partially because the gas stream analysis techniques used in EPA emission testing methods provide data directly in a volume percent format.
- **Partial Pressure**: Concentrations can also be expressed in partial pressures. This expression refers to the part of the total pressure exerted by one of the constituent gases.

  Gases composed of different chemical compounds, such as molecular nitrogen and oxygen, behave physically like gases of a single compound. At any given temperature, one mole of a gas exerts the same pressure as one mole of any other gas type. All the molecules move at a rate dependent on the absolute temperature and exert pressure. The total pressure is the sum of the pressures of each component. The equations below are often called Dalton's law of partial pressures.

$$P_{Total} = p_i + p_{ii} + p_{iii} \cdots p_n \tag{8.8}$$

$$P_{Total} = \sum_{i=1}^{n} p_i \tag{8.9}$$

$$\text{Partial pressure}(\text{gas}) = \left[\frac{\text{Volume \% (Gas)}}{100\ \%}\right] \times P_{Total} \tag{8.10}$$

Because the partial pressure value is related to the total pressure, concentration data expressed as partial pressure is not the same under actual and standard conditions. The partial pressure values also differ in American Engineering and CGS units.

Both Major and Minor Constituents

- **Mole Fraction**: This is simply an expression of the number of moles of a compound divided by the total number of moles of all the compounds in the gas.

Minor Constituents

- Parts per million (ppm)
- Milligrams per cubic meter (mg/m^3)
- Micrograms per cubic meter ($\mu g/m^3$)
- Nanograms per cubic meter (ng/m^3)

All concentration units above can be expressed in a dry format and corrected to a standard oxygen concentration. These corrections are necessary because moisture and oxygen concentrations can vary significantly in gas streams, causing variations in pollutant concentrations.

Gas Velocity

Gas velocity is one of the fundamental design variables for ventilation systems and air pollution control equipment. Gas streams containing particulate are usually maintained at velocities of 3,000–4,500 ft/min in ductwork, leading to particulate collectors to minimize particle deposition. The velocity of gas streams without particulate matter is 1,500–3,000 ft/min. The gas velocities in air pollution control equipment are usually low, allowing sufficient time to remove the contaminants. For example, gas velocities through electrostatic precipitators are generally 2.5–6 ft/s. The filtration velocities through pulse jet fabric filters are usually 2–10 ft/min. Variations in the gas velocity can directly impact the contaminant removal efficiency.

The average velocity of a gas stream in an emission testing probe, an industrial duct, or an air pollution control device is a function of the actual gas flow rate and the cross-sectional flow area.

$$v = \text{Gas velocity} = \frac{\text{Gas flow rate, actual}}{\text{Area}} \tag{8.11}$$

Gas Stream Treatment (Residence) Time

The flow rate of the gas stream through an air pollution control system determines the length of time that the pollutants can be removed from the stream. This is termed the *treatment time* or *residence time*. These standard equipment sizing parameters are defined mathematically in Equation 8.12.

$$\begin{aligned}\text{Treatment time} &= \text{Residence time}\\ &= \frac{\text{Volume of control device}}{\text{Gas flow rate, actual}}\end{aligned} \tag{8.12}$$

Gas Density

Gas density is essential primarily because it affects the flow characteristics of the moving gas streams. Gas density affects gas velocities through ductwork and air pollution control equipment. It determines the ability to move the gas stream using a fan. Gas density affects the velocities of gases emitted from the stack, influencing the dispersion of the pollutants remaining in the stack gases. It affects the ability of particles to move through gases. It also affects emission testing. Gas density data are needed in many of the calculations involved in air pollution control equipment evaluation, emission testing, and other air pollution control-related studies.

As discussed earlier, the volume of a gas increases as the temperature increases due to the motion of the gas molecules. As the volume occupied by the gas increases, its density decreases. Density is the mass per unit volume, as indicated in Equation 8.13.

$$P_{(T=i,\,P=j)} = \frac{m}{V_{(T=I,\,P=j)}} \tag{8.13}$$

where

$P_{(T=I,\,P=j)}$ = Density at T = I, P = j
M = Mass of a substance
$V_{(T=I,\,P=j)}$ = Volume at T = I, P = j
T = Absolute temperature
P = Absolute pressure

Heat Capacity and Enthalpy

The heat capacity of a gas is the amount of heat required to change the temperature of a unit mass of gas by one temperature degree.

Enthalpy represents the total quantity of internal energy, such as heat, measured for a specific amount of material at a given temperature. Enthalpy data are often described in power units (e.g., Btu, kcal, joule, etc.). The enthalpy content change is usually expressed in Btu/unit mass (Btu/lb_m) or Btu/unit time (Btu/SCF). The change in enthalpy of the total quantity of material in a system is expressed in units of Btu/unit time (Btu/min). The symbols H and ΔH denote enthalpy and the change in enthalpy, respectively.

Heat and Energy in the Atmosphere

In addition to the importance of heat on a particular air stream, it is essential to point out that heat also impacts Earth's atmosphere and thus atmospheric science. The Sun's energy is the prime source of Earth's climatic system. From the Sun, energy is reflected, scattered, absorbed, and reradiated within the system but without uniform distribution. Some areas receive more energy than they lose; in some places, the reverse occurs. If this situation continued for extended periods, the areas with an energy surplus would get hotter, and those with a deficit would get colder—too cold. This does not happen because the temperature differences produced help drive the world's wind and ocean currents. They carry heat with them, either in the sensible or latent forms, and help to counteract the radiation imbalance. Therefore, winds from the tropics are generally warm, carrying excess heat with them. Polar winds are blowing from areas with a deficit of heat, and so are cold. These energy transfer mechanisms help produce the present Earth's climate.

Adiabatic Lapse Rate

The atmosphere is restless, always in motion horizontally, vertically, or both. As air rises, pressure on it decreases, and in response, it expands. The act of expansion to encompass its new and larger dimensions requires an expenditure of energy; since temperature is a measure of internal energy, this use of energy makes its temperature drop—an important point—an essential process in physics (especially in air physics).

This phenomenon is known as the *adiabatic lapse rate*. Adiabatic refers to a process that occurs with or without loss of heat, especially the expansion or contraction of a gas in which a change takes place in the pressure or volume. However, no heat is allowed to enter or leave.

Lapse rate refers to the rate at which air temperature decreases with height. The average lapse rate in stationary air is about 3.5°F/1,000 ft (6.5°C/km). This value may vary with latitude and changing atmospheric conditions (e.g., seasonal changes). A parcel of air not immediately next to the Earth's surface is sufficiently well insulated by its surroundings that either expansion or compression of the parcel may be assumed to be adiabatic.

The air temperature may be calculated for any height using the general formula.

$$T = T_0 - Rh \tag{8.14}$$

where

T = Temperature of the air
h = Height of air
T_0 = Temperature of the air at the level from which the height is measured
R = Lapse rate

The formula states that for every 1,000 ft of altitude (height), 3.5° is subtracted from the initial air temperature.

Adiabatic lapse rates have an essential relationship with atmospheric stability and will be discussed in greater detail later in the text.

Viscosity

All fluids (gases included) resist flow. *Absolute viscosity* is a measure of this resistance to flow. The absolute viscosity of a gas for given conditions may be calculated from the following formula:

$$\begin{aligned}\mu &= 51.12 + 0.372(T) + 1.05\times10^{-4}(T)^2 \\ &\quad +53.147\left(\%O_2/100\%\right) - 74.143\left(\%\ H_2O/100\%\right)\end{aligned} \tag{8.15}$$

where

μ = Absolute viscosity of gas at the prevailing conditions, micropoise
T = Gas absolute temperature, °K
$\%\ O_2$ = Oxygen concentration, % by volume
$\%\ H_2O$ = Water vapor concentration, % by volume

As this equation indicates, the viscosity of a gas increases as the temperature increases. It's harder to push something (e.g., particles) through a hot gas stream than a cooler one due to increased molecular activity as the temperature rises, which results in increased momentum transfer between the molecules. For liquids, the opposite relationship between viscosity and temperature holds. The viscosity of a liquid decreases as temperature increases. It's harder to push something through a cold liquid than a hot one because hydrogen bonding increases with colder temperatures in liquids.

Gas viscosity increases slightly with pressure, but this variation is minimal in most air pollution-related engineering calculations.

A gas's absolute viscosity and density are occasionally combined into a single parameter since both are found in many common equations describing gas flow characteristics. The combined parameter is termed the kinematic viscosity, defined in Equation 8.16.

$$u = \mu / p \tag{8.15}$$

where

u = Kinematic viscosity, m^2/s
μ = Absolute viscosity, Pa•s
p = Gas density, gm/cm^3

Kinematic viscosity can be used in equations describing particle motion through gas streams. The expression for kinematic viscosity simplifies these calculations.

Flow Characteristics

When fluids, such as gases, move slowly, the bulk material moves as distinct layers in parallel paths. The molecular motion is the only movement across these layers, which creates viscosity. This is termed *laminar flow*. As the gas stream's velocity increases, the gas's bulk movement changes. Eddy currents develop, which cause mixing across the flow stream. T*urbulent flow* is essentially the only flow characteristic in air pollution control equipment and emission testing-related situations.

A dimensionless parameter called the *Reynolds number* is used to characterize fluid flow. It is the ratio of the inertial force, which causes gas movement, to the viscous force, which restricts movement. The Reynolds number is calculated using Equation 8.16. Consistent units must be used to ensure that the Reynolds number is dimensionless.

$$N_{Re(g)} = Lvp/\mu \tag{8.17}$$

where
L = Linear dimensions (usually duct diameter)
v = Fluid velocity
p = Fluid density
μ = Fluid viscosity

Reynolds numbers less than 2,000 are associated with laminar flow conditions. Due to the relatively low velocities associated with this type of flow, they are rarely encountered in air pollution field situations.

Reynolds numbers above 10,000 are associated with turbulent flow. In many field situations, they exceed 100,000.

All gas flow situations are turbulent in air pollution control systems, emission tests and monitoring equipment, and dispersion modeling studies. However, this does not mean that the gas stream is entirely well-mixed. In reality, the side-to-side mixing (and even mixing in the flow direction) can be limited. For this reason, it is possible to have different concentrations of pollutants at various positions in the duct. This is called *pollutant stratification*. Multiple factors can cause it: combining two separate gas streams into a single duct, temperature differences in the gas stream, and in-leakage of cold ambient air into the duct.

Stratification does not exist in most industrial gas handling systems. However, it is important to check for this condition before installing continuous emission monitors or other instruments located at a single sampling or measurement point in the gas stream. If stratification is severe, these measurements can be in error.

PARTICLE PHYSICS

Particulates constitute a significant level of air pollution. They have a variety of shapes; they include dust, fumes, mists, and smoke, and they have a wide range of physical and chemical properties. Important particulate matter characteristics include size, size distribution, shape, density, stickiness, corrosivity, reactivity, and toxicity. With a wide range of factors, it should be evident that the type of collection device used might be better suited than others for a particular particle. For example, particulate matter typically ranges from 0.005 to 100 µm in diameter. This wide range in size distribution must undoubtedly be considered.

Characteristics of Particles

Understanding the characteristics of particles is essential in air pollution control technology for the following reasons: the efficiency of the particle collection mechanisms strongly depends on particle size; the particle size distribution of flue gas dictates how air testing is performed; the particle size distribution of flue gas determines the operating conditions necessary to collect the particles; and particle characteristics are essential in determining the behavior of particles in the respiratory tract.

Surface Area and Volume

The size range of particles of concern in air pollution studies is extensive. Some of the droplets collected in the mist eliminators of wet scrubbers and the solid particles collected in large-diameter cyclones are as large as raindrops. Some small particles created in high-temperature incinerators and metallurgical processes can consist of less than 50 molecules clustered together. It is essential to recognize the categories since particles behave differently depending on their size. Most particles of primary interest in air pollution emission testing and control have diameters that range from 0.1 to 1,000 µm (microns).

Useful conversions for particle sizes:

$$\text{Micrometer} = \left(\frac{1}{1{,}000{,}000}\right)\text{m} \tag{8.18}$$

$$\text{Micrometer} = \left(\frac{1}{10{,}000}\right)\text{cm} \tag{8.19}$$

$$1{,}000\,\mu\text{m} = 1.0\,\text{mm} = 0.1\,\text{cm} \tag{8.19}$$

To appreciate particle size differences commonly found in emission testing and air pollution control, compare the diameters, volumes, and surface areas of particles in Table 8.12.

The data in Table 8.12 indicate that particles 1,000 µm in size are 1,000,000,000,000 (one trillion) times larger in volume than 0.1 µm particles.

As an analogy, assume that the 1,000-µm particle is a large domed sports stadium. A basketball in this "stadium" would be equivalent to a 5-µm particle. Approximately 100,000 spherical particles of 0.1-µm diameter would fit into this 5-µm "basketball." The entire 1,000-µm "stadium" is the size of a tiny raindrop. As previously stated,

TABLE 8.12
Spherical Particle Diameter, Volume, and Area

Particle Diameter (μm)	Particle Volume (cm^3)	Particle Area (cm^2)
0.1	5.23×10^{-16}	3.14×10^{-10}
1.0	5.23×10^{-13}	3.14×10^{-8}
10.0	5.23×10^{-10}	3.14×10^{-6}
100.0	5.23×10^{-7}	3.14×10^{-4}
1,000.0	5.23×10^{-4}	3.14×10^{2}

Source: EPA (2007).

particles over this entire size range of 0.1–1,000 μm are of interest in air pollution control (EPA, 2007).

Equations for calculating the volume and surface area of spheres are provided below.

$$\text{Surface area of a sphere} = 4\pi r^2 = \pi D^2 \quad (8.20)$$

$$\text{Volume of a sphere} = 4\pi r^3/3 = \pi D^3/6 \quad (8.21)$$

where

r = Radius of the sphere
D = Diameter of sphere

Aerodynamic Diameter

Particles emitted from air pollution sources and formed by natural processes have different shapes and densities. Defining particle size for spherical particles is easy; it is simply the diameter of the particle. The term "diameter" does not appear to be strictly applicable to non-spherical particles.

In air pollution control, particle size is based on particle behavior in the Earth's gravitational field. The *aerodynamic equivalent diameter* refers to a spherical particle of unit density that falls at a standard velocity. Particle size is important because it determines atmospheric lifetime, effects on light scattering, and deposition in human lungs. However, it is essential to note that we are not referring to particulate shape when we speak of particulate size. This is obvious but important because it is common practice to base assumptions on spherical particles for simplicity and theoretical and learning applications. Particles are not spherical; they are usually quite irregularly shaped. In pollution control technology design applications, the size of the particle is taken into consideration. Still, we are most concerned with the particle's behavior in a gaseous waste stream. To better understand this critical point, consider an analogy. If you drop a flat piece of regular typing paper from the top of a 6-ft ladder, the paper will settle irregularly to the floor. But if you crumple the same sheet of paper and drop it from the ladder, you will see that it falls rapidly. Thus, size is essential, but other factors, such as particle shape, are also important.

The aerodynamic diameter for all particles greater than 0.5 µm can be approximated using the following equation. For particles less than 0.5 µm, refer to aerosol textbooks to determine the aerodynamic diameter.

$$d_{pa} = d_{ps} \sqrt{P_p} \tag{8.22}$$

where

d_{pa} = Aerodynamic particle diameter, µm
d_{ps} = Stokes diameter, µm
P_p = Particle density, gm/cm^3

Along with particle size and shape, particle density must be considered. For example, a baseball and a whiffle ball have approximately the same diameter but behave quite differently when tossed into the air. Particle density affects the motion of a particle through a fluid and is considered in Equation 8.22. The Stokes diameter for a particle is the diameter of the sphere that has the same density and settling velocity as the particle. It is based on the aerodynamic drag force caused by the difference in velocity between the particle and the surrounding fluid. The Stokes diameter is identical to the physical or actual diameter for smooth, spherical particles.

Inertial sampling devices such as cascade impactors are used for particle sizing. These sampling devices determine the aerodynamic diameter. The term "aerodynamic diameter" is helpful for all particles, including fibers and particle clusters. It is not an actual size because "non-spherical" particles require more than one dimension to characterize their size.

Note: In this text, particle diameter and particle size refer to the aerodynamic diameter unless otherwise stated.

Particle Size Categories

Since the range of particle sizes of concern for air emission evaluation is quite broad, it is beneficial to divide this range into smaller categories. Defining different size categories is useful, as particles of various sizes behave differently in the atmosphere and the respiratory system.

The EPA has defined four terms for categorizing particles of different sizes. Table 8.13 displays the EPA terminology along with the corresponding particle sizes.

Regulated Particulate Matter Categories

In addition to the terminology provided in Table 8.13, the EPA also categorizes particles as follows:

- **Total Suspended Particulate Matter (TSP)**: Particles ranging in size from 0.1 to about 30 µm in diameter are called total suspended particulate matter (TSP). TSP includes many particle sizes, including fine, coarse, and super coarse particles.

TABLE 8.13
EPA Terminology for Particle Sizes

EPA Description	Particle Size
Super course	$d_{pa} > 10\,\mu m$
Coarse	$2.5\,\mu m < d_{pa} \leq 10\,\mu m$
Fine	$0.1\,\mu m < d_{pa} \leq 2.5\,\mu m$
Ultrafine	$d_{pa} \leq 0.1\,\mu m$

- **PM_{10}**: The U.S. EPA defines PM_{10} as particulate matter with a diameter of 10 μm collected with 50% efficiency by a PM_{10} sampling collection device. However, for convenience in this text, the term PM_{10} will be used to include all particles having an aerodynamic diameter of less than or equal to 10 μm.

  PM_{10} is regulated as a specific type of "pollutant" because this size range is considered respirable. In other words, particles less than 10 μm can penetrate the lower respiratory tract. The particle size range between 0.1 and 10 μm is especially important in air pollution studies. A significant fraction of the particulate matter generated in some industrial sources is in this size range.
- **$PM_{2.5}$**: As with PM10, the EPA defines PM2.5 as particulate matter with a diameter of 2.5 μm collected with 50% efficiency by a $PM_{2.5}$ sampling collection device. However, for convenience in this text, the term $PM_{2.5}$ will be used to include all particles having an aerodynamic diameter of less than or equal to 2.5 μm.

  The EPA chose 2.5 μm to divide fine and coarse particulate matter. Particles less than approximately 2.5 μm are regulated as **$PM_{2.5}$**. Air emission testing and pollution control methods for $PM_{2.5}$ particles differ from those for coarse and super-coarse particles.

  $PM_{2.5}$ particles settle slowly in the atmosphere relative to coarse and super-coarse particles. Typical weather patterns can keep $PM_{2.5}$ particles airborne for several hours to several days, enabling these particles to cover hundreds of miles. $PM_{2.5}$ particles can cause health problems due to their potentially long airborne retention time and the inability of the human respiratory system to defend itself against particles of this size.

  In addition, the chemical makeup of $PM_{2.5}$ particles is quite different from that of coarse and super-coarse particles. EPA data indicate that $PM_{2.5}$ particles comprise sulfates, nitrates, organic compounds, and ammonium compounds. The EPA also determined that PM2.5 particles often contain acidic materials, metals, and other contaminants believed to be associated with adverse health effects.

  Particles less than 1 μm in diameter are termed sub-micrometer particles and can be the most difficult size to collect. Particles in the 0.2–0.5 μm range are common in many types of combustion, waste incineration, and metallurgical sources. Particles in the range of 0.1–1.0 μm are essential because they can represent a significant fraction of the particulate emissions from some types of industrial sources and because they are relatively complex to collect.

- **Particles less than 0.1 µm**: Particles can be much smaller than 0.1 µm. Particles composed of as few as 20–50 molecules clustered together can exist in a stable form. Some industrial processes, such as combustion and metallurgical sources, generate particles in the 0.01–0.1 µm range. These sizes approach the size of individual gas molecules, which are in the range of 0.0002–0.001 µm. However, particles in the 0.01–0.1 µm size tend to agglomerate rapidly, forming particles greater than the 0.1-µm range. Accordingly, very little particulate matter entering an air pollution control device or leaving the stack remains in the tiny size range of 0.01–0.1 µm.
- **Condensable Particulate Matter**: Particulate matter that forms from condensing gases or vapors is called *condensable particulate matter.* Chemical reactions and physical phenomena form condensable particulate matter.

  Condensable particulate matter is usually formed from material that is not particulate matter under stack conditions but condenses and reacts upon cooling and dilution in the ambient air to become particulate matter. The formation of condensable particulate matter occurs within a few seconds after discharge from the stack.

  From a health standpoint, condensable particulate matter is crucial because it is almost entirely contained in the $PM_{2.5}$ classification.

These particle categories are essential because particulate matter is regulated and tested within these categories.

Size Distribution

Particulate emissions from both manmade and natural sources do not consist of particles of any one size. Instead, they are composed of particles over a relatively wide range of sizes. It is often necessary to describe this size range. Particulate matter for size distribution evaluation is measured in a variety of ways. The data must be measured in a manner that can be classified into successive particle diameter size categories.

Particle Formation

The range of particle sizes formed in a process largely depends on the types of particle formation mechanisms present. The general size range of particles can be estimated by simply recognizing which particle formation mechanisms are most important in the evaluation process. The most important particle formation mechanisms in air pollution sources include physical attrition/mechanical dispersion, combustion particle burnout, homogeneous and heterogeneous nucleation, and droplet evaporation. Several particle formation mechanisms can be present in an air pollution source. As a result, the particles created can have a wide range of sizes and chemical compositions. Particle formation mechanisms are described in detail below.

- **Physical Attrition**: Generates primarily moderate-to-large-sized particles and occurs when two surfaces rub together. For example, grinding a metal rod on a wheel yields small particles that break off from both surfaces. The compositions and densities of these particles are identical to the parent materials.

Collection Mechanisms

When sunlight streams into a quiet room, particles of many different shapes and sizes can be seen; some appear to float, while others slowly settle to the floor. These tiny particles are denser than the room air but do not settle quickly. The solid, liquid, and fibrous particles formed in air pollution sources behave in a manner very similar to that of standard household dust and other familiar particles. What we instinctively understand about these everyday particles can be, in many respects, applied to the particles from air pollution sources.

However, two major differences exist between industrially generated particles and those in more familiar settings. Industrial particles are much smaller than most household particles. Also, some industrial particles have complex chemical compositions and include compounds and elements known to be toxic.

Emission testing devices and air pollution control systems apply forces to particles to remove them from the gas stream. These forces include inertial impaction and interception, Brownian diffusion, gravitational settling, electrostatic attraction, thermophoresis, and diffusiophoresis. These forces are basically the "tools" for separating particles from the gas stream. All these collection mechanism forces are strongly dependent on particle size.

- **Inertial Impaction and Interception**: Due to inertia, a particle moving in a gas steam can strike slowly moving or stationary obstacles (targets) in its path. As the gas stream deflects around the obstacle, the particle continues toward the object and impacts it. The obstacle may be a solid particle, stationary object, or water droplet.

  Two primary factors affect the probability of an impaction: (1) aerodynamic particle size and (2) the difference in velocity between the particle and the obstacle. Larger particles are collected more quickly than smaller particles due to their greater inertia. Also, collection efficiency increases as the difference in velocity between the particle in the gas stream and the obstacle (or target) increases.

  Inertial impaction is analogous to a small car driving down an interstate highway at 65 mph and approaching a merge lane where a slowly moving truck is entering. If the vehicle cannot enter the passing lane to go around the merging truck, there could be an "impaction" incident. Larger cars will have more difficulty going around the truck than smaller cars. Also, the faster the vehicle is going relative to the truck, the more probable an impact is.

  The impaction efficiency is directly proportional to the impaction parameter shown in Equation 8.23. As the value of this parameter increases, the efficiency of inertial impaction increases. This parameter is related to the square of the Stokes particle diameter and the difference in velocity between the particle and the target droplet.

$$K_I = \frac{C_c \left(d_{ps}\right)^2 P_p}{18\mu\, D_c} \tag{8.23}$$

where

K_I = Impaction parameter (dimensionless)
C_c = Cunningham slip correction factor (dimensionless)
d_{ps} = Stokes particle diameter (μm)
v = Difference in velocity (cm/s)
P_p = Particle density (gm/cm³)
D_c = Diameter of a droplet (cm)
μ = Gas viscosity (gm/cm-s)

The Cunningham slip correction factor (also called Cunningham's correction factor) accounts for molecular slip. Molecular slip occurs when the size of the particle is of the same magnitude as the distance between gas molecules. The particle no longer moves as a continuum in the gas but as a particle among discrete gas molecules, thereby reducing the drag force. The Cunningham correction factor is significant for particles in air with an actual diameter of 1.0 μm and less.

Inertial impaction occurs when obstacles (e.g., water droplets) are directly in the path of the particle moving in the gas stream. Sometimes, the obstacle or target is offset slightly from the direct path of the moving particle. In this instance, as the particle approaches the obstacle's edge, the obstacle may collect the particle through *interception*.

- **Brownian Diffusion**: Becomes the dominant collection mechanism for particles less than 0.3 μm and is especially significant for particles in the 0.01–0.1-μm size range.

  Tiny particles in a gas stream deflect slightly when gas molecules strike them. The kinetic energy transfer from the rapidly moving gas molecule to the small particle causes this deflection, called Brownian diffusion. As a result of this random movement, these small particles are captured when they impact a target (e.g., a liquid droplet).

  Diffusivity measures the extent to which molecular collisions influence tiny particles, causing them to move randomly across the direction of gas flow. The diffusion coefficient in the equation below represents a particle's diffusivity under certain gas stream conditions.

$$D_p = \frac{C_c K\, T}{3\pi d_{pa}\mu} \tag{8.24}$$

where

D_p = Diffusion coefficient (cm²/s)
C_c = Cunningham slip correction factor (dimensionless)
K = Boltzmann constant (gm-cm²/s² °K)
T = Absolute temperature (°K)
d_{pa} = Particle aerodynamic diameter (micrometers)
μ = Gas viscosity (kg/m-sec)

- **Gravitational Settling**: Particles in still air have two forces acting on them: (1) a gravitational force downward and (2) the air resistance (or drag) force upward. When particles begin to fall, they quickly reach a terminal settling velocity, representing the constant velocity of a falling particle when the air resistance balances the gravitational force downward (or drag) upward. The terminal settling velocity can usually be expressed using Equation 8.25. [Note: Equation 8.25 applies to particles less than 80 μm in size (aerodynamic diameter) and having a Reynolds number less than 2.0 and a low velocity].

$$v_t = \frac{G\,P_p\left(d_{ps}\right)^2 C_c}{18\mu} \tag{8.25}$$

 where
 v_t = Terminal settling velocity (cm/s)
 g = Gravitational acceleration (cm/s^2)
 P_p = Density of particle (gm/cm^3)
 d_{ps} = Stokes particle diameter (cm)
 C_c = Cunningham slip correction factor (dimensionless)
 μ = Viscosity of air (gm/cm-s)

- **Electrostatic Attraction**: Electrostatic precipitators (ESPs) use electrostatic attraction for particulate collection in air pollution control. Electrostatic attraction of particles is accomplished by establishing a strong electrical field and creating unipolar ions. The ions driven along the electrical field lines charge the particles passing through the electrical field. Several parameters dictate the effectiveness of electrostatic attraction, including particle size, gas flow rate, and resistivity.

 The particles will eventually reach a maximum or saturation charge, a function of the particle area. The saturation charge occurs when the localized field created by the already captured ions is sufficiently strong to deflect the electrical field lines. Particles can also be charged by the diffusion of ions in the gas stream. The strength of the electrical charges imposed on the particles by both mechanisms is particle-size dependent.

 Resistivity is a measure of the ability of the particle to conduct electricity and is expressed in units of ohm-cm. Particles with low resistivity have a more remarkable ability to conduct electricity (and higher electrostatic attraction) than particles with high resistivity. The following factors influence resistivity:

 - Chemical composition of the gas stream
 - Chemical composition of the particle
 - Gas stream temperature

- **Thermophoresis**: Particle movement caused by thermal differences on two sides of the particle. Gas molecules at higher temperatures have more incredible kinetic energy than those at lower temperatures. Therefore, when the particle collides with a gas molecule from the hotter side, it receives more kinetic energy than when it collides with a gas molecule from the cooler side. Accordingly, particles tend to be deflected toward the colder area.

- **Diffusiophoresis**: Particle movement caused by concentration differences on two sides of the particle. When there is a big difference in the concentration of gas molecules on two sides of the particle, there is a difference in the number of molecular collisions, which causes an imbalance in the total kinetic energies of the gas molecules. Gas molecules in the high concentration area striking a particle transmit more kinetic energy to the particle than molecules in the lower concentration area. Therefore, particles tend to move toward the area of lower concentration.

ATMOSPHERIC CHEMISTRY AND RELATED PHENOMENA

Note: In the brief discussion that follows describing the chemistry of Earth's atmosphere, keep in mind that the atmosphere as it is at present is what is being referred to. The atmosphere before this period was chemically quite different. Also, note that "atmospheric chemistry" is a scientific discipline or entity that can stand independently. The "nuts and bolts" of atmospheric chemistry are beyond the scope of this text. Here, certain critical atmospheric chemistry phenomena are highlighted.

You should recall that the full range of chemistry (including slow and fast reactions, dissolving crystals, and precipitation of solids) all occur in the atmosphere, which is, as Graedel and Crutzen (1995) described it—a "flask without walls." We also pointed out that (excluding highly variable amounts of water vapor) more than 99% of the molecules constituting Earth's atmosphere are nitrogen, oxygen, and chemically inert gases (noble gases such as argon, etc.).

The chemistry (and thus the reactivity) of these natural gases (nitrogen, oxygen, carbon dioxide, argon, and others) is well known. The other chemically reactive chemicals (anthropogenically produced) that are part of Earth's atmosphere are also known. However, there are still differing opinions on their "exact" total effect on our environment. For example, methane is by far the most abundant reactive compound in the atmosphere and currently has a ground-level concentration (in the northern hemisphere) of about 1.7 ppmv. We know significant information about methane (its generation and fate when discharged) and its influence on the atmosphere; however, we are still researching to find out more—as we should.

Many different reactive molecules (other than methane) exist in the atmosphere. We may not be familiar with these reactants, but many of us certainly are familiar with their consequences: the ozone hole, the greenhouse effect and global warming, smog, acid rain, the rising tide, and so on. It may surprise you to know, however, that the total amount of all these reactants in the atmosphere is seldom more than 10 ppmv volume anywhere in the world at any given time. The significance should be obvious: the atmospheric problems on Earth result from less than one-thousandth of 1% of all the molecules in the atmosphere. This indicates that environmental damage (causing global atmospheric problems) can result from far less than the tremendous amounts of reactive substances we imagine are dangerous.

The quality of the air we breathe, visibility and atmospheric aesthetics, and climate (all of which depend upon chemical phenomena that occur in the atmosphere) are important to our health and quality of life. Global atmospheric problems, such as the nature and level of air pollutants, concern the air science practitioner the most because they affect our health and quality of life.

Let's look at some of the essential chemical species and their reactions within (primarily) the stratosphere of our atmosphere.

Photochemical Reaction—Smog-Production

A *photochemical reaction* is generally any chemical reaction in which light is produced or initiates the response. Light can initiate reactions by exciting atoms or molecules, making them more reactive. The light energy is converted into chemical energy.

The photochemical reaction we are concerned with here is the action or absorption of electromagnetic solar radiation (light) by chemical species, which causes the reactions. The ability of electromagnetic radiation to cause photochemical reactions is a function shown in the relationship.

$$E = hv \tag{8.26}$$

where

E = Energy of a photon

v = Frequency

h = Planck' constant, 6.62×10^{-27}

The primary photochemical reaction we are concerned with is the one that produces photochemical smog. *Photochemical smog* is initiated by nitrogen dioxide, which absorbs the visible or ultraviolet energy of sunlight, forming nitric oxide and freeing atoms of oxygen (O), which then combine with molecular oxygen (O_2) to form *ozone* (O_3). Various chemical reactions occur in the presence of hydrocarbons (other than methane) and certain other organic compounds. Some 80 separate reactions have been identified or postulated.

The photochemical reaction that produces the smog we are familiar with depends on two factors:

1. Smog concentration is linked to the amount of sunlight and hydrocarbons.
2. The amount of smog depends on the initial concentration of nitrogen oxides.

In producing photochemical smog, many different substances are formed in sequence, including acrolein, formaldehyde, and PAN (peroxyacetylnitrate). Photochemical smog (the characteristic haze of minute droplets) results from condensed low-volatility organic compounds. The organics irritate the eyes and, together with the ozone layer, can cause severe damage to leafy plants. Photochemical smog tends to be most intense in the early afternoon when sunlight intensity is most significant. It differs from traditional smog (Los Angeles-type smog), which is most intense in the early morning and dispersed by solar radiation.

In addition to the photochemical reactions that produce smog, many other chemical reactions occur in the Earth's atmosphere, including ozone production, free radicals, chain reactions, oxidation processes, acid-base reactions, the presence of electrons and positive ions, and many others.

AIR QUALITY

When undertaking a comprehensive discussion of air, the debate begins and ends with the Earth's atmosphere—and to a point, we have done just that. This chapter covers concepts that will enable us to better understand the anthropogenic impact of pollution on the atmosphere, allowing us to understand better the critical parameters used to measure air quality. Having a complete understanding of air quality is essential. We must review a few basic concepts to set the stage for information to follow in subsequent sections related to air pollution and air pollution control.

EARTH'S HEAT BALANCE

The energy expended in virtually all atmospheric processes is originally derived from the sun. This energy is transferred by heat radiation in the form of electromagnetic waves. The radiation from the sun has its peak energy transmission in the visible wavelength range [038–0.78 micrometers (μm)] of the electromagnetic spectrum. However, the sun also releases considerable energy in the ultraviolet and infrared regions. Ninety-nine percent of the sun's energy is emitted in wavelengths between 0.15 and 40 μm. Furthermore, wavelengths longer than 2.5 μm are strongly absorbed by water vapor and carbon dioxide in the atmosphere. Radiation at wavelengths less than 0.29 μm is absorbed high in the atmosphere by nitrogen and oxygen. Therefore, solar radiation striking the Earth generally has a wavelength between 0.29 and 2.5 μm (EPA, 2005).

Since energy from the sun constantly enters the atmosphere, the Earth would overheat if all this energy were stored in the Earth-atmosphere system. So, energy must eventually be released back into space. This is what happens—approximately 50% of the solar radiation entering the atmosphere reaches Earth's surface, either directly or after being scattered by clouds, particulate matter, or atmospheric gases. The other 50% is either reflected directly or absorbed in the atmosphere, and its energy is reradiated into space later as infrared radiation. Most solar energy reaching the surface is absorbed and must be returned to space to maintain *heat balance (radiational balance)*. The energy produced within the Earth's interior (from the hot mantle area via convection and conduction), which reaches the Earth's surface (about 1% of that received from the sun), must also be lost.

Reradiation of energy from the Earth is accomplished by three energy transport mechanisms: radiation, conduction, and convection. As stated earlier, radiation occurs through electromagnetic radiation in the infrared region of the spectrum. The crucial importance of the radiation mechanism is that it carries energy away from Earth on a much longer wavelength than the solar energy (sunlight) that brings energy to the Earth and, in turn, works to maintain the Earth's heat balance. The Earth's heat balance is fascinating to us in this text because it is susceptible to human activity upsetting it.

A comparatively smaller but significant amount of heat energy is transferred to the atmosphere by conduction from the Earth's surface. *Conduction* of energy occurs through the interaction of adjacent molecules with no visible motion accompanying the transfer of heat; for example, the whole length of a metal rod will become hot when one end is held in a fire. Because air is a poor heat conductor, conduction is restricted to the layer of air in direct contact with the Earth's surface. The heated air is then transferred aloft by *convection*, the movement of whole air masses, which may be either relatively warm or cold. Convection is the mechanism by which abrupt temperature variations occur when large air masses move across an area. Air temperature tends to be greater near the surface of the Earth and decreases gradually with altitude. A large amount of the Earth's surface heat is transported to clouds in the atmosphere by conduction and convection before ultimately being lost by radiation, and this redistribution of heat energy plays a vital role in weather and climate conditions.

The Earth's average surface temperature is maintained at about 15°C because of the atmospheric greenhouse effect. The greenhouse effect occurs when the gases of the lower atmosphere transmit most of the visible portion of incident sunlight in the same way as the glass of a garden greenhouse. The warmed Earth emits radiation in the infrared region, which is selectively absorbed by the atmospheric gases whose absorption spectrum is similar to that of glass. This absorbed energy heats the atmosphere and helps maintain the Earth's temperature. Without this greenhouse effect, the surface temperature would average around −18°C. Water molecules perform most infrared energy absorption in the atmosphere. In addition to water molecules' vital role, carbon dioxide, although to a lesser extent, is also essential in maintaining the heat balance. Environmentalists and others concerned with environmental issues are concerned that an increase in the carbon dioxide level in the atmosphere could prevent sufficient energy loss, causing damaging increases in the Earth's temperature. This phenomenon, commonly known as the anthropogenic greenhouse effect, may occur from elevated carbon dioxide levels caused by increased use of fossil fuels and reduced carbon dioxide absorption because of the destruction of the rainforest and other forest areas.

Insolation (EPA, 2005)

The amount of **in**coming **sol**ar radi**ation** received at a particular time and location in the Earth-atmosphere system is called insolation. Four factors govern insolation:

- Solar constant
- Transparency of the atmosphere
- Daily sunlight duration
- Angle at which the sun's rays strike the Earth

Solar Constant

The *solar constant* is the average amount of radiation received at a point, perpendicular to the sun's rays, that is located outside the Earth's atmosphere at the Earth's mean distance from the sun. The average amount of solar radiation received at the outer edge of the atmosphere would vary slightly depending on the energy output of the sun and the distance of the Earth relative to the sun. Due to the eccentricity of the Earth's orbit around the sun, the Earth is closer to the sun in January than in July. Also, the radiation emitted from the sun varies slightly, probably less than a few percent. These slight variations affecting the solar constant are trivial considering the atmospheric properties that deplete the overall amount of solar radiation reaching the Earth's surface. Transparency of the atmosphere, duration of daylight, and the angle at which the sun's rays strike the Earth are much more important in influencing the amount of radiation received, including the weather.

Transparency

The transparency of the atmosphere does have an essential bearing on the amount of insolation that reaches the Earth's surface. The emitted radiation is depleted as it passes through the atmosphere. Different atmospheric constituents absorb or reflect energy in various ways and varying amounts. Transparency refers to how much radiation penetrates the atmosphere and reaches the Earth's surface without being depleted.

The general reflectivity of the various surfaces of the Earth is referred to as the albedo. *Albedo* is the fraction (or percentage) of incoming solar energy reflected into space. Different surfaces (water, snow, sand, etc.) have different albedo values. For the Earth and atmosphere, the average albedo is 30% for average cloudiness conditions over the Earth. This reflectivity is most significant in the visible range of wavelengths.

Some of the gases in the atmosphere (notably water vapor) absorb solar radiation, causing less radiation to reach the Earth's surface. Water vapor, although comprising only about 3% of the atmosphere, on average, absorbs about six times as much solar radiation as all other gases combined. The radiation received at the Earth's surface is considerably less than that received outside the atmosphere, as represented by the solar constant.

Daylight Duration

The duration of daylight also affects the amount of insolation received: the longer the period of sunlight, the greater the total possible insolation. Daylight duration varies with latitude and the seasons. At the equator, day and night are always equal. In the Polar Regions, the daylight period reaches a maximum of 24 hours in summer and a minimum of zero hours in winter.

Angle of Sun's Rays

The angle at which the sun's rays strike the Earth varies considerably as the sun "shifts" back and forth across the equator. A relatively flat surface perpendicular to an incoming vertical sun ray receives the most significant insolation. Therefore, areas where the sun's rays are oblique receive less insolation because the oblique rays must pass through a thicker layer of reflecting and absorbing the atmosphere and are spread over a greater surface area. This same principle also applies to the daily shift of the sun's rays. At solar noon, the intensity of insolation is most incredible. The amount of insolation is small when the sun is at a low angle in the morning and evening hours.

Heat Distribution

The Earth, as a whole, experiences great contrasts in heat and cold at any particular time. Warm, tropical breezes blow at the equator while ice caps form in the Polar Regions. In fact, due to the extreme temperature differences at the equator and the poles, the Earth-atmosphere system resembles a giant "heat engine." Heat engines depend on hot-cold contrasts to generate power. As you will see, this global "heat engine" influences the major atmospheric circulation patterns as warm air is transferred to cooler areas. Different parts of the Earth receive various amounts of insolation, accounting for much of this heat imbalance. As discussed earlier, latitude, the seasons, and daylight duration cause different locations to receive varying amounts of insolation.

Differential Heating

Different amounts of solar radiation reach the Earth's surface, and Earth's surfaces absorb heat energy at different rates. For example, land masses absorb and store heat differently than water masses. Also, various types of land surfaces vary in their ability to absorb and store heat. The color, shape, surface texture, vegetation, and presence of buildings can all influence the heating and cooling of the ground. Generally, dry surfaces heat and cool faster than moist surfaces. Plowed fields, sandy beaches, and paved roads become hotter than surrounding meadows and wooded areas. During the day, the air over a plowed field is warmer than over a forest or swamp; during the night, the situation is reversed. The property of different surfaces, which causes them to heat and cool at different rates, is referred to as *differential heating.*

Absorption of heat energy from the sun is confined to a shallow layer of land surface. Consequently, land surfaces heat rapidly during the day and cool quickly at night. Water surfaces, on the other hand, heat and cool more slowly than land surfaces for the following reasons:

- Water movement distributes heat
- The sun's rays can penetrate the water surface
- More heat is required to change the water temperature due to its higher specific heat. (It takes more energy to raise the temperature of water than it does to change the temperature of the same amount of soil.)
- Evaporation of water occurs, which is a cooling process

Transport of Heat

Earlier, it was pointed out that heat is transferred by conduction, convection, and advection in addition to radiation. These processes affect the temperature of the atmosphere near the Earth's surface. *Conduction* is the process by which heat is transferred through matter without the transfer of matter itself. For example, the handle of an iron skillet becomes hot due to the conduction of heat from the stove burner. Heat is transmitted from a warmer object to a cooler one. Heat transfer occurs when the matter is in motion. Air warmed by a heated land surface (by conduction) will rise because it is lighter than the surrounding air. This heated air rises, transferring heat vertically. Likewise, cooler air will sink because it is heavier than the surrounding air. This goes hand in hand with increasing air and is part of heat transfer by convection. Meteorologists also use the term *advection* to denote heat transfer mainly by horizontal motion rather than vertical movement of air (convection).

Global Distribution of Heat

As mentioned, the world distribution of insolation is closely related to latitude. Total annual insolation is greatest at the equator and decreases toward the poles. The amount of insolation received annually at the equator is over four times that received at either of the poles. As the sun's rays shift seasonally from one hemisphere to another, the maximum possible daily insolation zone moves with them. For the Earth, the gains in solar energy equal the energy losses back into space (heat balance). However, since the equatorial region gains more heat than it loses and the poles lose more heat than they gain, something must happen to distribute heat more evenly around the Earth. Otherwise, the equatorial regions would continue to heat, and the poles would continue to cool. Therefore, atmospheric and oceanic circulations carry out a continuous large-scale transfer of heat (from low to high altitudes) to reach equilibrium.

The atmosphere drives warm air poleward and brings cold air toward the equator. Heat transfer from the tropics poleward occurs throughout the year but at a much slower rate in summer than in winter. The temperature difference between low and high latitudes is considerably smaller in summer than in winter (only about half as significant in the Northern Hemisphere). As expected, the winter hemisphere has a net energy loss, and the summer hemisphere has a net gain. Most of the summertime gain is stored in the land and ocean surface layers, mainly in the sea.

The oceans also play a role in heat exchange. Warm water flows poleward along the western side of an ocean basin, and cold water flows toward the equator on the eastern side. At higher latitudes, warm water moves poleward on the east side of the ocean basin, and cold water flows toward the equator on the western side. Oceanic currents are responsible for about 40% of energy transport from the equator to the poles. The remaining 60% is attributed to the movement of air.

AIR QUALITY MANAGEMENT

Proper air quality management includes several areas related to air pollutants and their control. For example, we can mathematically predict where pollutants emitted from a source will disperse in the atmosphere, eventually falling to the ground and at what concentration. We have found that pollution control equipment can be added to various sources to reduce pollutants before they are emitted into the air. We have also found that certain phenomena, such as acid rain, the greenhouse effect, and global warming, are indicators of adverse effects on the air and other environmental mediums resulting from the excessive release of pollutants into the air. We must concern ourselves with ambient air quality in our local outdoor environment and the issue of indoor air quality.

To accomplish air quality management, we have found that managing is one thing—and achieving significant improvement is another. We must add regulatory authority, regulations, and regulatory enforcement authority to the air quality management scheme; strict voluntary compliance is ineffective.

We cannot maintain a quality air supply without proper management, regulation, and regulatory enforcement. This section presents the regulatory framework governing air quality management. It provides an overview of the environmental air quality laws and rules used to protect human health and the environment from the potential hazards of air pollution. New legislation, reauthorizations of acts, and new National Ambient Air Quality Standards (NAAQS) have changed how government and industry manage their business. Fortunately for our environment and us, they are practical management tools that work to manage air quality.

CLEAN AIR ACT (CAA)

Looking at a historical overview of air quality regulations, you might be surprised to discover that most are recent. For example, in the United States, the first attempt at regulating air quality came about through the passage of the Air Pollution Control Act of 1955 (Public Law 84–159). This act was a step forward, but that's all; it did little more than move us toward effective legislation. Revised in 1960 and again in 1962, the act was supplanted by the Clean Air Act (CAA) of 1963 (Public Law 88–206). The CAA of 1963 encouraged state, local, and regional programs for air pollution control but reserved the right of federal intervention should pollution from one state endanger the health and welfare of citizens residing in another state. In addition, in 1963, the CAA initiated the development of air quality criteria on which the air quality and emissions standards of the 1970s were based.

The move toward air pollution control gained momentum in 1970, first by creating the Environmental Protection Agency (EPA) and second by the passage of the Clean Air Act of 1970 (Public Law 91-604), which the EPA was responsible for implementing. The act was important because it set primary and secondary ambient air quality standards. Primary standards (based on air quality criteria) allowed for an extra margin of safety to protect public health, while secondary standards (also based on air quality criteria) were established to protect public welfare—animals, property, plants, and materials.

The Clean Air Act of 1977 (Public Law 95-95) further strengthened the existing laws and set the nation on a course toward cleaning our atmosphere.

In 1990, the President signed the Clean Air Act Amendments of 1990. Specifically, the new law:

- Encourages market-based principles and innovative approaches like performance-based standards, emission banking, and trading.
- Promotes using clean, low-sulfur coal and natural gas and innovative technologies to clean high-sulfur coal through the acid rain program.
- Reduces enough energy waste and creates enough of a market for clean fuels derived from grain and natural gas to cut dependency on oil imports by 1 million barrels/day.
- Promotes energy conservation through an acid rain program that allows utilities to obtain needed emission reductions through programs encouraging customers to conserve energy.

Under CAA 1990, several Titles are listed with specific requirements. For example:

- **Title 1**: Specifies provisions for attaining and maintaining National Ambient Air Quality Standards (NAAQS).
- **Title 2**: Specifies provisions relating to mobile sources of pollutants.
- **Title 3**: Covers air toxics.
- **Title 4**: Covers specifications for acid rain control.
- **Title 5**: Addresses permits.
- **Title 6**: Specifies stratospheric ozone and global protection measures.
- **Title 7**: Discusses provisions relating to enforcement.

Title 1: Attainment & Maintenance of NAAQS

The Clean Air Act of 1977 has significantly improved U.S. air quality. Still, urban air pollution problems, such as smog (ozone), carbon monoxide (CO), and particulate matter (PM-10), persist. For example, over 100 million Americans live in cities that do not meet the public health standards for ozone.

A new, balanced strategy for attacking the urban smog problem was needed. The Clean Air Act of 1990 created this new strategy. Under these latest amendments, states are given more time to meet the air quality standard (e.g., up to 20 years for the ozone layer in Los Angeles), but they must make steady, impressive progress in reducing emissions. Specifically, it requires the federal government to reduce emissions from (1) cars, buses, and trucks; (2) consumer products such as window-washing compounds and hairspray; and (3) ships and barges during loading and unloading of petroleum products. In addition, the federal government must develop the technical guidance that states need to control stationary sources. In urban air pollution problems of smog (ozone), carbon monoxide (CO), and particulate matter (PM-10), the new law clarifies how areas are designated and redesignated "attainment." The EPA is also allowed to define the boundaries of "nonattainment" areas (geographical areas whose air quality does not meet federal air quality standards designed to protect public health). CAA 1990 also establishes provisions defining when and how the federal government can impose sanctions on areas of the country that have not met certain conditions.

For ozone specifically, the new law established nonattainment area classifications ranked according to the severity of the area's air pollution problem. These classifications are

- Marginal
- Moderate
- Serious
- Severe
- Extreme

The EPA assigns each nonattainment area one of these categories, prompting varying requirements the areas must comply with to meet the ozone standard.

Again, nonattainment areas must implement different control measures depending on their classifications. For example, those closest to meeting the standard are the marginal areas, which must conduct an inventory of their ozone-causing emissions and institute a permit program. Nonattainment areas with more severe air quality problems must implement various control measures; the worse the air quality, the more control areas will have to implement.

For carbon monoxide and particulate matter, CAA 1990 also establishes similar programs for areas that do not meet the federal health standard. Areas exceeding the standards for these pollutants are divided into "moderate" and "serious" classifications. Areas that exceed the carbon monoxide standard (i.e., to the degree to which they exceed it) are required primarily to implement programs introducing oxygenated rules and enhanced emission inspection programs. Likewise, areas exceeding the particulate matter standard have to (among other requirements) implement either reasonably available control measures (RACMs) or best available control measures (BACMs).

Title 1 attainment and maintenance of NAAQS requirements have gone a long way toward improving air quality in most locations throughout the United States. However, on November 27, 1996, to upgrade NAAQS for ozone and particulate matter, the USEPA proposed the National Ambient Air Quality Standards. Carol Browner, EPA Administrator, later signed a Notice of rulemaking putting into effect the two new NAAQS for ozone and particulate matter smaller than 2.5 μm diameter ($PM_{2.5}$). These rules appear at 62 FR 38651 for particulate matter and 62 FR 38855 for ozone. They are the first update in 20 years for ozone (smog) and the first in 10 years for particulate matter (soot).

Table 8.14 lists the National Ambient Air Quality Standards, including the new requirements (updated as of January 2007). Note that NAAQS is essential but not

TABLE 8.14
National Ambient Air Quality Standards (NAAQS)

Pollutant	**Standard Value**	
Carbon Monoxide (CO)		
8-hour average	9 ppm	10 mg/m^3
1-hour average	35 ppm	40 mg/m^3
Lead (Pb)		
Quarterly average	1.5 μg/m^3	
Nitrogen dioxide (NO_2)		
Annual arithmetic mean	0.053 ppm	100 μg/m^3
Ozone (O_3)		
1-hour average	0.12 ppm	235 μg/m^3
8-hour average	0.08 ppm	157 μg/m^3

(Continued)

TABLE 8.14 (*Continued*)
National Ambient Air Quality Standards (NAAQS)

Pollutant	Standard Value	
Particulate matter (PM-10)		
Annual arithmetic mean	50 μg/m³	
24-hour average	150 μg/m³	
Particulate matter (PM-2.5)		
Annual arithmetic mean	15 μg/m³	
24-hour average	65 μg/m³	
Sulfur dioxide (SO_2)		
Annual arithmetic mean	0.03 ppm	80 μg/m³
24-hour average	0.14 ppm	365 μg/m³

Source: USEPA, NAAQS (2007).

enforceable by itself. The standards set ambient concentration limits to protect human health and environment-related values. However, it is necessary to remember that it is an infrequent case where any one source of air pollutants is responsible for the concentrations in an entire area.

Title 2: Mobile Sources

Cars, trucks, and buses account for almost half of the emissions (even though great strides have been made since the 1960s in reducing the amounts) of the ozone precursors, volatile organic carbons (VOCs), and nitrogen oxides, and up to 90% of the CO emissions in urban areas. A large portion of the emission reductions gained from motor vehicle emission controls has been offset by the rapid growth in the number of vehicles on the highways and the total miles driven.

Due to the unforeseen growth in automobile emissions in urban areas, further compounded by the severe air pollution problems in many metropolitan regions, Congress made significant changes to the motor vehicle provisions of the 1977 Clean Air Act. The Clean Air Act of 1990 established even tighter pollution standards for emissions from motor vehicles. These standards were designed to reduce tailpipe emissions of hydrocarbons, nitrogen oxides, and carbon monoxide on a phased-in basis, starting with model year 1994. Automobile manufacturers must also reduce vehicle emissions from gasoline's evaporation during refueling.

The latest Clean Air Act (1990 w/1997 amendments for ozone and particulate matter) also requires controlling fuel quality. New programs for cleaner or reformulated gasoline were needed, initiated in 1995 for the cities (nine total) with the worst ozone problems. Other towns were allowed to "buy in" to the reformulated gasoline program. In addition, a clean fuel car pilot program was established in California, requiring the phasing in of tighter emission limits for several thousand vehicles in 1996 and up to

300,000 by 1999. The law allows these standards to be met with vehicle technology and cleaner fuels. Note that the standards became even stricter in 2001.

Title 3: Air Toxics

Toxic air pollutants (those hazardous to human health and the environment—carcinogens, mutagens, and reproductive toxins) were not explicitly covered under the Clean Air Act of 1977. This situation is quite surprising and alarming when you consider that information generated from SARA Title III (Superfund Section 313) indicates that in the U.S., more than 2 billion pounds of toxic air pollutants are emitted annually.

The Clean Air Act of 1990 offered a comprehensive plan for significantly reducing emissions of hazardous air pollutants from primary sources. The new law improved the EPA's ability to address this problem effectively and dramatically accelerated progress in controlling major toxic air pollutants.

The 1990 law includes a list of 189 toxic air pollutants whose emissions must be reduced. The EPA was required to publish a list of source categories that emit certain levels of these pollutants and issue maximum achievable control technology (MACT) standards for each listed source category. The law also established a Chemical Safety Board to investigate accidental releases of highly hazardous chemicals.

Title 4: Acid Deposition

The purity of rainfall is a significant concern for many people, especially regarding the acidity of the precipitation. Most rainfall is slightly acidic because of decomposing organic matter, the movement of the sea, and volcanic eruptions, but the principal factor is atmospheric carbon dioxide, which causes carbonic acid to form. *Acid rain* ($pH < 5.6$) (in the pollution sense) is produced by the conversion of the primary pollutants sulfur dioxide and nitrogen oxides to sulfuric acid and nitric acid, respectively. These processes are complex, depending on the physical dispersion processes and the rates of chemical conversions.

Contrary to popular belief, acid rain is not a new phenomenon, nor does it result solely from industrial pollution. For example, natural processes—volcanic eruptions and forest fires—produce and release acid particles into the air. Burning forest areas to clear land in Brazil, Africa, and other countries also contribute to acid rain. However, the rise in manufacturing, which began with the Industrial Revolution, literally dwarfs all other contributions to the problem.

The main culprits are sulfur dioxide emissions from burning fossil fuels, such as oil and coal, and nitrogen oxides, formed chiefly from internal combustion engine emissions, which are readily transformed into nitrogen dioxide. These mix in the atmosphere to form sulfuric acid and nitric acid.

The Earth's ecosystems are not entirely defenseless against atmospheric acid deposition; they can deal with a certain amount of acid through natural alkaline substances in soil or rocks that buffer and neutralize acid. The American Midwest and southern England have highly alkaline soil (limestone and sandstone), providing some natural neutralization. However, areas with thin soil and those laid on granite bedrock cannot neutralize acid rain.

Scientists continue to study how acid rain damages and kills living beings. This complex subject has many variables. We know from various episodes of acid rain that pollution can travel very long distances. Lakes in Canada and New York feel the effects of coal burning in the Ohio Valley. Most scientific studies have occurred in the world's lakes for this and other reasons. In lakes, the smaller organisms often die off first, leaving the larger animals starving. Sometimes, the larger animals (fish) are killed directly; as lake water becomes more acidic, it dissolves heavy metals, leading to concentrations at toxic and often lethal levels. Have you ever wandered to the lake shore and observed thousands of belly-up fish? Not a pleasant sight or smell. Loss of life in lakes also disrupts the life system on the land and the air around them.

In some parts of the United States, rainfall acidity has fallen below 5.6. In the northeastern U.S., for example, the average pH of rainfall is 4.6, and rainfall with a pH of 4.0 is 1,000 times more acidic than distilled water.

Despite intensive research into most aspects of acid rain, scientists still have many areas of uncertainty and disagreement. Progressive, forward-thinking countries emphasize the importance of further research into acid rain. That is why the 1990 Clean Air Act was strengthened to initiate a permanent reduction in SO_2 levels.

One of the exciting features of the 1990 Act is that it allowed utilities to trade allowances within their systems and buy or sell allowances to and from other affected sources. Each source must have sufficient allowances to cover its annual emissions. If not, the source is subject to excess emissions fees and a requirement to offset the excess in the following year. The 1990 law also included specific requirements for reducing emissions of nitrogen oxides from certain boilers.

Title 5: Permits

The 1990 law also introduced an operating permit system similar to the *National Pollution Discharge Elimination System* (NPDES). The permit system has a two-fold purpose: (1) to ensure compliance with all applicable requirements of the CAA and (2) to enhance the EPA's ability to enforce the Act. Under the Act, air pollution sources must develop and implement the program, and the EPA must issue permit program regulations, review each state's proposed program, and oversee the state's effort to implement any approved program. The EPA must also develop and implement a federal permit program when a state fails to adopt and enforce its program.

Title 6: Ozone and Global Climate Protection

Ozone is formed in the stratosphere by sun radiation and helps shield life on Earth from some of the sun's potentially destructive ultraviolet (UV) radiation.

In the early 1970s, scientists suspected that the ozone layer was being depleted. By the 1980s, it became clear that the ozone shield was indeed thinning in some places and, at times, even had a seasonal hole, notably over Antarctica. The exact causes and extent of the depletion are not yet fully known, but most scientists believe that various chemicals in the air are responsible.

Most scientists identify the family of chlorine-based compounds, most notably chlorofluorocarbons (CFCs) and chlorinated solvents (carbon tetrachloride and methyl chloroform), as the primary culprits involved in ozone depletion. It has been hypothesized that CFCs containing chlorine are responsible for ozone depletion. Scientists pointed out that chlorine molecules are highly active. They continually break apart the three-atom ozone into the two-atom form of oxygen generally found close to Earth in the lower atmosphere.

The Interdepartmental Committee for Atmospheric Sciences (1975) estimated that a 5% reduction in the ozone layer could result in nearly a 10% increase in cancer. This already frightening scenario became even more terrifying by 1987 when evidence showed that CFCs destroy the ozone layer in the stratosphere above Antarctica every spring. The ozone hole had become more extensive, with more than half of the total ozone column wiped out, and essentially all ozone disappeared from some regions of the stratosphere (Davis & Cornwell, 1991).

In 1988, Zurer reported that the ozone layer had shrunk approximately 2.5% worldwide in the preceding decade. This obvious thinning of the ozone layer, with its increased chances of skin cancer and cataracts, is also implicated in suppressing the human immune system and damaging other animals and plants, especially aquatic life and soybean crops. The urgency of the problem spurred the 1987 signing of the Montreal Protocol by 24 countries, which required signatory countries to reduce their CFC consumption by 20% by 1993 and by 50% by 1998, marking a significant achievement in solving a global environmental problem.

The Clean Air Act of 1990 borrowed from EPA requirements in other regulations and mandated the phase-out of the production of substances that deplete the ozone layer. Under these provisions, the EPA was required to list all regulated substances along with their ozone-depletion potential, atmospheric lifetime, and global warming potential.

Title 7: Enforcement

The Clean Air Act contains many authorities that make the law more readily enforceable. The EPA was given new authority to issue administrative penalties and field citations (with fines) for smaller infractions. In addition, sources must certify their compliance, and the EPA has the authority to issue administrative subpoenas for compliance data.

CLEAN AIR ACT AMENDMENTS (EPA, 2005)

The Clean Air Act amendments require that State Implementation Plans (SIPs) be developed, the impact of new sources on the atmosphere evaluated, and air quality modeling analyses performed. These regulatory programs require knowledge of the air quality in the region around a source, air quality modeling procedures, and the fate and transport of pollutants in the atmosphere. Implicit in air pollution programs is knowledge of the climatology of the area in question.

State Implementation Plans (SIPs)

State Implementation Plans (SIPs) are federally approved plans developed by state (or local) air quality management authorities to attain and maintain the National Ambient Air Quality Standards (NAAQS). Generally, these SIPs are a state's (local) air quality rules and regulations, considered an acceptable control strategy once the EPA approves them. The purpose of SIPs is to control the amount and types of pollution in any given area or region of the United States.

In these control strategies, emission limits should be based on ambient pollutant concentration estimates for the averaging time that results in the most stringent control requirements. In all cases, these concentration estimates are assumed to be the sum of the pollutant concentrations contributed by the source and an appropriate background concentration. An air quality model determines which averaging time (e.g., annual, 24-hour, 8-hour, 3-hour, 1-hour) results in the highest ambient impact. For example, if the average yearly air quality standard is approached by a greater degree (percentage) than standards for other averaging times, the annual average is considered the restrictive standard. In this case, the sum of the highest estimated yearly average concentration and the annual average background concentration provides the concentration that should be used to specify emission limits. However, suppose a short-term standard is approached by a greater degree and is thus identified as the restrictive standard. In that case, other considerations are required because the frequency of occurrence must also be considered.

New Source Review

New primary stationary sources or major modifications to existing sources of air pollution are required by the Clean Air Act to obtain an air quality permit before construction starts. This process is called New Source Review (NSR), and it is required for any new major stationary source or significant modification to an existing source, regardless of whether the NAAQS are exceeded. Sources located in areas that exceed the NAAQS (nonattainment areas) undergo nonattainment New Source Review. New Source Review for major sources in areas where the NAAQS are not violated (attainment areas) involves the preparation of a Prevention of Significant Deterioration (PSD) permit. Some sources will have the potential to emit pollutants for which their area is in attainment (or unclassifiable) and the potential to emit pollutants for which their area is in nonattainment. When this is the case, the source's permit will contain terms and conditions to meet both the PSD and nonattainment requirements, which are major NSR requirements because these requirements are pollutant-specific.

In most cases, any new source must obtain a nonattainment NSR permit if it will emit or has the potential to emit 100 tons per year or more of any regulated NSR pollutant for which that area is in nonattainment, from "marginal" to "extreme." In areas where air quality problems are more severe, the EPA has established lower thresholds for three criteria pollutants: ozone (VOCs), particulate matter (PM_{10}), and carbon monoxide. The "significance levels" are lower for modifications to existing sources.

In general, a new source located in attainment or unclassifiable must get a PSD permit if it will emit, or has the potential to emit, 250 tons per year (tpy) or more of any criteria or NSR-regulated pollutant. If the source is on the EPA's list of 28 PSD source categories, a PSD permit is required if it will or may emit 100 tpy or more of any NSR-regulated pollutant. The "significant levels" are lower for modifications to existing sources. In addition, PSD review would be triggered, with respect to a particular pollutant, if a new source or major modification is constructed within 10 km of a Class 1 area (see below) and would have an impact on such an area equal to or greater than 1 mg/m^3, (24-hour average) for the pollutant, even though the emissions of such a pollutant would not otherwise be considered "significant."

Some new sources or modifications to those in attainment areas may be required to perform an air quality modeling analysis. This *air quality impact analysis* should determine if the source will cause a violation of the NAAQS or cause air quality deterioration more significantly than the available PSD increments. PSD requirements provide an area classification system based on land use for areas within the United States. These three areas are Class 1, Class 2, and Class III, and each class has an established set of increments that cannot be exceeded. Class I areas consist of national parks and wilderness areas that only allow a small amount of air quality deterioration. Due to the pristine nature of these areas, the most stringent limits on air pollution are enforced in Class I areas. Class II areas consist of normal, well-managed industrial development. Moderate levels of air quality deterioration are permitted in these regions. Class III areas allow the most significant amount of air quality deterioration. When a PSD analysis is performed, the PSD increments set forth a maximum allowable increase in pollutant concentrations, which limits the allowable amount of air quality deterioration in an area. This limits the amount of pollution that enters the atmosphere for a given region. To determine if a source of sulfur dioxide, for example, will cause an air quality violation, the air quality analysis uses the highest estimated concentration for annual averaging periods and the second-highest for averaging periods of 24 hours or less. The new NAAQS for PM and ozone contain specific procedures for determining modeled air quality violations.

For reviews of new or modified sources, the air quality impact analysis should generally be limited to the area where the source's impact is "significant," as defined by regulations. In addition, due to the uncertainties in making concentration estimates over large downwind distances, the air quality impact analysis should generally be limited to a downwind distance of 50 km unless adverse impacts in a Class I area may occur at greater distances.

AIR QUALITY MONITORING

As mentioned in the previous two sections, air quality modeling must ensure that a source follows the requirements of SIP and New Source Review. When air quality modeling is required, the selection of a model depends on the source characteristics, pollutants emitted, terrain, and meteorological parameters. The EPA has compiled the *Guideline on Air Quality Modeling* (40 CFR 1 Appendix W), which summarizes the available models, techniques, and guidance in conducting air quality modeling analyses used in regulatory programs. This document was written to promote consistency among modelers so that all air quality modeling activities would be based on the same procedures and recommendations.

When air quality modeling is required, the specific model (from a simple screening tool to a refined analysis) will need meteorological data. The data can vary from a few factors, such as average wind speed and Pasquill-Gifford stability categories, to a mathematical representation of turbulence. Whatever model is chosen to estimate air quality, the meteorological data must match the quality of the model used. For example, the average wind speed used in a simple screening model will not be sufficient for a complex, refined model. An air quality modeling analysis incorporates the evaluation of terrain, building dimensions, ambient monitoring data, relevant emissions from nearby sources, and the aforementioned meteorological data.

For a dispersion model to provide useful and valid results, the meteorological data used in the model must be representative of the transport, and the dispersions used in the model must be representative of the transport and dispersion characteristics in the vicinity of the source that the model is trying to simulate. The representativeness of the meteorological data is dependent on the following:

- The proximity of the meteorological monitoring site to the area under consideration
- The complexity of the terrain in the area
- The exposure of the meteorological monitoring site
- The period during which the data are collected

In addition, large distances between the source and the receptor of interest can adversely affect the representativeness of the data. Similarly, valley/mountain, land/water, and urban/rural characteristics affect the accuracy of the meteorological data for the source under consideration.

For control strategy evaluations and New Source Review, the minimum meteorological data required to describe the transport and dispersion of air pollutants in the atmosphere are wind direction, wind speed, mixing height, and atmospheric stability (or related indicators of atmospheric turbulence and mixing). Because of the question of representativeness of meteorological data, site-specific data are preferable to data collected off-site. Typically, 1 year of on-site data is required. If an off-site database is used (from a nearby airport, for example), 5 years of data are normally required. With 5 years of data, the model can incorporate most of the possible variations in the meteorological conditions at the site.

Visibility

Visibility is the distance an observer can see along the horizon. Air pollutants in the atmosphere impair visibility by scattering and absorpting light.

There are generally two types of air pollution that impair visibility. The first type consists of smoke, dust, or gaseous plumes that obscure the sky or horizon and are emitted from a single source or small group of sources. The second type is a widespread area of haze that impairs visibility in every direction over a large area and originates from many sources. Regardless of the type of air pollution that impairs visibility at a particular location, any change in the meteorology or source emissions that would increase the pollutant concentration in the atmosphere will result in increased visibility impairment.

Therefore, PSD Class I areas have the most stringent PSD increments and must be protected from high pollutant concentrations and the additional problems pollutants in the atmosphere can cause. Under the Clean Air Act, PSD Class I areas must be evaluated for visibility impairment. This may involve a visibility impairment analysis. According to EPA regulations, visibility impairment is defined as any humanly perceptible change in visibility (visual range, contrast, or coloration) from natural conditions. Therefore, any location is susceptible to visibility impairment due to air pollution sources. Since PSD Class I areas (national parks and wilderness areas) are known for their aesthetic quality, any change or alteration in the area's visibility must be analyzed.

Pollutant Dispersion

Pollutant dispersion is the process of pollutants being removed from the atmosphere and deposited onto the Earth's surface. Stack plumes contain gases and small particles not removed from the gas stream. When the plume emerges from the stack, these particles are carried with it. Once airborne, the particles settle and become deposited on the ground and surface objects. There are basically two ways the particles can be deposited: dry deposition (gravitational settling) or wet deposition (precipitation scavenging). Depending on the meteorological conditions during the time of pollutant emission, these may:

1. Settle out quickly due to their weight and the effect of gravity
2. Be transported further downwind of the source due to buoyancy and wind conditions, or
3. Be washed out of the atmosphere by precipitation or clouds (wet deposition).

In any case, the deposition of these pollution particles is important to understand and quantify since pollutants deposited upon the ground can impact human health, vegetation, and wildlife.

Pollutant deposition concentrations must be predicted to minimize the risk to human health. Air quality models can quantify the pollutant deposition that occurs from stack emissions. These models determine pollution deposition based on the chemical reactivity and solubility of various gases and detailed data on precipitation for the areas in question.

Vapor Plume-Induced Icing

Vapor plumes are emitted from cooling towers and stacks and consist mainly of water vapor. Although pollutant concentrations are not a significant concern with vapor plumes, other problems arise when vapor plume sources are close to frequently traveled roads and populated areas. The vapor emitted from a stack is warm and misty. When meteorological conditions are favorable, the moisture in the vapor plume condenses out and settles on cooler objects (e.g., road surfaces). This phenomenon is similar to the moisture that collects on the sides of a glass of water on a warm day. Road surfaces can freeze rapidly if temperatures are at or below freezing when the moisture condenses, creating hazardous driving conditions. In addition, light winds can cause the plume to remain stagnant, creating a form of ground fog that can also cause low visibility. Water vapor plumes that lower visibility can create hazards for aircraft, especially during critical phases of flight, including landings and takeoffs.

AIR POLLUTION MECHANICS

In the past, belching smokestacks were a comforting sight to many people: more smoke meant more business, indicating that the economy was healthy. But many of us are now troubled by evidence that suggests that polluted air adversely affects our health. Many toxic gases and fine particles entering the air pose health hazards: cancer, genetic defects, and respiratory disease. Nitrogen and sulfur oxides, ozone, and other air pollutants from fossil fuels are inflicting damage on our forests, crops, soils, lakes, rivers, coastal waters, and buildings. Chlorofluorocarbons (CFCs) and other atmospheric pollutants are depleting the Earth's protective ozone layer, allowing more harmful ultraviolet radiation to reach the Earth's surface. Fossil fuel combustion increases the amount of carbon dioxide in the atmosphere, which can have severe long-term environmental impacts.

Interestingly, the composition of "unpolluted" air is unknown when ambient air is considered. Humans have lived on the planet for thousands of years and influenced the composition of the air through their many activities before it was possible to measure its constituents.

EPA (2006) points out that, in theory, the air has always been polluted to some degree. Natural phenomena such as volcanoes, wind storms, the decomposition of plants and animals, and even the aerosols emitted by the ocean "pollute" the air. However, the pollutants we usually refer to when discussing air pollution are those generated due to human activity. An *air pollutant* can be considered a substance in the air that, in high enough concentrations, produces a detrimental environmental effect. These effects can be either health effects or welfare effects. A pollutant can affect the health of humans, as well as the health of plants and animals. Pollutants can also affect non-living materials such as paints, metals, and fabrics. An *environmental effect* is a measurable or perceivable detrimental change from contact with an air pollutant.

Human activities have had a detrimental effect on the makeup of air. Activities such as driving cars and trucks, burning coal, oil, and other fossil fuels, and manufacturing chemicals have changed the composition of air by introducing many pollutants. There are hundreds of contaminants in the ambient air. Ambient air is the air the general public has access to, i.e., any unconfined portion of the atmosphere. The two primary physical forms of air pollutants are particulate matter and gases. Particulate matter includes small solid and liquid particles such as dust, smoke, sand, pollen, mist, and fly ash. Gases include substances such as carbon monoxide (CO), sulfur dioxide (SO_2), nitrogen oxides (NO_2), and volatile organic compounds (VOCs).

Historically, many felt that the air renewed itself, through interaction with vegetation and the oceans, in sufficient quantities to make up for the influx of anthropogenic pollutants into our atmosphere. Today, however, this kind of thinking is being challenged by evidence that indicates that the increased use of fossil fuels, expanding industrial production, and growing use of motor vehicles are detrimental to the atmosphere, air, and the environment. In this section, we discuss air pollution mechanics:

pollutant dispersal, transformation, and deposition mechanisms, and we examine the types and sources of air pollutants related to these concerns.

Atmospheric Dispersion, Transportation & Deposition (EPA, 2005)

A source of air pollution is any activity that causes pollutants to be emitted into the air.

There have always been natural sources of air pollution, also known as biogenic sources. For example, volcanoes have spewed particulate matter and gases into our atmosphere for millions of years. Lightning strikes have caused forest fires, with their resulting contribution of gases and particles, for as long as storms and forests have existed. Organic matter in swamps decays, and wind storms whip up dust. Trees and other vegetation contribute significant amounts of pollen and spores to our atmosphere. These natural pollutants can be problematically generated pollutants or anthropogenic sources.

The quality of daily life depends on many modern conveniences. People enjoy the freedom to drive cars and travel in airplanes for business and pleasure. They expect their homes to have electricity and water heated for bathing and cooking. They use various products such as clothing, pharmaceuticals, and furniture made of synthetic materials. They sometimes rely on services that use chemical solvents, such as the local dry cleaner and print shop. Yet the availability of these everyday conveniences comes at a price because they all contribute to air pollution.

Air pollutants are released from both stationary and mobile sources. Scientists have gathered much information that is available on the sources, quantity, and toxicity levels of these pollutants. Air pollution measurement is an essential scientific skill, and practitioners of this skill are usually well-grounded in the related sciences and in modeling aspects applicable to their studies and analyses of air pollutants in the ambient atmosphere. However, to get at the heart of air pollution, the practitioner must also be well-versed in determining the pollutants' origin and understanding the pollutant dispersal, transport, and deposition mechanics.

Air pollution practitioners must constantly deal with one basic fact: air pollutants rarely stay at their release location. Instead, wind flow conditions and turbulence, local topographic features, and other physical conditions work to disperse these pollutants. So, along with a thorough knowledge and understanding of the pollutants in question, the air pollution practitioner needs detailed knowledge of the atmospheric processes that govern their subsequent dispersal and fate.

The conversion of precursor substances to secondary pollutants such as ozone is an example of chemical transformation in the atmosphere. Transformations (both physical and chemical) affect the ultimate impact of originally emitted air pollutants.

Pollutants emitted into the atmosphere do not remain there forever. Two common deposition (depletion) mechanisms are *dry deposition* (the removal of both particles and gases as they come into contact with the Earth's surface) and *washout* (the uptake of particles and gases by water droplets and snow and their removal from the atmosphere as precipitation that falls to the ground). Acid deposition (acid rain) is a form of pollution depletion from the atmosphere.

The following sections discuss the atmospheric dispersion of air pollutants in greater detail and the main factors associated with this phenomenon, including weather, turbulence, air parcels, buoyancy factors, lapse rates, mixing, topography, inversions, plume behavior, and transport.

Weather

The air contained in Earth's atmosphere is not still. Constantly in motion, air masses warmed by solar radiation rise at the equator and spread toward the colder poles, where they sink and flow downward, eventually returning to the equator. Near the Earth's surface, as a result of the Earth's rotation, major wind patterns develop. During the day, the land warms more quickly than the sea; at night, the land cools more quickly. Local wind patterns are driven by this differential warming and cooling between the land and adjacent water bodies. Normally, onshore breezes bring cooler, denser air from over the water bodies out over the lands during the night. Precipitation is also affected by wind patterns. Warm, moisture-laden air rising from the oceans is carried inland, where the air masses eventually cool, causing the moisture to fall as rain, hail, sleet, or snow.

Even though pollutant emissions may remain relatively constant, air quality varies daily. The determining factor is the weather.

Weather conditions significantly impact both favorable and unfavorable air quality and air pollution, especially local conditions. For example, on hot, sun-filled days, when the weather is calm with stagnating high-pressure cells, air quality suffers because these conditions allow the buildup of pollutants at ground level. When local weather conditions include cool, windy, stormy weather with turbulent low-pressure cells and cold fronts, these conditions allow the upward mixing and dispersal of air pollutants.

Weather directly impacts pollution levels in both mechanical and chemical ways. Mechanically, precipitation cleanses the air of pollutants (transferring the pollutants to rivers, streams, lakes, or the soil). Winds transport contaminants from one place to another. Winds and storms often dilute pollutants with cleaner air, making pollution levels less severe in the area of their release. Air (and its accompanying pollution in a low-pressure cell) is also carried aloft by air heated by the sun. The pollutants are diluted with fresh air when wind accompanies this rising air mass. In a high-pressure cell, the opposite occurs, with air and the contaminants it carries sinking toward the ground. With no wind, these pollutants are trapped and concentrated near the ground, where severe air pollution may occur.

Chemically, weather can also affect pollution levels. Winds and turbulence mix pollutants in a giant chemical broth in the atmosphere. Energy from the sun, moisture in the clouds, and the proximity of highly reactive chemicals may cause chemical reactions, forming secondary pollutants. Many of these secondary pollutants may be more dangerous than the original pollutants.

Turbulence

In the atmosphere, the degree of turbulence (which results from wind speed and convective conditions related to the change of temperature with height above the Earth's surface) is directly related to stability (a function of the vertical distribution of atmospheric temperature). The atmosphere's stability refers to the susceptibility of rising air parcels to vertical motion (attributed to high- and low-pressure systems, airlifting over terrain or fronts, and convection). Consideration of atmospheric stability or instability is essential in establishing the dispersion rate of pollutants. When discussing the atmosphere's stability, we refer to the Earth's lower boundary where air pollutants are emitted. The degree of turbulence in the atmosphere is usually classified by stability class. Ambient and adiabatic lapse rates are a measure of atmospheric stability.

Stability is divided into three classes: stable, unstable, and neutral. A *stable atmosphere* is marked by air that is cooler at the ground than aloft, by low wind speeds, and consequently, by a low degree of turbulence. A plume of pollutants released into a stable lower layer of the atmosphere can remain relatively intact for long distances. Thus, we can say that stable air discourages the dispersion and dilution of pollutants. An *unstable atmosphere* is marked by a high degree of turbulence. A plume of contaminants released into an unstable atmosphere may exhibit a characteristic looping appearance produced by turbulent eddies. A *neutrally stable atmosphere* is an intermediate class between stable and unstable conditions. A plume of pollutants released into a neutral stability condition is often characterized by a coning appearance as the edges of the plume spread out in a V-shape.

The importance of the "state of the atmosphere" and its effects cannot be overstated. The ease with which pollutants can disperse vertically into the atmosphere is mainly determined by the rate of air temperature change with height (altitude). Therefore, air stability is a primary factor in deciding where pollutants will travel and how long they will remain aloft. Stable air discourages the dispersion and dilution of contaminants. Conversely, in unstable air conditions, rapid vertical mixing takes place, encouraging pollutant dispersal, which increases air quality.

Air Parcels

Concerning air parcels described in this section, think of air inside a balloon as an analogy for the air parcel. This theoretically infinitesimal parcel is a relatively well-defined body of air (a constant number of molecules) that acts as a whole. Self-contained, it does not readily mix with the surrounding air. The heat exchange between the parcel and its surroundings is minimal, and the temperature within the parcel is generally uniform.

Buoyancy Factors

Atmospheric temperature and pressure influence the buoyancy of air parcels. When holding other conditions constant, the temperature of air (a fluid) increases as atmospheric pressure increases and conversely decreases as pressure decreases. For the atmosphere, where air pressure decreases with rising altitude, the typical temperature profile of the troposphere is one where temperature decreases with height.

An air parcel that becomes warmer than the surrounding air (for example, by heat radiating from the Earth's surface) begins to expand and cool. As long as the parcel's temperature exceeds the surrounding air, the parcel is less dense than the cooler surrounding air. Therefore, it rises or is buoyant. As the parcel rises, it expands, decreasing its pressure and, thus, its temperature decreases. The initial cooling of an air parcel has the opposite effect. In short, warm air rises and cools while cool air descends and warms.

The extent to which an air parcel rises or falls depends on the relationship of its temperature to that of the surrounding air. As long as the parcel's temperature is more relaxed, it will descend. When the temperatures of the parcel and the surrounding air are the same, the parcel will neither rise nor descend unless influenced by wind flow.

Lapse Rates

The *lapse rate* is defined as the rate of temperature change with height. The ambient air temperature usually decreases with an increase in altitude in the troposphere. On average, temperature decreases from −6 to −7°C/km. This is the regular lapse rate, but it varies widely depending on location and time of day. We define a temperature *decrease* with height as a negative lapse rate and a temperature *increase* with height as a positive lapse rate.

In a dry environment, when a parcel of warm, dry air is lifted in the atmosphere, it undergoes adiabatic expansion and cooling. Primarily, a parcel of air does not exchange heat across its boundaries. Therefore, an air parcel warmer than the surrounding air does not transfer heat to the atmosphere. Increases or decreases in molecular activity cause any temperature changes within the parcel. Such changes occur adiabatically and are due only to the change in atmospheric pressure as a parcel moves vertically. The term "adiabatic" literally means impassable from, corresponding to an absence of heat transfer. That is, an adiabatic process is one in which there is no transfer of heat or mass across the boundaries of the air parcel. In an adiabatic process, compression results in heating, and expansion results in cooling. A dry air parcel rising in the atmosphere cools at the dry adiabatic rate of 9.8°C/1,000 m and has a lapse rate of −9.8°C/1,000 m. Likewise, a dry air parcel sinking in the atmosphere heats up at the dry adiabatic rate of 9.8°C/1,000 m and has a lapse rate of 9.8°C/1,000 m. Air is considered dry, in this context, as long as water remains in a gaseous state.

The *dry adiabatic lapse rate* is fixed and independent of ambient air temperature. Thus, a parcel of dry air moving upward in the atmosphere will always cool at the rate of 9.8°C/1,000 m, regardless of its initial temperature or the surrounding air temperature.

When the ambient lapse rate exceeds the adiabatic lapse rate, it is said to be superadiabatic, and the atmosphere is highly unstable. When the two lapse rates are exactly equal, the atmosphere is said to be neutral. When the ambient lapse rate is less than the dry adiabatic lapse rate, it is termed *subadiabatic*, and the atmosphere is stable.

The cooling process within a rising air parcel is assumed to be adiabatic (occurring without adding or losing heat). An increasing parcel of air (under adiabatic conditions) behaves like a rising balloon, with the air in that distinct parcel expanding as it encounters air of lesser density until its density is equal to that of the atmosphere surrounding it. This process is assumed to occur without heat exchange between the rising parcel and the ambient air (Peavy et al., 1985).

A rising parcel of dry air containing water vapor will continue to cool at the dry adiabatic lapse rate until it reaches its condensation temperature or dew point. At this point, the water vapor's pressure equals the air's saturation vapor pressure, and some water vapor condenses. Condensation releases latent heat in the parcel, and thus, the cooling rate of the parcel slows. This new rate is called the *wet adiabatic lapse rate*. Unlike the dry adiabatic lapse rate, the wet adiabatic lapse rate is not constant but depends on temperature and pressure. In the middle troposphere, however, it is assumed to be approximately −6 to −7°C/1,000 m.

Mixing

Turbulent mixing is essential for effective pollutant dispersal within the atmosphere. It results from air movement in the vertical dimension and is enhanced by vertical temperature differences. The steeper the temperature gradient and the larger the vertical air column in which the mixing takes place, the more vigorous the convective and turbulent mixing of the atmosphere.

Topography

On a local scale, topography may affect air motion. In the United States, most significant urban centers are located along sea and lake coastal areas. Contained within these large urban centers is much heavy industry. Local air flow patterns in these urban centers significantly impact pollution dispersion processes. Topographic features also affect regional weather patterns, especially in large urban centers near lakes, seas, and open land. Breezes from these features affect vertical mixing and pollutant dispersal. Seasonal differences in heating and cooling land and water surfaces may also precipitate the formation of inversions near the sea or lake shore.

River valley areas are also geographical locations that routinely suffer from industry-related pollution. Many early settlements began in river valleys because of the readily available water supply and the ease of transportation afforded to settlers by river systems within such valleys. Along with settlers came industry—the industry that invariably produces air pollutants. Because of the terrain and physical configuration of the valley, these air pollutants are not easily removed from the valley.

Winds that move through a typical river valley are called slope winds. Slope winds, like water, flow downhill into the valley floor. At the valley floor level, slope winds transform into valley winds, which flow down the valley with the river's flow. Down-valley winds are lighter than slope winds. The valley floor becomes flooded with a large volume of air, intensifying the surface inversion typically produced by radiative cooling. As the inversion deepens throughout the night, it often reaches its maximum depth just before sunrise, with the height of the inversion layer dependent on the depth of the valley and the intensity of the radiative cooling process.

Hills and mountains can also affect local airflow. These natural topographical features tend to decrease wind speed (because of their roughness) and form physical barriers that prevent air movement.

Inversions

An inversion occurs when air temperature increases with altitude. Temperature inversions (extreme cases of atmospheric stability) create a virtual lid on the upward movement of atmospheric pollution. This situation occurs frequently but is generally confined to a relatively shallow layer. Plumes emitted into air layers experiencing an inversion (inverted layer) do not disperse much as they are transported with the wind. Plumes emitted above or below an inverted layer do not penetrate that layer; instead, these plumes are trapped either above or below the inverted layer. High concentrations of air pollutants are often associated with inversions since they inhibit plume dispersion. Two inversions are essential from an air quality standpoint: radiation and subsidence inversions.

Radiation inversions are the most common form of surface inversion and occur when the Earth's surface cools rapidly. They prompt fog formation and trap gases and particulates, creating a concentration of pollutants. They are characteristically a nocturnal phenomenon caused by the cooling of the Earth's surface. On a cloudy night, water vapor in the atmosphere tends to absorb the Earth's radiant heat. Some of this is re-radiated back to the surface. However, on clear winter nights, the surface more readily radiates energy to the atmosphere and beyond, allowing the ground to cool more rapidly. The air in contact with the cooler ground also cools, and the air just above the ground becomes more stable than the air above it, creating an inversion close to the ground that lasts only hours. These radiation inversions usually begin to form at the worst time of the day for human concerns in large urban areas—during the late afternoon rush hour, trapping automobile exhaust at ground level and causing elevated pollution concentrations for commuters. During evening hours, photochemical reactions cannot occur, so the biggest problem can be the accumulation of carbon monoxide. The sun warms the ground at sunrise, and the inversion begins to break up. Pollutants trapped in the stable air mass are suddenly brought back to Earth in a process known as *fumigation,* which can cause a short-lived, high concentration of pollution at ground level (Masters, 1991).

The second type of inversion is the subsidence inversion, usually associated with anticyclones (high-pressure systems). These inversions may significantly affect the dispersion of pollutants over large regions. A subsidence inversion is caused by the characteristic sinking motion of air in a high-pressure cell. Air in the middle of a high-pressure zone descends slowly. As the air descends, it is compressed and heated. It forms a blanket of warm air over the cooler air below, thus creating an inversion (anywhere from several hundred meters above the surface to several thousand meters) that prevents further vertical air movement.

Plume Behavior

One way to quickly determine the stability of the lower atmosphere is to observe the shape of a smoke trail or *plume* from a tall stack located on flat terrain. Visible plumes usually contain pollutants emitted from a smoke stack into the atmosphere. The formation and fate of the plume itself depend on several related factors: (1) the nature of the pollutants, (2) meteorological factors (combination of vertical air movement and horizontal airflow), (3) source obstructions, and (4) local topography, especially downwind. Overall, maximum ground-level concentrations will occur in a range from the vicinity of the smokestack to some distance downwind.

An already contaminated environment compounds air quality problems associated with the dispersion of city plumes. Even though conventional processes typically work to disperse emissions from point sources within the city plume, because of micro-climates within the city and the volume of pollutants they must handle, these conventional processes often cannot disperse effectively. Additional compounding conditions present in areas where city plumes are generated (topographical barriers, surface inversions, and stagnating anticyclones) work to intensify the city plume, resulting in high pollutant concentrations.

Plume Rise Equation

Plume rise has been studied over the years. The most common plume rise formulas are those developed by Briggs, which have been extensively validated with stack plume observations (EPA, 2005). One of these, which applies to buoyancy-dominated plumes, is included in Equation 8.27. Plume rise formulas will be used on plumes with temperatures greater than the ambient air temperature. The *Briggs' plume rise formula* is as follows:

$$\Delta h = \frac{1.6\,F^{1/3} \times^{2/3}}{u} \tag{8.27}$$

where Δh = Plume rise (above stack)
F = Buoyancy Flux (see below; Equation 8.28)
u = Average wind speed
x = Downwind distance from the stack/source
g = Acceleration due to gravity (9.8 m/s^2)
V = Volumetric flow rate of tack gas
T_s = Temperature of stack gas
T_a = Temperature of ambient air

$$\text{Buoyancy flux} = F = g/\pi V\left(Ts - T_a/T_s\right) \tag{8.28}$$

Transport

Those living east of the Mississippi River would be surprised to learn that they breathe air contaminated by pollutants from various sources many miles from their location. Most people view pollution under the cliché "out of sight, out of mind." As far as they are concerned, if they don't see it, it doesn't exist. For example, assume that a person on a farm heaps a massive pile of assorted rubbish to be burned. The person preparing this huge bonfire probably gives little thought to the long-range transport (and consequences) of any contaminants that might be generated from that bonfire. This person has trash he or she no longer wants, and an easy solution is to burn it.

This pile of rubbish contains various elements: discarded rubber tires, old compressed gas bottles, assorted plastic containers, paper, oils and greases, wood, and old paint cans. The person burning it doesn't consider this hazardous material, just household trash. The truth is when the pile of rubbish burns, a huge plume of smoke forms and is carried away by a westerly wind. The fire starter looks downwind and notices that the smoke disappears just a few miles over the property line. The dilution processes and the enormity of the atmosphere work together to dissipate and move away the smoke plume; the fire starter doesn't give it a second thought. However, elevated levels of pollutants from many such fires may occur hundreds to thousands of miles downwind of the combination point sources producing such plumes. The result is that people living many miles from such pollution generators breathe contaminated air that is transported over distance to their location.

Transport or dispersion estimates are determined by using distribution equations and air quality models. These dispersion estimates are typically valid for the layer of the atmosphere closest to the ground, where frequent changes occur in the temperature and distribution of the winds. These two variables have an enormous effect on how plumes are dispersed.

DISPERSION MODELS

Air quality dispersion models are mathematical equations that interpret and predict pollutant concentrations due to plume dispersal and impaction. They indicate or describe the fate of airborne gases, particulate matter, and ground-level concentrations downwind of point sources. To determine the significance of air quality impact on a particular area, the first consideration is average background concentrations, those pollutant concentrations from natural sources and distant, unidentified man-made sources. Each specific geographical area has a "signature" or background level of contamination, which is considered an annual mean background concentration level of certain pollutants. An area, for example, might typically have a particulate matter reading of 30–40 $\mu g/m^3$. If particulate matter readings are significantly higher than the background level, this suggests an additional source. Air quality data about that site and its vicinity must be collected and analyzed to establish background contamination for a particular source.

The USEPA recognized that in calculating the atmospheric dispersion of air pollutants, some means by which consistency could be maintained in air quality analysis had to be established. Thus, the USEPA promulgated two guidebooks to assist in modeling for air quality analyses: *Guidelines on Air Quality Models (Revised) (1986)* and *Industrial Source Complex (ISC) Dispersion Models User's Guide (1986)*.

In performing dispersion calculations, particularly for health effect studies, the USEPA and other recognized experts in the field recommend following a four-step procedure:

1. Estimate the release rate, duration, and location in the environment.
2. Select the best available model to perform the calculations.
3. Perform the calculations and generate downstream concentrations, including lines of constant concentration (isopleths) resulting from the source emission(s).
4. Determine what effect, if any, the resulting discharge has on the environment, including humans, animals, vegetation, and materials of construction. These calculations often include estimates of the so-called vulnerability zones—regions that may be adversely affected by the emissions (Holmes et al., 1993).

Before beginning any dispersion determination activity, you must determine the acceptable ground-level concentration of the waste pollutant(s). Local meteorological conduits and topography must be considered, and accurate knowledge of the constituents of the waste gas and its chemical and physical properties is paramount.

Air quality models provide a relatively inexpensive means of determining compliance and predicting the degree of emission reduction necessary to attain ambient air quality standards. Under the 1977 Clean Air Act Amendments, the use of models is required for the evaluation of permit applications associated with permissible increments under the so-called Prevention of

Significant Degradation (PSD) requirements, which require localities to "protect and enhance" uncontaminated air (Godish, 1997).

Several dispersion models have been developed. Really equations, these models are mathematical descriptions of the meteorological transport and dispersion of air contaminants in a particular area, which permit estimates of contaminant concentrations, either in the plume from a ground level or an elevated source (Carson & Moses, 1969). User-friendly modeling programs that produce quick, accurate results from the operator's pertinent data are now available.

There are four generic types of models: Gaussian, numerical, statistical, and physical. The *Gaussian* models use the Gaussian distribution equation and are widely used to estimate the impact of nonreactive pollutants. *Numerical* models are more appropriate than Gaussian models for area sources in urban locations involving reactive pollutants, but numerical models require extreme measures. Detailed source and pollutant information are not widely used. *Statistical* models are used when scientific details on a source's chemical and physical processes are incomplete or vague, making the use of either Gaussian or numerical models impractical. Lastly, *physical* models require fluid modeling studies or wind tunneling. This approach involves the construction of a scaled model and observing fluid flow around these models. This type of modeling is very complex and requires expert technical support. However, this type of modeling may be the best choice for large areas with complex terrain, stack downwash, complex flow conditions, or large buildings.

As mentioned, selecting an air quality model for a particular air quality analysis depends on the type of pollutants being emitted, the complexity of the source, and the kind of topography surrounding the facility. The combination of precursor pollutants forms some contaminants. For example, ground-level ozone is formed when volatile organic compounds (VOCs) and nitrogen oxides (NO_x) react in sunlight. Models to predict ground-level ozone concentrations would use the emission rate of VOCs and NO_x as inputs. Also, some pollutants readily react when emitted into the atmosphere. These reactions deplete the concentrations of these pollutants and may need to be accounted for in the model. Source complexity also plays a role in model selection. Some pollutants may be emitted from short stacks subject to aerodynamic downwash. If this is the case, a model capable of accounting for this phenomenon must be used. Again, topography plays a significant role in the dispersal of plumes and their air pollutants and must be considered when selecting an air quality model. Elevated plumes may impact areas of high terrain. Elevated terrain heights may experience higher pollutant concentrations since they are closer to the plume centerline. A model that considers terrain heights should be used when elevated terrain exists.

This handbook does not develop each dispersion model in detail but recommends the one with the most excellent applicability today. Probably the best atmospheric dispersion workbook for modeling published to date is D.B. Turner's *Workbook of Atmospheric Dispersion Estimates* for the EPA, and most of the air dispersion models used today are based on the Pasquill-Gifford Model.

MAJOR AIR POLLUTANTS

The most common and widespread anthropogenic pollutants currently emitted are sulfur dioxide (SO_2), nitrogen oxides (NO_x), carbon monoxide (CO), carbon dioxide (CO_2), volatile organic compounds (hydrocarbons), particulates, lead, and several toxic chemicals—Table 8.15 lists important air pollutants and their sources.

Recall that in the United States, the Environmental Protection Agency (EPA) regulates air quality under the Clean Air Act (CAA) and amendments that charge the federal government with developing uniform National Ambient Air-Quality Standards (NAAQS). These include a dual standard requirement of primary standards (covering criteria pollutants) designed to protect health and secondary standards to protect public welfare. Primary standards were to be achieved by July 1975, and secondary standards in "a reasonable period." Pollutant levels for the protection of public welfare take priority (and are more stringent) over those for public health; achievement of the primary health standard had immediate priority. In 1971, the USEPA promulgated NAAQS for six classes of air pollutants. Later, in 1978, an air-quality standard was also enacted for lead, and the photochemical oxidant standard was revised to an ozone (O_3) standard (the ozone permissible level was increased). The PM standard was revised and redesignated as the PM_{10} standard in 1987. This revision reflected the need for a PM standard based on particle sizes (≤10 μm) that can potentially enter the respiratory tract and affect human health.

Thus, air pollutants were categorized into two groups: primary and secondary. Primary pollutants are emitted directly into the atmosphere, adversely affecting human health and the environment. Of particular concern are primary pollutants emitted in large quantities: carbon dioxide, carbon monoxide, sulfur dioxide, nitrogen oxides, hydrocarbons, and particulate matter (PM). Once in the atmosphere, primary pollutants may react with other primary pollutants or atmospheric compounds,

TABLE 8.15
Common Air Pollutants

Pollutant	Source
Sulfur and nitrogen oxides	From fossil fuel combustion
Carbon monoxide	Mostly from motor vehicles
Volatile organic compounds	From vehicles and industry
Ozone	From atmospheric reactions between nitrogen oxides and organic compounds

Source: USEPA, *Environmental Progress & Challenges*, p. 13 (1988a).

such as water vapor, to form secondary pollutants. A secondary pollutant that has received a lot of press and attention is acid precipitation, formed when sulfur or nitrogen oxides react with water vapor in the atmosphere.

Sulfur Dioxide (SO_2)

Sulfur enters the atmosphere as corrosive *sulfur dioxide* (SO_2) gas. Sulfur dioxide is a colorless gas possessing burning rubber's sharp, pungent odor. On a global basis, nature and anthropogenic activities produce sulfur dioxide in roughly equivalent amounts. Its natural sources include volcanoes, decaying organic matter, and sea spray, while anthropogenic sources include the combustion of sulfur-containing coal and petroleum products and the smelting of nonferrous ores. In industrial areas, much more sulfur dioxide comes from human activities than natural sources. Sulfur-containing substances are often present in fossil fuels; SO_2 is a product of combustion from burning sulfur-containing materials. The largest single source (65%) of sulfur dioxide is burning fossil fuels to generate electricity. Thus, it is often encountered as an air pollutant near major industrialized areas.

In the air, sulfur dioxide converts to sulfur trioxide (SO^3) and sulfate particles (SO_4). Sulfate particles restrict visibility and form sulfuric acid (H_2SO_4), a highly corrosive substance that also lowers visibility in the presence of water. According to McKenzie & El-Ashry (1988), the global sulfur dioxide output has increased sixfold since 1900. Most industrial nations, however, have since (1975–1985) lowered sulfur dioxide levels by 20%–60% by shifting away from heavy industry and imposing stricter emission standards. Significant sulfur dioxide reductions have come from burning coal with lower sulfur content and using less coal to generate electricity.

Two major environmental problems have developed in highly industrialized regions of the world, where the atmospheric sulfur dioxide concentration has been relatively high: sulfurous smog and acid rain. Sulfurous smog is the haze that develops in the atmosphere when sulfuric acid molecules accumulate, growing in size as droplets until they become sufficiently large to serve as light scatterers. The second problem, acid rain, is precipitation contaminated with dissolved acids like sulfuric acid. Acid rain has threatened the environment by causing certain lakes to become devoid of aquatic life.

Nitrogen Oxides (NO_x)

There are seven oxides of nitrogen that are known to occur—NO, NO_2, NO_3, N_2O, N_2O_3, N_2O_4, and N_2O_5—but only two are essential in the study of air pollution: nitric oxide (NO) and nitrogen dioxide (NO_2). Both natural and human actions produce nitric oxide. Soil bacteria produce most nitric oxide naturally and release it into the atmosphere. Within the atmosphere, nitric oxide readily combines with oxygen to form nitrogen dioxide, and together, those two oxides of nitrogen are usually referred to as NO_x (nitrogen oxides). NO_x is formed naturally by lightning and by decomposing organic matter. Approximately 50% of anthropogenic NO_x is emitted by motor vehicles, about 30% comes from power plants, and industrial processes produce the other 20%.

Scientists distinguish between two types of NO_x—thermal and fuel—depending on their mode of formation. Thermal NO_x is created when nitrogen and oxygen in the combustion air, such as those within internal combustion engines, are heated to a high enough temperature (above 1,000 K) to cause nitrogen (N_2) and oxygen (O_2) in the air to combine. Fuel NO_x results from the oxidation (i.e., combining with oxygen in the air) of nitrogen contained within a fuel such as coal. Both types of NO_x generate nitric oxide first, and then, when vented and cooled, a portion of nitric oxide is converted to nitrogen dioxide. Although thermal NO_x can significantly contribute to total NOx emissions, fuel NO_x is usually the dominant source, with approximately 50% coming from power plants (stationary sources) and the other half released by automobiles (mobile sources).

Nitrogen dioxide is more toxic than nitric oxide and a more serious air pollutant. At high concentrations, it is believed to contribute to heart, lung, liver, and kidney damage. In addition, it reduces visibility because it occurs as a brownish haze (giving smog its reddish-brown color). When nitrogen dioxide combines with water vapor in the atmosphere, it forms nitric acid (HNO_3), a corrosive substance precipitated as acid rain, causing damage to plants and corrosion of metal surfaces.

NO_x levels rose in several countries and then leveled off or declined during the 1970s. During this same timeframe, nitrogen oxide levels did not drop as dramatically as sulfur dioxide, primarily because a large part of total NO_x emissions comes from millions of motor vehicles. In contrast, most sulfur dioxide is released by a relatively small number of emission-controlled, coal-burning power plants.

Carbon Monoxide (CO)

Carbon monoxide is a colorless, odorless, tasteless gas formed when carbon in fuel is not burned completely; it is by far the most abundant of the primary pollutants, as Table 8.16 indicates. When inhaled, carbon monoxide gas restricts the blood's ability to absorb oxygen, causing angina, impaired vision, and poor coordination. Carbon monoxide has little direct effect on ecosystems but has an indirect environmental impact by contributing to the greenhouse effect and depleting the Earth's protective ozone layer.

TABLE 8.16
United States Emission Estimates, 1986 (10^{12} g/year)

Source	SO_x	NO_x	VOC	CO	Lead	PM
Transportation	0.9	8.5	6.5	42.6	0.0035	1.4
Stationary source fuel	17.2	10.0	2.3	7.2	0.0005	1.8
Industrial processes	3.1	0.6	7.9	4.5	0.0019	2.5
Solid waste disposal	0.0	0.1	0.6	1.7	0.0027	0.3
Miscellaneous	0.0	0.1	2.2	5.0	0.0000	0.8
Total	28.4	18.1	19.5	60.9	0.0086	6.8

Source: USEPA (1988). *National Air Pollutant Emission Estimates 1940–1986*, Washington, DC, Environmental Protection Agency.

The most important natural source of atmospheric carbon monoxide is the combination of oxygen with methane (CH_4), a product of vegetation's anaerobic decay. (Anaerobic decay occurs in the absence of oxygen.) At the same time, however, carbon monoxide is removed from the atmosphere by the activities of certain soil microorganisms. Hence, the net result is a harmless average concentration of less than 0.12–15 ppm in the northern hemisphere. Because stationary source combustion facilities are under much tighter environmental control than mobile sources, the principal source of carbon monoxide caused by human activities is motor vehicle exhaust, contributing to about 70% of all CO emissions in the United States.

Volatile Organic Compounds (VOCs—Hydrocarbons)

Volatile Organic Compounds (VOCs) (also listed under the general heading of hydrocarbons) encompass various chemicals that exclusively contain hydrogen and carbon. Emissions of volatile hydrocarbons from human sources are primarily the result of incomplete combustion of fossil fuels. Fires and the decomposition of matter are the natural sources. Of the VOCs that occur naturally in the atmosphere, methane (CH_4) is present at the highest concentrations (approximately 1.5 ppm). However, even at relatively high concentrations, methane does not interact chemically with other substances and causes no ill health effects. However, in the lower atmosphere, sunlight causes VOCs to combine with other gases, such as NO_2, oxygen, and CO, to form secondary pollutants such as formaldehyde, ketones, ozone, peroxyacetyl nitrate (PAN), and other types of photochemical oxidants. These active chemicals can irritate the eyes, respiratory system, and vegetation.

Ozone and Photochemical Smog

The most damaging photochemical air pollutant is ozone (each ozone molecule contains three atoms of oxygen and thus is written O_3). Other photochemical oxidants [peroxyacetyl nitrate (PAN), hydrogen peroxide (H_2O_2), and aldehydes] play minor roles. These are secondary pollutants because they are not emitted but are formed in the atmosphere by photochemical reactions involving sunlight and gases, especially NO_x and hydrocarbons.

Ozone is a bluish gas about 1.6 times heavier than air and relatively reactive as an oxidant. It is present in a relatively large concentration in the stratosphere and is formed naturally by ultraviolet radiation. At ground level, ozone is a serious air pollutant; it has caused severe air pollution problems throughout the industrialized world, posing threats to human health and damaging foliage and building materials.

According to MacKenzie and El-Ashry (1988), ozone concentrations in industrialized countries of North America and Europe are up to three times higher than the level at which damage to crops and vegetation begins. Ozone harms vegetation by damaging plant tissues, inhibiting photosynthesis, and increasing susceptibility to disease, drought, and other air pollutants.

In the upper atmosphere, where "good" (vital) ozone is produced, ozone is depleted by anthropogenic emissions of ozone-depleting chemicals, which have increased on the ground. With this increase, concern has been raised over a potential upset of the dynamic equilibria among stratospheric ozone reactions, with a consequent reduction in ozone concentration. This situation is serious because stratospheric ozone absorbs much incoming solar ultraviolet (UV) radiation. As a UV shield, ozone helps to protect organisms on the Earth's surface from some of the harmful effects of this high-energy radiation. If not interrupted, UV radiation could cause serious damage, such as disruption of genetic material, which could lead to increased rates of skin cancers and heritable problems.

In the mid-1980s, a serious problem with ozone depletion became apparent. A springtime decrease in stratospheric ozone concentration (ozone holes) occurred at high latitudes, most notably over Antarctica between September and November. Scientists strongly suspected that chlorine atoms or simple chlorine compounds may play a key role in this ozone depletion problem.

On rare occasions, upper stratospheric ozone (good ozone) can enter the lower atmosphere (troposphere). Generally, this phenomenon only occurs during significant turbulence in the upper atmosphere. On rare incursions, atmospheric ozone reaches ground level for a short period. Most of the tropospheric ozone is formed and consumed by endogenous photochemical reactions, which result from the interaction of hydrocarbons, oxides of nitrogen, and sunlight, producing a yellowish-brown haze commonly called smog (Los Angeles-type smog).

Although the incursion of stratospheric ozone into the troposphere can cause smog formation, the actual formation of Los Angeles-type smog involves a complex group of photochemical interactions. These interactions are between anthropogenically emitted pollutants (NO and hydrocarbons) and secondarily produced chemicals (PAN, aldehydes, NO_2, and ozone). Note that the concentrations of these chemicals exhibit a pronounced diurnal pattern depending on their rate of emission, the intensity of solar radiation, and atmospheric stability at different times of the day (Freedman, 1989).

Historical records show that various air pollutants in the atmosphere of Los Angeles include NO (emitted as NO_x), which has a morning peak of concentration at 0600–0700, mainly due to emissions from morning rush-hour vehicles. Hydrocarbons are emitted from vehicles and refineries; they display a similar pattern to NO, except their peak concentration is slightly later. In bright sunlight, the NO is photochemically oxidized to NO_2, decreasing NO concentration and causing a peak of NO_2 at 0700–0900. Photochemical reactions involving NO_2 produce O atoms, which react with O_2 to form O_3. These result in a net decrease in NO_2 concentration and an increase in O_3 concentration, peaking between 1,200 and 1,500. Aldehydes are also formed photochemically and peak earlier than O_3. As the day proceeds, the various gases decrease in concentration as they are diluted by fresh air masses or are consumed by photochemical reactions. This cycle is typical of an area that experiences photochemical smog and is repeated daily (Urone, 1976).

Table 8.17 shows a tropospheric ozone budget for the northern hemisphere. The considerable range of the estimates reflects uncertainty in calculating the ozone fluxes. On average, stratospheric incursions account for about 18% of the total ozone influx to the troposphere, while endogenous photochemical production accounts for the remaining 82%. About 31% of the tropospheric ozone is consumed by oxidative reactions at vegetative and inorganic surfaces at ground level, while the other 69% is consumed by photochemical reactions in the atmosphere (Freedman, 1989).

TABLE 8.17
Tropospheric Ozone Budget (Northern Hemisphere) (kg/ha-year)

Transport from stratosphere	13–20
Photochemical production	48–78
Destruction at ground	18–35
Photochemical destruction	48–55

Source: Adapted from Hov, O. (1984). Ozone in the troposphere: High-level pollution. *Ambio* **13**, 73–79.

Carbon Dioxide

Carbon-laden fuels, when burned, release carbon dioxide (CO_2) into the atmosphere. Much of this carbon dioxide is dissipated and then absorbed by ocean water; some is taken up by vegetation through photosynthesis, and some remains in the atmosphere. Today, the concentration of carbon dioxide in the atmosphere is approximately 350 ppm, rising to approximately 20 ppm every decade. The increasing rate of coal and oil combustion has been primarily responsible for this occurrence, which may eventually impact global climate.

Particulate Matter

Atmospheric particulate matter is defined as any dispersed matter, solid or liquid, in which the individual aggregates are larger than single small molecules but smaller than about 500 μm. Particulate matter is highly diverse and complex since size, chemical composition, and atmospheric concentrations are important characteristics (Masters, 1991).

Depending on their size and phase (liquid or solid), particulates are categorized using several terms listed and described in Table 8.18.

Dust, spray, forest fires, and the burning of certain types of fuels are among the sources of particulates in the atmosphere. Even with the implementation of stringent emission controls, which have worked to reduce particulates in the atmosphere, the U.S. Office of Technology Assessment (Postel, 1987) estimates that current levels of particulates and sulfates in ambient air may cause the premature death of 50,000 Americans every year.

Lead

Lead is emitted into the atmosphere primarily from human sources, such as burning leaded gasoline, in the form of inorganic particulates. In high concentrations, lead can damage human health and the environment. Once lead enters an ecosystem, it remains there permanently. Lead can affect the neurological system in humans and animals and cause kidney disease. Lead can inhibit respiration and photosynthesis in plants and block microorganisms' decomposition. Since the 1970s, stricter emission standards have caused a dramatic reduction in lead output.

TABLE 8.18
Atmospheric Particulates

Term	Description
Aerosol	Is a general term for particles suspended in air
Mist	Aerosol consists of liquid droplets
Dust	Aerosol consists of solid particles blown into the air or produced from larger particles by grinding them down
Smoke	Aerosol consists of solid particles or a mixture of solid and liquid particles produced by chemical reactions, such as those occurring during fires
Fume	Generally means the same as smoke but often applies specifically to aerosols produced by the condensation of hot vapors, especially from metals
Plume	The geometrical shape or form of the smoke coming out of a stack or chimney
Fog	Aerosol consisting of water droplets
Haze	Any aerosol other than fog obscures the view through the atmosphere
Smog	A popular term originating in England that describes a mixture of smoke and fog; it implies photochemical pollution

AIR POLLUTION CONTROL TECHNOLOGY

There are two primary motivations behind using industrial air pollution control technologies. These are:

1. They must be used because of legal or regulatory requirements.
2. They are integral to the economical operation of an industrial process.

Although economists would point out that both motivations are the same, it is less expensive for an industrial user to operate with air pollution control than without the distinction in application type, which is an important one. In general, air pollution control is used to describe those applications that are driven by regulations and health considerations. In contrast, applications that deal with product recovery are considered process applications. Nevertheless, the technical issues, equipment design, operation, etc., will be similar, if not identical. These uses differ because the economics that affect the decision-making process will often vary to some degree (W.L. Heumann, p. xv, 1997).

Air pollution control begins with regulation. Regulations (for example, to clean up, reduce, or eliminate a pollutant emission source) are generated because of specific community concerns. Buonicore et al. (1992) point out that regulations usually evolve around three considerations:

1. Legal limitations imposed to protect public health and welfare.
2. Social limitations imposed by the community where the pollution source is or is to be located.
3. Economic limitations imposed by marketplace constraints.

The environmental engineer assigned to mitigate an air pollution problem must ensure that the design control methodology used will bring the source into full compliance with applicable regulations. To accomplish this feat, environmental engineers must first understand the problem(s) and then rely heavily on technology to correct the situation. Various air pollution control technologies are available to environmental engineers or air pollution control practitioners working to mitigate air pollution source problems. By analyzing the problem carefully and applying the most effective method, the engineer or practitioner can ensure that a particular pollution source is controlled and that the responsible parties fully comply with regulations.

This section discusses the various air pollution control technologies available to environmental engineers and practitioners to mitigate air pollution source problems.

Air Pollution Control: Choices

Assuming that the design engineer has complete knowledge of the contaminant and the source, all available physical and chemical data on the effluent from the source, and the control agencies' regulations, he or she must decide which control methodology to employ. Since only a few control methods exist, the choice is limited. Control of atmospheric emissions from a process will generally consist of one of four methods depending on the process, types, fuels, availability of control equipment, etc. The four general control methods are (1) elimination of the process entirely or in part, (2) modification of the operation to a fuel that will give the desired level of emission, (3) installation of control equipment between the pollutant source and the receptor, and (4) relocation of the operation.

Tremendous costs are involved in eliminating or relocating a complete process, making either choice the last resort. Let's examine the first and last control methods first. Eliminating a process is no easy undertaking, especially when the "process" to be eliminated is the process for which the facility exists. Relocation is not always an answer, either.

The second pollution control method—modifying the operation to a fuel that will give the desired emission level—often looks favorable to those who have weighed the high costs associated with air pollution control systems. Modifying the process to eliminate as much of the pollution problem as possible at the source is generally the first approach to be examined.

Again, often, the easiest way to modify a process for air pollution control is to change the fuel. If a power plant, for example, emits large quantities of sulfur dioxide and fly ash, conversion to cleaner-burning natural gas is cheaper than installing the necessary control equipment to reduce the pollutant emissions to permitted values.

However, changing from one fuel to another causes problems related to costs, availability, and competition. Today's fuel prices are high, and no one counts on the trend to reverse. Finding a low-sulfur fuel isn't easy, especially since many industries own dedicated supplies (which are not available for use in other sectors). With regulation compliance threatening everyone, everyone wants their share of any available low-cost, low-sulfur fuel. With limited supplies available, the law of supply and demand takes over, and prices increase.

Some industries employ other process modification techniques. These may include evaluating alternative manufacturing and production techniques, substituting raw materials, and improving process control (Buonicore et al., 1992).

When elimination of the process entirely or in part, relocation of the operation, or modification of the operation to a fuel that will give the desired level of emission is not possible, the only alternative control method left is the installation of control equipment between the pollutant source and the receptor (the purpose of installing pollution control equipment or a control system is to remove the pollution from the polluted carrier gas). To accomplish this, the polluted carrier gas must pass through a control device or system, which collects or destroys the pollutant and releases the cleaned carrier gas into the atmosphere (Boubel et al., 1994). The rest of this section will focus on these air pollution control equipment devices and systems.

AIR POLLUTION CONTROL EQUIPMENT AND SYSTEMS

Several considerations must be factored into selecting air pollution control equipment or systems. Careful consideration must be given to costs. No one ever said air pollution equipment/systems were inexpensive—they are not. The equipment/system must be designed to comply with applicable regulatory emission limitations. The operational and maintenance history/record (costs of energy, labor, and repair parts should also be factored in) of each equipment/system must be evaluated. Remember, emission control equipment must be operated continually without interruptions. Any interruption could be subject to severe regulatory penalties, which could be costly again.

The major factor to consider in the equipment/system selection process is what type of pollutant or pollutant stream is under consideration. Suppose the pollutant is conveyed in a carrier gas, for example. In that case, factors such as carrier gas pressure, temperature, viscosity, toxicity, density, humidity, corrosiveness, and inflammability must all be considered before any selection is made. Other important factors must also be considered when selecting air pollution control equipment. Many of the general factors are listed in Table 8.19.

In addition to those factors listed in Table 8.19, process considerations dealing with gas flow rate and velocity, pollutant concentration, allowable pressure drop, and the variability of gas and pollutant flow rates, including temperature, must all be considered.

TABLE 8.19
Factors in Selecting Air Pollution Control Equipment/Systems

1.	Best available technology (BAT)
2.	Reliability
3.	Lifetime and salvage value
4.	Power requirements
5.	Collection efficiency
6.	Capital cost, including operation and maintenance costs
7.	Track record of equipment/system and manufacturer
8.	Space requirements and weight
9.	Power requirements
10.	Availability of spare parts and manufacturer's representatives

The type of pollutant is also an essential factor that must be considered: gaseous or particulate. Specific pertinent questions must be asked and answered. If the pollutant, for example, is gaseous, how corrosive, flammable, reactive, and toxic is it? After these factors have been evaluated, the focus shifts to selecting the best air pollution control equipment/system—affordable, practical, and permitted by regulatory requirements—depending, of course, on the type of pollutant to be removed.

The following sections discuss two types of pollutants (dry particulates and gaseous pollutants) and the various air pollution control equipment/processes available for their removal.

Removal of Dry Particulate Matter

Constituting a significant class of air pollutants, particulates have a variety of shapes and sizes. As either liquid droplets or dry dust, they have a wide range of physical and chemical characteristics. Dry particulates are emitted from various sources, including combustion and non-combustion sources in industry, mining, construction activities, incinerators, and internal combustion engines. Dry particulates are also emitted from natural sources—volcanoes, forest fires, pollen, and windstorms.

All particles and particulate matter exhibit specific essential characteristics, which, along with process conditions, must be considered in any engineering strategy to separate and remove them from a stream of carrier gas. Particulate size range and distribution, particle shape, corrosiveness, agglomeration tendencies, abrasiveness, toxicity, reactivity, inflammability, and hygroscopic tendencies must all be examined in light of equipment limitations.

In an air pollution control system, particulates are separated from the gas stream by applying one or more forces in gravity settlers, centrifugal settlers, fabric filters, electrostatic precipitators, or wet scrubbers. The particles are then collected and removed from the system.

As mentioned earlier in the text, when a flowing fluid (engineering and science applications consider both liquid and gaseous states as a fluid) approaches a stationary object such as a metal plate, a fabric thread, or a large water droplet, the fluid flow will diverge around that object. Particles in the fluid (because of inertia) will not follow stream flow strictly but will tend to continue in their original directions. If the particles have enough inertia and are located close enough to the stationary object, they will collide with the object and can be collected by it. This is an important phenomenon.

Particles are collected by impaction, interception, and diffusion. *Impaction* occurs when the center of mass of a particle diverging from the fluid strikes a stationary object. *Interception* occurs when the particle's center of mass closely misses the object, but because of its finite size, the particle strikes the object. *Diffusion* occurs when small particulates happen to "diffuse" toward the object while passing near it. Particles that strike the object by any means are collected—if short-range forces (chemical, electrostatic, and so forth) are strong enough to hold them to the surface (Cooper & Alley, 1990).

Control technologies for particles focus on capturing the particles emitted by a pollution source. Several factors must be considered before choosing a particulate control device. Typically, particles are collected and channeled through a duct or stack. The particulate exhaust stream's characteristics affect the control device's choice. These characteristics include the range of particle sizes, the exhaust flow rate, the temperature, the moisture content, and various chemical properties such as explosiveness, acidity, alkalinity, and flammability.

The most commonly used control devices for controlling particulate emissions include

gravity settlers, cyclones, electrostatic precipitators, wet (Venturi) scrubbers, and baghouse (fabric filters). In many cases, more than one of these devices is used in a series to obtain the desired removal efficiencies. For example, a settling chamber can remove larger particles before a pollutant stream enters an electrostatic precipitator. This section will briefly introduce each of the major types of particulate control equipment and point out their advantages and disadvantages.

Gravity Settlers

Gravity settlers (or settling chambers) have long been used by industry to remove solid and liquid waste materials from gaseous streams. Constructed, a gravity settler is nothing more than an enlarged chamber in which the horizontal gas velocity is reduced, allowing large particles to settle out of the gas by gravity and be recollected in hoppers. Gravity settlers have the advantage of having low initial cost and are relatively inexpensive to operate—there's not much to go wrong. However, because settling chambers effectively remove only larger particles, they have relatively low efficiency, especially for removing small particles (<50 μm). Thus, gravity settlers are used with a more efficient control device. In addition, although simple in design, gravity settlers require a large space for installation.

Cyclone Collectors

The cyclone (or centrifugal) collector provides a low-cost, low-maintenance method of removing larger particulates from a gas stream. The cyclone removes particles by inertia separation, causing the entire gas stream to flow in a spiral pattern inside a tube, and is the collector of choice for removing particles greater than 10 μm in diameter. By centrifugal force, the larger particles move outward and collide with the narrowing wall of the tube. The particles slide down the wall and fall to the bottom of the cone, where they are removed. The cleaned gas flows out the top of the cyclone. Along with their relatively low construction costs, cyclones need relatively small space requirements for installation. However, cyclones are efficient in removing large particles. Still, they are not as efficient with smaller particles, especially those below 10 μm, and they do not handle sticky materials well. For this reason, they are used with other particulate control devices. The most severe problems with cyclones are airflow equalization and their tendency to plug. Cyclones have been used successfully at feed and grain mills, cement plants, fertilizer plants, petroleum refineries, and other applications involving large quantities of gas containing relatively large particles.

Electrostatic Precipitators (ESPs)

An electrostatic precipitator (ESP) is a particle control device that uses electrical forces to move the particles out of the flowing gas stream and onto collector plates.

ESPs usually remove small particles from moving gas streams at high collection efficiencies. Widely used in power plants for removing fly ash from the gases before discharge, an electrostatic precipitator applies electrical force to separate particles from the gas stream. A high voltage drop is established between electrodes, and particles passing through the resulting electrical field acquire a charge. The charged particles are attracted to and collected on an oppositely charged plate, and the cleaned gas flows through the device. Periodically, the plates are cleaned by rapping to shake off the accumulated dust layer, and the dust is collected in hoppers at the bottom of the device. Although electrostatic precipitators have the advantages of low operating costs, capability for operating in high-temperature applications (up to 1,300°F), low-pressure drop—and high particulate (coarse and fine) collection efficiencies, they have the disadvantages of high capital costs and space requirements. The removal efficiencies for ESPs are highly variable; however, for tiny particles alone, the removal efficiency is about 99%. Typical ESP applications include industrial and utility boilers, cement plants, steel mills, petroleum refineries, municipal waste incinerators, hazardous waste incinerators, Kraft pulp and paper mills, and lead, zinc, and copper smelters.

Wet (Venturi) Scrubbers

Wet scrubbers (or collectors) have been widely used to clean contaminated gas streams (e.g., foundry dust emissions, acid mists, and furnace fumes) because of their ability to remove particulate and gaseous pollutants. Wet scrubbers vary in complexity, from simple spray chambers to remove coarse particles to high-efficiency systems (venturi types) to remove fine particles. Whichever system is used, the operation employs the same basic principles of inertial impingement or impaction and interception of dust particles by water droplets. The larger, heavier water droplets are easily separated from the gas by gravity. The solid particles can then be independently separated from the water, or the water can be otherwise treated before reuse or discharge. Increasing either the gas velocity or the liquid droplet velocity in a scrubber increases the efficiency because of the greater number of collisions per unit of time. The venturi scrubber is used for the ultimate wet scrubbing, where high collection efficiency is desired. The venturi operates at extremely high gas and liquid velocities with a high-pressure drop across the throat. The reduced velocity at the expanded section of the throat allows the droplets of water containing the particles to drop out of the gas stream. Venturi scrubbers are most efficient for removing particulate matter in the size range of 0.5–5 μm, with removal efficiencies of up to 99%, making them especially effective for removing sub-micron particulates associated with smoke and fumes.

Although wet scrubbers require relatively small space, can remove both gases and particles, can neutralize corrosive gases, have low capital cost, and can handle high-temperature, high-humidity gas streams, their power and maintenance costs are relatively high, they may create wastewater disposal problems, their corrosion problems are more severe than those of dry systems, and the final product they produce is collected wet. Table 8.20 summarizes these advantages and disadvantages. Wet scrubbers have been used in various industries such as acid plants, fertilizer plants, steel mills, asphalt plants, and larger power plants.

TABLE 8.20
Advantages and Disadvantages of Wet Scrubbers

Advantages	Disadvantages
Small space requirements	Corrosion problems
No secondary dust sources	High power requirements
Handles high-temperature, high-humidity gas streams	Water-disposal problems
Minimal fire and explosion hazards	Difficult product recovery
Ability to collect both gases and particles	Meteorological problems (Plume = fog)

Source: EPA (2006).

Baghouse (Fabric) Filters

Baghouse filters (or fabric filters) are the most commonly used air pollution control filtration system. In much the same manner as a standard vacuum cleaner, filter fabric material, capable of removing most particles as small as 0.5 μm and substantial quantities of particles as small as 0.1 μm, is formed into cylindrical or envelope bags and suspended in the baghouse. The particulate-laden gas stream is forced through the porous fabric filter, and as the air passes through the fabric, particulates accumulate on the cloth, providing a clean air stream. As particulates build up on the inside surfaces of the bags, the pressure drop increases. Before the pressure drop becomes too severe, the bags must be relieved of some particulate layer. The particulates are periodically removed from the cloth by shaking or reversing the airflow.

The selection of the fiber material and fabric construction is vital to baghouse performance. The fiber material from which the fabric is made must have adequate strength characteristics at the maximum gas temperature expected and proper chemical compatibility with both the gas and the collected dust. One disadvantage of the fabric filter is that high-temperature gases often must be cooled before contacting the filter medium.

Fabric filters are used in the power generation, incineration, chemical, steel, cement, food, pharmaceutical, metalworking, aggregate, and carbon black industries.

Fabric filters are relatively simple to operate, provide high overall collection efficiencies of up to 99+%, and are very effective in controlling sub-micrometer particles. However, they do have limitations. These include relatively high capital costs, high maintenance requirements (bag replacement, etc.), high space requirements, and flammability hazards for some dust.

Removal of Gaseous Pollutants: Stationary Sources

The principal gases of concern in removing gaseous air pollutants are sulfur oxides (SO_x), carbon oxides (CO_x), nitrogen oxides (NO_x), organic and inorganic acid gases, and hydrocarbons (HC). The most common method for controlling gaseous pollutants is the addition of add-on control devices to recover or destroy a pollutant. Four major treatment processes (add-ons) are currently available to prevent these and other gaseous emissions: absorption, adsorption, condensation, and combustion (incineration).

TABLE 8.21
Comparison of Air Pollution Control Technologies

Treatment Technology	Concentration & Efficiency	Comments
Incineration	(<100 ppmv) 90%–95% efficiency (>100 ppmv) 95%–99% efficient	Incomplete combustion may require additional controls
Carbon adsorption	(>200 ppmv) 90+% efficiency (>1,000 ppmv) 95+% efficiency	Recovered organics may need additional treatment—can increase the costs
Absorption	(<200 ppmv) 90%–95% efficiency (>200 ppmv) 95+% efficiency	Can the blowdown stream be accommodated at the site?
Condensation	(>2,000 ppmv 80+% efficiency	Must have low temperature or high pressure for efficiency

Note: Typically, only incineration and absorption technologies can consistently achieve greater than 99% gaseous pollutant removal.

Deciding which single or combined air pollution control technique for stationary sources to use is not always easy. A wide variety of devices can control gaseous pollutants, and choosing the most cost-effective, most efficient units requires careful attention to the particular operation for which the control devices are intended. Specifically, the choice of control technology depends on the pollutant(s) to be removed, the removal efficiency required, pollutant and gas stream characteristics, and specific characteristics of the site (EPA, 2006). Absorption, adsorption, and condensation are all recovery techniques, while incineration involves the destruction of the pollutant.

When making the complicated and often complex decision on which air pollution control technology to employ, it is helpful to follow guidelines based on experience set forth by Buonicore and Davis (1992) in the prestigious engineering text, the *Air Pollution Engineering Manual*. Table 8.21 summarizes these guidelines.

Absorption

Absorption (or scrubbing) is a significant chemical engineering unit operation that involves bringing contaminated effluent gas into contact with a liquid absorbent so that one or more constituents of the flowing gas are selectively dissolved into a relatively nonvolatile liquid. Absorption units are designed to transfer the pollutant from a gas to a liquid phase (water is the most commonly used absorbent liquid). The absorption unit accomplishes this by providing intimate contact between the gas and the liquid, providing optimum gas diffusion into the solution. The actual removal of a pollutant from the gas stream takes place in three steps: (1) diffusion of the pollutant gas to the surface of the liquid; (2) transfer across the gas/liquid interface; and (3) diffusion of the dissolved gas away from the interface into the liquid (Davis & Cornwell, 1991). Absorption is commonly used to recover products or purify gas streams with high concentrations of organic compounds. Absorption equipment is designed to achieve as much mixing between the gas and liquid as possible.

Several types of absorbers are available, including spray chambers (and towers or columns), plate or tray towers, packed towers, and venturi scrubbers. Pollutant gases commonly controlled by absorption include sulfur dioxide, hydrogen sulfide, hydrogen chloride, chlorine, ammonia, and oxides of nitrogen.

Absorbers are often referred to as scrubbers, and various types of absorption equipment exist. The principal types of gas absorption equipment include spray towers, packed columns, spray chambers, and venturi scrubbers. The plate and packed tower systems are today's two most common absorbent units. Plate towers contain perforated horizontal plates or trays designed to provide large liquid-gas interfacial areas. The polluted air stream is usually introduced at one side of the bottom of the tower or column and rises through the perforations in each plate; the rising gas prevents the liquid from draining through the openings rather than through a downpipe. During continuous operation, contact is maintained between air and fluid, allowing gaseous contaminants to be removed, with clean air emerging from the top of the tower.

The packed tower scrubbing system is by far the most commonly used to control gaseous pollutants in industrial applications, where it typically demonstrates a removal efficiency of 90%–95%. Usually configured vertically, the packed tower is literally "packed" with devices having a large surface-to-volume ratio and a large void ratio, offering minimum resistance to gas flow. In addition, packing should provide even distribution of both fluid phases, be sturdy enough to support them in the tower, and be cheap, available, and easily handled (Hesketh, 1991).

The flow through a packed tower is typically countercurrent, with gas entering at the bottom and liquid entering at the top. The fluid flows over the surface of the packing in a thin film, affording continuous contact with the gases.

Though highly efficient for removing gaseous contaminants, packed towers may create wastewater disposal problems (converting an air pollution problem into a water pollution problem), become easily clogged when gases with high particulate loads are introduced, and have relatively high maintenance costs.

Adsorption

Adsorption is a mass transfer process that involves passing a stream of effluent gas through the surface of prepared porous solids (adsorbents). The surfaces of the porous solid substance attract and hold (bind) the gas (the adsorbate) by either physical or chemical adsorption. In physical adsorption (a readily reversible process), a gas molecule adheres to the surface of the solid because of an imbalance in electron distribution. In chemical adsorption (not readily reversible), once the gas molecule adheres to the surface, it reacts chemically with it.

Several materials possess adsorptive properties. These materials include activated carbon, alumina, bone char, magnesia, silica gel, molecular sieves, and strontium sulfate. Activated charcoal is the most important adsorbent for air pollution control. Activated carbon is the universal standard for purifying and removing trace organic contaminants from liquid and vapor streams. The surface area of activated charcoal will preferentially adsorb hydrocarbon vapors and odorous organic compounds from an airstream.

In an adsorption system (in contrast to the absorption system where the collected contaminant is continuously removed by flowing liquid), the collected contaminant remains in the adsorption bed. The most common adsorption system is the fixed bed adsorber, which can be contained in a vertical or horizontal cylindrical shell. The adsorbent (usually activated carbon) is arranged in beds or trays in layers about 0.5 inch thick. One or more beds absorb vapors in multiple-bed systems while the other regenerates them.

Most adsorbers' efficiency is near 100% at the beginning of the operation and remains high until a breakpoint or breakthrough occurs. When the adsorbent becomes saturated with adsorbate, contaminants leak out of the bed, signaling that the adsorbent should be renewed or regenerated. The identical activated carbon particles can be used repeatedly by regenerating the carbon bed.

Although adsorption systems are high-efficiency devices that may allow recovery of product, have excellent control and response to process changes, and have the capability of being operated unattended, they also have some disadvantages, including the need for exotic, expensive extraction schemes if product recovery is required, relatively high capital cost, and gas stream prefiltering needs (to remove any particulates capable of plugging the adsorbent bed).

Condensation

Condensation is when volatile gases are removed from the contaminant stream and changed into a liquid. In air pollution control, a condenser can be used in two ways: pretreatment to reduce the load on other air pollution control equipment or effectively controlling contaminants in the form of gases and vapors.

Condensers condense vapors to the liquid phase by either increasing the system pressure without a change in temperature or decreasing the system temperature to its saturation temperature without a pressure change. The composition of the contaminant gas stream affects condensation. When gases condense under different conditions, condensation is hindered.

Condensers are widely used to recover valuable products in a waste stream. Condensers are simple, inexpensive devices that use water or air to cool and condense a vapor stream. Condensers are typically utilized as pretreatment devices. They can be used ahead of adsorbers, absorbers, and incinerators to reduce the total gas volume, which can be treated by more expensive control equipment.

Two basic types of condensation equipment are used for pollution control—surface and contact condensers. A surface condenser is normally a shell-and-tube heat exchanger. A surface condenser uses a cooling medium of air or water where a metal wall separates the vapor to be condensed from the cooling medium. Coolant flows through the tubes while the vapor is passed over and condenses on the outside of the tubes and drains off to storage (EPA, 2006).

In a contact condenser (which resembles a simple spray scrubber), the vapor is cooled by spraying liquid directly on the vapor stream. The cooled vapor condenses, and the water and condensate mixture are removed, treated, and disposed of.

Contact condensers are generally less expensive, more flexible, and more straightforward than surface condensers. Still, surface condensers require much less water and produce much less wastewater that must be treated than contact condensers. Depending on design and applications, removal efficiencies of condensers typically range from 50% to more than 95%. Condensers are used in various industrial applications, including petroleum refining, petrochemical manufacturing, basic chemical manufacturing, dry cleaning, and degreasing.

Combustion (Incineration)

Even though combustion (or incineration) is a significant source of air pollution, it is also, if properly operated, a beneficial air pollution control system in which the objective is to convert specific air contaminants (usually CO and hydrocarbons) to innocuous substances such as carbon dioxide and water (EPA, 2006).

Combustion is a chemical process defined as rapid, high-temperature gas-phase oxidation. The combustion equipment that controls air pollution emissions is designed to push these oxidation reactions as close to complete combustion as possible, leaving minimal unburned residue. Any combustion operation is governed by four variables: oxygen, temperature, turbulence, and time. For complete combustion to occur, oxygen must be available, put into contact with sufficient temperature (turbulence), and held at this temperature for an adequate time. These four variables are not independent—changing one affects the entire process.

Depending on the contaminant being oxidized, equipment used to control waste gases by combustion can be divided into three categories: direct-flame combustion (or flaring), thermal combustion (afterburners), or catalytic combustion. Choosing the proper device depends on many factors, including the type of hazardous contaminants in the waste stream, the concentration of combustibles in the stream, the process flow rate, control requirements, and an economic evaluation.

Direct-Flame Combustion (Flaring)

Direct-flame combustion devices (flares) are the most commonly used air pollution control devices by which waste gases are burned directly (with or without the addition of a supplementary fuel). Standard flares include steam-assisted, air-assisted, and pressure head types. EPA studies have shown that a flare's destruction efficiency is about 98%. Flares are typically elevated from 100 to 400 ft to protect the surroundings from heat and flames. Often designed for steam injection at the flare top, steam is used in this application because it provides sufficient turbulence to ensure complete combustion, preventing visible smoke or soot production. Flares are also noisy, which can cause problems for adjacent neighborhoods, and some flares produce nitrogen oxides, thus creating a new air pollutant.

Thermal Combustion (Afterburners)

The thermal incinerator or afterburner is usually the unit of choice in cases where the concentration of combustible gaseous pollutants is too low to make flaring practical. Widely used in industry, the thermal combustion system typically operates at high temperatures. Within the thermal incinerator, the contaminant airstream passes around or through a burner and into a refractory-lined residence chamber where oxidation occurs. Residence time is the time the fuel mixture remains in the combustion chamber. Flue gas from a thermal incinerator (which is relatively clean) is at high temperature and contains recoverable heat energy. When operated correctly, thermal incinerators can destroy gaseous pollutants at efficiencies greater than 99%.

Catalytic Combustion

Catalytic combustion passes a preheated contaminant-laden gas stream through a catalyst bed (usually a thinly-coated platinum mesh mat, a honeycomb, or other configurations designed to increase surface area), promoting the oxidization reaction at lower temperatures. The metal catalyst is used to initiate and encourage combustion at much lower temperatures than those required for thermal combustion (metals in the platinum family are recognized for their ability to promote combustion at low temperatures). Catalytic incineration may require 20–50 times less residence time than thermal incineration (see Table 8.22 for other advantages of catalytic incinerators over thermal incinerators). Catalytic incinerators typically operate at 700°F–900°F. At this reduced temperature range, fuel usage and cost savings are realized; however, this may be offset by the cost of the catalytic incinerator itself. Destruction efficiencies greater than 95% are possible using a catalytic incinerator. Higher efficiencies are possible if larger catalyst volumes or higher temperatures are used.

A heat exchanger is an option for systems with heat transfer between two gas streams (recuperative heat exchange). The need for dilution air, combustion air, and flue gas treatment is based on site-specific conditions. Catalysts are subject to physical and chemical deterioration, and sulfur-containing compounds suppress their usefulness. For best performance, catalyst surfaces must be clean and active.

Catalytic incineration is used in various industries to treat effluent gases, including emissions from paint and enamel bake ovens, asphalt oxidation, coke ovens, formaldehyde manufacturing, and varnish cooking. Catalytic incinerators are best suited for emission streams with low VOC content.

TABLE 8.22
Advantages of Catalytic Over Thermal Incinerators

1.	Catalytic incinerators have lower fuel requirements
2.	Catalytic incinerators have lower operating temperatures
3.	Catalytic incinerators have little or no insulation requirements
4.	Catalytic incinerators have reduced fire hazards
5.	Catalytic incinerators have reduced flashback problems

Source: Adaptation from Buonicore & Davis, *Air Pollution Engineering Manual,* (1992).

INDOOR AIR QUALITY

Various factors influence the quality of the air we breathe and its attendant consequences for human health. These include hazardous material discharges indoors and outdoors, meteorological and ventilation conditions, and pollutant decay and removal processes. Over 80% of our time is spent in indoor environments, so the influence of building structures, surfaces, and ventilation is an important consideration when evaluating air pollution exposures (Wadden & Scheff, p. 1, 1983).

For those familiar with *Star Trek, Trekkies* (and for those who are not), consider a quotable quote: "The air is the air." However, regarding the air we breathe, according to USEPA (2001), few of us realize that we all face a variety of risks to our health as we go about our day-to-day lives. Driving our cars, flying in planes, engaging in recreational activities, and being exposed to environmental pollutants pose varying risks. Some risks are simply unavoidable. Some we choose to accept because doing otherwise would restrict our ability to lead our lives the way we want. We might decide to avoid some risks if we had the opportunity to make informed choices. Indoor air pollution is one risk that we can act to address.

In the last several years, scientific evidence has indicated that the air within homes and other buildings can be more seriously polluted than the outdoor air in even the largest and most industrialized cities. Other research suggests that people spend approximately 90% of their time indoors. A type of climate we don't often think about (if at all) is the micro-climates we spend 80% of our time in the office and the home (indoors) (Wadden & Scheff, 1983). Thus, the health risks may be more significant for many people due to exposure to indoor air pollution than outdoors (USEPA, 2001).

Not much attention was given to indoor microclimates until a few years ago after two events took place: the first involved Legionnaires' disease, and the second involved Sick Building Syndrome. In addition, people who may be exposed to indoor air pollutants for the longest periods are often those most susceptible to the effects of indoor air pollution. Such groups include the young, the elderly, and the chronically ill, especially those suffering from respiratory or cardiovascular disease.

The impact of energy conservation on indoor environments may be substantial, particularly concerning decreases in ventilation rates (Hollowell et al., 1979a) and "tight" buildings constructed to minimize infiltration of outdoor air (Woods, 1980; Hollowell et al., 1979b). The purpose of creating "tight buildings" is to save energy—to keep the heat or air conditioning inside the structure. The problem is that indoor air contaminants within these tight structures are trapped and can be concentrated, exposing inhabitants to even more exposure.

These topics and others, along with causal factors leading to indoor air pollution, are covered in this section. What about indoor air quality problems in the workplace? In this section, we also discuss this pervasive but often overlooked issue. In this regard, we discuss the basics of IAQ (related to the workplace environment) and the major contaminants that currently contribute to this problem. Moreover, mold and mold remediation, although not new to the workplace, are the latest buzzwords attracting attention. Contaminants such as asbestos, silica, lead, and formaldehyde are also discussed. Various related remediation practices are also covered.

Legionnaires' Disease

Since the infamous event that occurred in Philadelphia in 1976 at the Bellevue-Stratford Hotel during a convention of American Legion members, which included 182 cases and 29 deaths, *Legionella pneumophila* (the deadly bacterium) has become synonymous with the term *Legionnaires' disease*. The deaths were attributed to colonized bacteria in the air conditioning system cooling tower.

Let's take a look at this deadly killer—a killer that inhabits the micro-climate we call office, hotel, and other indoor spaces.

Organisms of the genus *Legionella* are ubiquitous in the environment. They are found in natural fresh water and potable water, as well as in closed-circuit systems such as evaporative condensers, humidifiers, recreational whirlpools, air handling systems, and, of course, cooling tower water.

Legionella bacteria's potential presence is dependent on certain environmental factors: moisture, temperature (50°F–140°F), oxygen, and a source of nourishment such as slime or algae.

We do not know how Legionnaires' disease can be spread at this time; however, we do know that it can be spread through the air. The Centers for Disease Control (CDC) states (in its Questions and Answers on Legionnaires' Disease, CDC No. 28L0343779) that there is no evidence that it is spread person-to-person.

Air-conditioning cooling towers and evaporative condensers have been the source of most outbreaks to date, and the bacterium is commonly found in both. Unfortunately, we do not know if this is an essential means of spreading Legionnaires' disease because other outbreaks have occurred in buildings that did not have air conditioning.

Not all people are at risk of contracting Legionnaires' disease. The people most at risk include those:

1. with lowered immunological capacity
2. who smoke cigarettes and abuse alcohol
3. who are exposed to high concentrations of *Legionella pneumophila*

Most commonly recognized as a form of pneumonia, the symptoms of Legionnaires' disease usually become apparent 2–10 days after known or presumed exposure to airborne Legionnaires' disease bacteria. A sputum-free cough is common, but sputum production is sometimes associated with the disease. Within less than a day, the victim can experience a rapidly rising fever and the onset of chills. Mental confusion, chest pain, abdominal pain, impaired kidney function, and diarrhea are associated manifestations of the disease. The CDC estimates that around 25,000 people develop Legionnaires' disease annually.

The obvious question becomes: How do we prevent or control Legionnaires' disease? Good question.

The controls presently being used are targeted at cooling towers and air handling units (condensate drain pans).

Cooling tower procedures used to control bacterial growth vary somewhat in the various regions of a cooling tower system. However, control procedures usually include a good maintenance program, including repair/replacement of damaged components, routine cleaning, and sterilization.

A typical sterilization protocol calls for using chlorine in a residual solution at about 50 ppm combined with a compatible detergent to produce the desired sterilization effect. Ensuring that even those inaccessible spaces are adequately cleaned of slime and algae accumulations is essential.

Control measures for air handling units: Condensate drain pans typically require cleaning and checking for proper fluid drainage. This is important to prevent stagnation and the buildup of slime, algae, and bacteria. A cleaning and sterilization program is required anytime algae or slime are found in the unit.

Sick Building Syndrome

The second event that got the public's attention regarding micro-climates and the possibility of unhealthy environments contained therein was spawned by the first, the Legionnaires' event, and other incidents or complaints that followed. What we are referring to here is *Sick Building Syndrome.*

Sick Building Syndrome was coined by an international working group under the *World Health Organization* (WHO) in 1982. The WHO working group studied the literature about indoor climate problems and found that these micro-climates in buildings are characterized by the same frequently appearing complaints and symptoms. WHO identified five categories of symptoms, exemplified by some complaints reported by occupants who were supposed to suffer from sick building syndrome (SBS). These categories are listed as follows:

1. **Sensory Irritation in Eyes, Nose, and Throat**: Pain, sensation of dryness, smarting feeling, stinging, irritation, hoarseness, and voice problems.
2. **Neurological or General Health Symptoms**: Headache, sluggishness, mental fatigue, reduced memory, reduced capability to concentrate, dizziness, intoxication, nausea and vomiting, tiredness.
3. **Skin Irritation**: Pain, reddening, smarting or itching sensations, dry skin.
4. **Nonspecific Hypersensitivity Reactions**: Running nose and eyes, asthma-like symptoms among non-asthmatics, sounds from the respiratory system.
5. **Odor and Taste Symptoms**: Changed sensitivity of olfactory or gustatory sense, unpleasant olfactory or gustatory perceptions.

In the past, similar symptoms had been used to define other syndromes such as the building disease, building illness syndrome, building-related illness, or tight-fitting office syndrome, which in many cases appear to be synonyms for the sick building syndrome; thus, the WHO definition of SBS worked to combine these syndromes into one general definition or summary. A summary compiled by WHO (1982, 1984) and Molhave (1986) of this combined definition includes the five categories of symptoms listed earlier and also:

1. Irritation of mucous membranes in the eye, nose, and throat is among the most frequent symptoms.
2. Other symptoms, e.g., from lower airways or internal organs, should be infrequent.
3. A large majority of occupants report symptoms.
4. The symptoms appear especially frequently in one building or part of it.
5. No evident causality about exposures or occupant sensitivity can be identified.

The WHO group suggested the possibility that the SBS symptoms have a common causality and mechanism (WHO, 1982). However, the existence of SBS is still a postulate because the descriptions of the symptoms in the literature are anecdotal and unsystematic (Molhave, 1992).

Why is indoor air pollution a problem? As indicated above, recognizing that the indoor air environment may be a health problem is a relatively recent emergence. The most significant indoor air quality is the impact of cigarette smoking, stove and oven operation, and emanations from certain types of particleboard, cement, and other building materials (Wadden & Scheff, 1983).

The significance of the indoor air quality problem became apparent not only because of the Legionnaires' Incident of 1976 and the WHO study of 1982 but also because of another factor that came to the forefront in the mid-1970s: the need to conserve energy. In the early 1970s, when hundreds of thousands of people were standing in line to obtain gasoline for their automobiles, it was not difficult to drive home the need to conserve energy supplies.

The resulting impact of energy conservation on indoor environments has been substantial. This is especially the case for building modifications that decreased ventilation rates and new construction practices that were incorporated to ensure "tight" buildings to minimize outdoor air infiltration.

There is some irony in this development, of course. While proper building design, construction, and ventilation guidelines are necessary to avoid exposing inhabitants to unhealthy environments, this dash to reduce ventilation rates and "tighten" buildings from infiltration was a tradeoff: energy economics versus air quality.

According to Byrd (2003), Indoor Air Quality (IAQ) refers to the effect, good or bad, of the contents of the air inside a structure on its occupants. Stated differently, in this text, indoor air quality (IAQ) refers to the air quality inside workplaces as represented by concentrations of pollutants and thermal (temperature and relative humidity) conditions that affect employees' health, comfort, and performance. Usually, temperature (too hot or too cold), humidity (too dry or too damp), and air velocity (draftiness or motionlessness) are considered "comfort" rather than indoor air quality issues. They may make someone uncomfortable but won't make someone ill unless extreme. Other factors affecting employees, such as light and noise, are essential indoor environmental quality considerations but not core elements of indoor air quality problems. Nevertheless, most industrial hygienists must consider these factors when investigating environmental quality situations.

Byrd (2003) further points out that "good IAQ is the quality of air, which has no unwanted gases or particles at concentrations, which will adversely affect someone. Poor IAQ occurs when gases or particles are present at an excessive concentration, affecting the satisfaction and health of occupants."

Poor IAQ in the workplace may only be annoying to one person; however, at the extreme, it could be fatal to all the occupants.

The concentration of the contaminant is crucial. Potentially infectious, toxic, allergenic, or irritating substances are always in the air. Note that there is nearly always a threshold level below which no effect occurs.

Common Indoor Air Pollutants in the Home

This section takes a brief, source-by-source look at the most common indoor air pollutants, their potential health effects, and ways to reduce their levels in the home.

Radon

Radon is a noble, nontoxic, colorless, odorless gas produced in the decay of radium-226 and is found everywhere at deficient levels. Radon is ubiquitous in the soil and air near the Earth's surface. As radon undergoes radioactive decay, it releases an alpha particle, gamma ray, and progeny that quickly decay to release alpha and beta particles and gamma rays. Because radon progeny are electrically charged, they readily attach to particles, producing a radioactive aerosol. It is when radon becomes trapped in buildings and concentrations build up in indoor air that radon exposure becomes a concern. This is the case because aerosol radon-contaminated particles may be inhaled and deposited in the bifurcations of respiratory airways. Irradiation of tissue at these sites poses a significant risk of lung cancer (depending on exposure dose).

How does radon enter a house?

The most common way radon enters a house is through the soil or rock upon which the house is built. The most common source of indoor radon is uranium, which is expected in many soils and rocks. As uranium breaks down, it releases soil or radon gas, and radon gas breaks down into radon decay products or progeny (commonly called *radon daughters*). Radon gas is transported into buildings by pressure-induced convective flows.

There are other sources of radon, such as water and masonry materials.

Radon levels in a house vary in response to temperature-dependent and wind-dependent pressure differentials and to changes in barometric pressures. When the base of a home is under significant negative pressure, radon transport is enhanced.

Studies by the USEPA (1987, 1988) indicate that as many as 10% of all American homes, or about 9 million homes, may have elevated levels of radon, and the percentage may be higher in geographic areas with certain soils and bedrock formations.

According to the USEPA's booklets, *A Citizen's Guide to Radon*, *Radon Reduction Methods: A Homeowner's Guide*, and *Radon Measurement Proficiency Report* (for each state), exposure to radon in the home can be reduced by the following steps:

1. Measure levels of radon in the home.
2. The state radiation protection office can provide information on the availability of detection devices or services.

3. Refer to EPA guidelines when deciding whether and how quickly to act based on test results.
4. Learn about control measures.
5. Avoid drawing more significant amounts of radon into the house.
6. Select a qualified contractor to draw up and implement a radon mitigation plan.
7. Stop smoking and discourage smoking in your home.
8. Treat radon-contaminated well water by aerating or filtering through granulated activated charcoal.

Environmental Tobacco Smoke

The use of tobacco products by approximately 45 million smokers in the United States results in significant indoor contamination from combustion by-products that pose significant exposures to millions of others who do not smoke but must breathe contaminated indoor air. Composed of side-stream smoke (smoke that comes from the burning end of a cigarette) and smoke exhaled by the smoker, it contains a complex mixture of over 4,700 compounds, including gases and particles.

According to reports issued in 1986 by the Surgeon General and the National Academy of Sciences, environmental tobacco smoke is a cause of disease, including lung cancer, in both smokers and healthy nonsmokers. Environmental tobacco smoke may also increase the lung cancer risk associated with radon exposure.

The following steps can reduce exposure to environmental tobacco smoke in the office and home:

1. Give up smoking and discourage smoking in your home and place of work or require smokers to smoke outdoors.
2. A standard method of reducing exposure to indoor air pollutants such as environmental tobacco smoke is ventilation, which reduces but does not eliminate exposure.

Biological Contaminants

A variety of biological contaminants can cause significant illness and health risks. These include mold and mildew, viruses, animal dander and cat saliva, mites, cockroaches, pollen, and infections from airborne exposures to viruses that cause colds and influenza and bacteria that cause Legionnaires' disease and tuberculosis (TB).

The following steps can reduce exposure to biological contaminants in the home and office:

1. Install and use exhaust fans vented to the outdoors in kitchens, bathrooms, and vent clothes dryers outdoors.
2. Ventilate the attic and crawl spaces to prevent moisture buildup.
3. Keep water trays in cool mist or ultrasonic humidifiers clean and filled with fresh distilled water daily.
4. Water-damaged carpets and building materials should be thoroughly dried and cleaned within 24 hours.
5. Maintain good housekeeping practices both in the home and office.

Combustion By-Products

Combustion byproducts are released into indoor air from various sources, including unvented kerosene and gas space heaters, woodstoves, fireplaces, gas stoves, and water heaters. The major pollutants released from these sources are carbon monoxide, nitrogen dioxide, and particles.

The following steps can reduce exposure to combustion products in the home (and office):

1. Fuel-burning unvented space heaters should only be operated using great care and special safety precautions.
2. Install and use exhaust fans over gas cooking stoves and ranges, and adjust the burners properly.
3. Furnaces, flues, and chimneys should be inspected annually, and any needed repairs should be made promptly.
4. Woodstove emissions should be kept to a minimum.

Household Products

Various organic compounds are widely used in household products because of their useful characteristics, such as their ability to dissolve substances and evaporate quickly. Cleaning, disinfecting, cosmetic, degreasing, and hobby products all contain organic solvents, as do paints, varnishes, and waxes. All of these products can release organic compounds while being used and when they are stored.

The following steps can reduce exposure to household organic compounds:

1. Always follow label instructions carefully.
2. Throw away partially full containers of chemicals safely.
3. Limit the amount you buy.

Pesticides

Pesticides represent a case of chemical contamination of buildings where the EPA estimates 80%–90% of most people's exposure to the air occurs. These products are hazardous if not used properly.

The following steps can reduce exposure to pesticides in the home:

1. Read the label and follow directions.
2. Use pesticides only in well-ventilated areas.
3. Dispose of unwanted pesticides safely.

Asbestos in the Home

Asbestos became a primary indoor air-quality concern in the U.S. in the late 1970s. Asbestos is a mineral fiber commonly used in a variety of building materials and has been identified as having the potential (when friable) to cause cancer in humans.

The following steps can reduce exposure to asbestos in the home (and/or office):

1. Do not cut, rip, or sand asbestos-containing materials.
2. Use a professional, trained contractor when removing or cleaning up asbestos.

Building Factors Affecting Indoor Air Quality

Building factors affecting indoor air quality can be grouped into two factors: factors affecting indoor climate and factors affecting indoor air pollution.

- **Factors Affecting Indoor Climate**: For several reasons, the thermal environment (temperature, relative humidity, and airflow) is an essential dimension of indoor air quality. First, many complaints of poor indoor air may be resolved by altering the temperature or relative humidity. Second, thermally uncomfortable people will be less tolerant of other building discomforts. Third, the rate at which chemicals are released from building materials is usually higher at higher building temperatures. Thus, if occupants are too warm, it is also likely that they are being exposed to higher pollutant levels.
- **Factors Affecting Indoor Air Pollution**: Much of the building fabric, furnishings, equipment, occupants, and activities produce pollution. In a well-functioning building, some of these pollutants will be directly exhausted to the outdoors, and some will be removed as outdoor air enters the building and replaces the air inside. The air outside may also contain contaminants brought inside in this process. This air exchange is brought about by the mechanical introduction of outdoor air (outdoor air ventilation rate), the mechanical exhaust of indoor air, and the air exchanged through the building envelope (infiltration and exfiltration).

Pollutants inside can travel through the building as air flows from areas of higher atmospheric pressure to areas of lower atmospheric pressure. Some of these pathways are planned and deliberate to draw pollutants away from occupants, but problems arise when unintended flows draw contaminants into occupied areas. In addition, some pollutants may be removed from the air through natural processes, such as the adsorption of chemicals by surfaces or the settling of particles onto surfaces. Removal processes may also be deliberately incorporated into the building systems. Air filtration devices, for example, are commonly incorporated into building ventilation systems.

Thus, the factors most important for understanding indoor pollution are (a) indoor sources of pollution, (b) outdoor sources of pollution, (c) ventilation parameters, (d) airflow patterns and pressure relationships, and (e) air filtration systems.

Types of Workplace Air Pollutants

Table 8.23 provides common pollutants or pollutant classes of concern in commercial buildings and common sources of these pollutants.

Sources of Workplace Air Pollutants

Air quality is affected by various types of contaminants in the air. Some are in the form of gases. These would be generally classified as toxic chemicals. The types of interest are combustion products (carbon monoxide, nitrogen dioxide), volatile organic compounds (formaldehyde, solvents, perfumes, fragrances, etc.), and semi-volatile organic compounds (pesticides). Other pollutants include animal dander, soot;

TABLE 8.23
Indoor Pollutants and Potential Sources

Pollutant or Pollutant Class	Potential Sources
Environmental tobacco smoke	Lighted cigarettes, cigars, pipes
Combustion contaminants	In furnaces, generators, gas or kerosene space heaters, tobacco products, outdoor air, and vehicles
Biological contaminants	Wet materials, cooling towers, humidifiers, cooling coils or drain pans, damp duct insulation or filters, condensation, re-entrained sanitary exhausts, bird droppings, cockroaches or rodents, dust mites on upholstered furniture or carpeting, body odors
Volatile organic compounds (VOCs)	Paints, stains, varnishes, solvents, pesticides, adhesives, wood preservatives, waxes, polishes, cleansers, lubricants, sealants, dyes, air fresheners, fuels, plastics, copy machines, printers, tobacco products, perfumes, dry-cleaned clothing
Formaldehyde	Particle board, plywood, cabinetry, furniture, fabrics
Soil gases (radon, sewer gas, VOCs,	Soil, rock (radon), sewer drain leak, dry drain methane) traps, leaking underground storage tanks, landfill
Pesticides	Termiticides, insecticides, rodenticides, fungicides, disinfectants, and herbicides.

particles from buildings, furnishings, and occupants, such as fiberglass, gypsum powder, paper dust, lint from clothing, carpet fibers, etc.; and dirt (sandy and earthy material), etc.

Burge and Hoyer (1998) point out many specific sources of contaminants that result in adverse health effects in the workplace, including workers (contagious diseases, carriage of allergens, and other agents on clothing); building compounds (VOCs, particles, fibers); contamination of building components (allergens, microbial agents, pesticides); and outdoor air (microorganisms, allergens, and chemical air pollutants).

When workers complain of IAQ problems, the industrial hygienist is called upon to determine if the problem is an IAQ issues. If he or she determines that some form of contaminant is present in the workplace, proper remedial action is required. This usually includes removing the source of the contamination.

Tables 8.24 and 8.25 identify the indoor and outdoor sources (respectively) of contaminants commonly found in the workplace and offer some measures for controlling them.

Table 8.25 identifies common sources of contaminants introduced from outside buildings. These contaminants frequently find their way inside through the building shell, openings, or other pathways.

TABLE 8.24
Indoor Sources of Contaminants

Category/Common Sources	Mitigation and Control
Housekeeping and Maintenance	
• Cleanser	• Use low-emitting products
• Waxes and polishes	• Avoid aerosols and sprays
• Disinfectants	• Dilute to proper strength
• Air fresheners	• Do not overuse; use during unoccupied hours
• Adhesives	
• Janitor's/storage closets	• Use proper protocol when diluting and mixing
• Wet mops	
• Drain cleaners	• Store properly with containers closed and lid tight
• Vacuuming	
• Paints and coatings	• Use exhaust ventilation for storage spaces (eliminate return air)
• Solvents	
• Pesticides	
• Lubricants	• Clean mops, store mop top-up to dry
	• Avoid "air fresheners"— clean and exhaust instead
	• Use high-efficiency vacuum bags/filters
	• Use integrated pest management
Occupant-Related Sources	
• Tobacco products	• Smoking policy
• Office equipment (printers/copiers)	• Use exhaust ventilation with pressure control for major from local sources
• Cooking/microwave	
• Art supplies	
• Marking pens	• Low emitting art supplies/marking pens
• Paper products	
• Personal products (e.g., perfume)	• Avoid paper clutter
• Tracked in dirt/pollen	• Education material for occupants and staff
Building Uses as Major Sources	
• Print/photocopy shop	• Use exhaust ventilation and pressure control
• Dry cleaning	
• Science laboratory	• Use exhaust hoods where appropriate; check hood airflows
• Medical office	
• Hair/nail salon	
• Cafeteria	
• Pet store	
Building-Related Sources	
• Plywood/compressed wood	• Use low-emitting sources
• Construction adhesives	• Air out in the open/ventilated area before installing

(*Continued*)

TABLE 8.24 (*Continued*)
Indoor Sources of Contaminants

Category/Common Sources	Mitigation and Control
• Asbestos products • Insulation • Wall/floor coverings (vinyl/plastic) • Carpets/carpet adhesives • Wet building products • Transformers • Upholstered furniture • Renovation/remodeling	• Increase ventilation rates during and after installing • Keep material dry prior to enclosing
HVAC System	
• Contaminated filters • Contaminated duct lining • Dirty drain pans • Humidifiers	• Perform HVAC preventive maintenance • Change filter • Clean drain pans; proper slope and drainage
• Lubricants • Refrigerants • Mechanical room • Maintenance activities • Combustion appliances boilers/furnaces/stoves/generators	• Use portable water for humidification • Keep duct lining dry; move lining outside of the duct if possible • Fix leaks/clean spills • Maintain a spotless mechanical room (not a storage area) • Avoid back-drafting • Check/maintain flues from the boiler to outside • Keep combustion appliances properly tuned • Disallow unvented combustion appliances • Perform polluting activities during unoccupied hours
Moisture	
• Mold	• Keep building dry
Vehicles	
• Underground/attached garage	• Use exhaust ventilation • Maintain the garage under negative pressure relative to the building • Check airflow patterns frequently • Monitor CO

Indoor Air Contaminant Transport

Air contaminants reach worker breathing zones by traveling from the source to the worker through various pathways. Usually, the contaminants travel with the flow of air, which moves from areas of high pressure to areas of low pressure. That is why controlling workplace air pressure prevents pollution and enhances building IAQ performance.

TABLE 8.25
Outdoor Sources of Contaminants

Category/Common Sources	Mitigation and Control
Ambient Outdoor Air	
• Air quality in the general area	• Filtration or air cleaning of intake air
Vehicular Sources	
• Local vehicular traffic	• Locate air intake away from the source
• Vehicle idling areas	• Require engines shot off at loading dock
• Loading dock	• Pressurize building/zone
	• Add vestibules/sealed doors near the source
Commercial/Manufacturing Sources	
• Laundry or dry cleaning	• Locate air intake away from the source
• Paint shop	
• Restaurant	• Pressurize building relative to outdoors
• Photo-processing	
• Automotive shop/gas station	• Consider air cleaning options for outdoor air intake
• Electronics manufacture/assembly	
• Various industrial operations	• Use landscaping to block or Redirect flow of contaminants
Utilities/Public Works	
• Utility power plant	
• Incinerator	
• Water treatment plant	
Agricultural	
• Pesticide spraying	
• Processing or packing plants	
• Ponds	
Construction/Demolitions	
	• Pressurize building
	• Use walk-off mats
Building Exhaust	
• Bathroom exhaust	• Separate exhaust or relief from air intake
• Restaurant exhaust	• Pressurize building
• Air handler relief vent	
• Exhaust from the major tenant (e.g., dry cleaner)	
Water Sources	
• Pools of water on the roof	• Proper roof drainage
• Cooling tower mist	• Separate air intake from the source of water
	• Treat and maintain cooling tower water

(Continued)

TABLE 8.25 (*Continued*)
Outdoor Sources of Contaminants

Category/Common Sources	Mitigation and Control
Birds and Rodents	
• Fecal contaminants	• Bird-proof intake grills
• Bird nesting	• Consider vertical grills
Building Operations and Maintenance	
• Trash and refuse area	• Separate source from air intake
• Chemical/fertilizer/groundskeeping storage	• Keep source area clean/lids on tight
• Painting/roofing/sanding	• Isolate storage area from occupied areas
Ground Sources	
• Soil gas	• Depressurize soil
• Sewer gas	• Seal foundation and penetrations to foundations
• Underground fuel storage tank	• Keep air ducts away from ground sources

Air movements should flow from occupants toward a source and then out of the building rather than from the source to the occupants and out of the building. Pressure differences will control the direction of air motion and the extent of occupant exposure.

Indoor Air Dispersion Parameters

Several parameters (some characterizing observed patterns of contaminant distribution and others characterizing flow features, such as stability) describe the dispersion of a contaminant inside a room.

Parameters

- **Contaminant Concentration**: An indicator of contaminant distribution in a room, i.e., the contaminant mass per unit air volume (measured in kg/m^3).
- **Local Mean Age of Air**: The average time air travels from the inlet to any point P in the room (Di Tommaso et al., 1999).
- **Purging Effectiveness of Inlets**: A quantity that can be used to identify each inlet's relative performance in a room with multiple inlets.
- **Local Specific Contaminant-Accumulating Index**: A general index that reflects the interaction between the ventilation flow and a specific contaminant source.
- **Air Change Efficiency (ACE)**: Measures how effectively the air in a room is replaced by fresh air from the ventilation system (Di Tommaso et al., 1999). The ratio of room mean age would exist if the air in the room were utterly mixed to the average time of room replacement.

- **Ventilation Effectiveness Factor (VEF)**: Defined as the ratio of two contaminant concentration differentials: the contaminant concentration in the supply air (typically zero) and the contaminant concentration in the room under complete mixing conditions (Zhang et al., 2001).
- **Relative Ventilation Efficiency**: The ratio of the local mean age that would exist if the air in the room were thoroughly mixed with the local mean age, measured at a point.
- **Air Diffusion Performance Index (ADPI)**: Primarily a measure of occupant comfort rather than an indicator of contaminant concentrations. It expresses the percentage of locations in an occupied zone that meet air movement and temperature specifications for comfort.
- **Temperature Effectiveness**: Similar in concept to ventilation effectiveness and reflects the ability of a ventilation system to remove heat.
- **Effective Draft Temperature**: Indicates the feeling of coolness due to air motion.
- **Reynolds Number**: Expresses the ratio of the inertial forces to viscous forces.
- **Rayleigh Number**: Characterizes natural convection flows.
- **Grashof Number**: Equivalent to the Rayleigh number divided by the Prandtl number (a dimensionless number approximating the ratio of viscosity and thermal diffusivity).
- **Froude Number**: A dimensionless number used to characterize flow through corridors and doorways and in combined displacement and wind ventilation cases.
- **Richardson Number**: Characterizes the importance of buoyancy.
- **Flux Richardson Number**: Used to characterize the stabilizing effect of stratification on turbulence.
- **Buoyancy Flux**: Used to characterize buoyancy-driven flows.
- **Archimedes Number**: Conditions of the supplied air are often characterized by the discharge Archimedes number, which expresses the ratio of the buoyancy forces to momentum forces or the strength of natural convection to forced convection.

COMMON AIRFLOW PATHWAYS

Contaminants travel along pathways—sometimes over great distances. Pathways may lead from an indoor source to an indoor location or from an outdoor source to an indoor location.

The location experiencing a pollution problem may be nearby, in the same or an adjacent area, but it may be a great distance from and on a different floor than the contaminant source.

Knowledge of common pathways helps track down the source and prevent contaminants from reaching building occupants (see Table 8.26).

TABLE 8.26
Common Airflow Pathways for Contaminants

Common Pathway	Comment Indoors
	Indoors
Stairwell/elevator shaft	The stack effect brings about airflow by drawing air toward these chases on the lower floors and the elevator shaft away from these chases on the higher floors, affecting the flow of contaminants
Vertical electrical or plumbing chases	
Receptacles, outlets, openings	Contaminants can easily enter and exit building cavities, moving from space to space
Duct or plenum	The HVAC system commonly carries contaminants throughout the occupied spaces.
Duct or plenum leakage	Duct leakage accounts for significant unplanned air flow and energy loss in buildings
Flue or exhaust leakage	Leaks from sanitary exhausts or combustion flues can cause serious health problems
Room spaces	Air and contaminants move within or through doors and corridors to adjoining spaces
	Outdoors to Indoors
Indoor air intake	Polluted outdoor air or exhaust air can enter the building through the air intake
Windows/doors	A negatively pressurized building will draw air and outside pollutants into the building through any cracks and crevices available opening
Substructures/slab penetrations	Radon and other soil gases, moisture-laden air, and microbial-contaminated air often travel through crawlspaces and other substructures into the building

MAJOR INDOOR AIR QUALITY CONTAMINANTS

Environmental engineers and industrial hygienists spend much of their time working with and mitigating air contaminant problems in the workplace. The list of potential contaminants workers might be exposed to while working is extensive. However, a few major chemical-/material-derived air contaminants (other than those poisonous gases and materials that are automatically top priorities for the engineer and industrial hygienist to investigate and mitigate) are considered highly hazardous. These, too, garner the industrial hygienist's immediate attention and remedial action(s). In this text, we focus on asbestos, silica, formaldehyde, and lead as "those" hazardous contaminants (keeping in mind that there are others) requiring the immediate attention of the industrial hygienist.

Asbestos Exposure (OSHA, 2002)

Asbestos is the name given to a group of naturally occurring minerals widely used in certain products, such as building materials and vehicle brakes, to resist heat and corrosion. Asbestos includes chrysotile, amosite, crocidolite, tremolite asbestos, anthophyllite asbestos, actinolite asbestos, and any materials that have been chemically treated and altered. Typically, asbestos appears as a whitish, fibrous material that may release texture fibers ranging from coarse to silky; however, airborne fibers that can cause health damage may be too small to be seen with the naked eye.

An estimated 1.3 million construction and general industry employees face significant asbestos exposure. The heaviest exposures occur in the construction industry, particularly during asbestos removal during renovation or demolition (abatement). Employees are also likely to be exposed during the manufacturing of asbestos products (such as textiles, friction products, insulation, and other building materials) and automotive brake and clutch repair work.

The inhalation of asbestos fibers by workers can cause severe diseases of the lungs and other organs that may not appear or manifest until years after the exposure has occurred. For instance, asbestosis can cause a buildup of scar-like tissue in the lungs and result in a loss of lung function that often progresses to disability and death. As mentioned, asbestos fibers associated with these health risks are too small to be seen with the naked eye, and smokers are at higher risk of developing some asbestos-related diseases. For example, asbestos exposure can cause asbestosis (scarring of the lungs resulting in a loss of lung function that often progresses to disability and death), mesothelioma (cancer affecting the membranes lining the lungs and abdomen), lung cancer, and cancers of the esophagus, stomach, colon, and rectum.

Occupational Safety and Health Administration (OSHA) has issued the following three standards to assist industrial hygienists with compliance and to protect workers from exposure to asbestos in the workplace:

- 29 CFR 1926.1101 covers construction work, including alteration, repair, renovation, and demolition of structures containing asbestos.
- 29 CFR 1915.1001 covers asbestos exposure during work in shipyards.
- 29 CFR 1910.1001 applies to asbestos exposure in general industry, such as exposure during brake and clutch repair, custodial work, and the manufacture of asbestos-containing products.

The standards for the construction and shipyard industries classify the hazards of asbestos work activities and prescribe particular requirements for each classification:

- **Class I**: The most potentially hazardous class of asbestos jobs and involves the removal of thermal system insulation and sprayed-on or troweled-on surfacing asbestos-containing materials or presumed asbestos-containing materials.
- **Class II**: Includes removing other types of asbestos-containing materials that are not thermal systems insulation, such as resilient flooring and roofing materials containing asbestos.
- **Class III**: Focuses on repair and maintenance operations where asbestos-containing materials are disturbed.
- **Class IV**: Pertains to custodial activities where employees clean up asbestos-containing waste and debris.

There are equivalent regulations in states with OSHA-approved state plans.

Permissible Exposure Limits

Employee exposure to asbestos must not exceed 0.1 fibers per cubic centimeter (f/cc) of air, averaged over an 8-hour work shift. Short-term exposure must also be limited to not more than 1 f/cc, averaged over 30 minutes. Rotation of employees to achieve compliance with either permissible exposure limit (PEL) is prohibited.

Exposure Monitoring

In construction and shipyard work, unless the industrial hygienist can demonstrate that employee exposures will be below the PELs (a "negative exposure assessment"), it is generally a requirement that monitoring for workers in Class I and II regulated areas be conducted. For workers in other operations where exposures are expected to exceed one of the PELs, periodic monitoring must be performed. In general industry, initial monitoring must be conducted for workers who may be exposed above a PEL or the excursion limit. Subsequent monitoring at reasonable intervals must be performed, and in no case at intervals greater than 6 months for employees exposed above a PEL.

Competent Person

In all operations involving asbestos removal (abatement), employers must name a "competent person" qualified and authorized to ensure worker safety and health, as required by Subpart C., "General Safety and Health Provisions for Construction" (29 CFR 1926.20). Under the requirements for safety and health prevention programs, the competent person must frequently inspect job sites, materials, and equipment. A fully trained and licensed industrial hygienist often fills this role.

In addition, for Class I jobs, the competent person must inspect onsite at least once during each work shift and upon employee request. For Class II and III jobs, the competent person must inspect often enough to assess changing conditions and upon employee request.

Regulated Areas

In general industry and construction, regulated areas must be established where the 8-hour TWA or 30-minute excursion values for airborne asbestos exceed the PELs. Only authorized persons wearing appropriate respirators can enter a regulated area. Eating, smoking, drinking, chewing tobacco or gum, and applying cosmetics are prohibited in regulated areas. Warning signs must be displayed at each regulated area and posted at all approaches to regulated areas.

Methods of Compliance

In general industry and construction, employers must control exposures using engineering controls to the extent feasible. Where engineering controls are not possible to meet the exposure limit, they must be used to reduce employee exposures to the lowest attainable levels and supplemented by respiratory protection.

Respirators

In general industry and construction, the level of exposure determines what type of respirator is required; the standards specify the respirator to be used. Remember that respirators must be used during all Class I asbestos jobs. Refer to 29 CFR 1926.103 for further guidance on when respirators must be worn.

Labels

Caution labels must be placed on all raw materials, mixtures, scrap, waste, debris, and other products containing asbestos fibers.

Protective Clothing

For any employee exposed to airborne concentrations of asbestos that exceed the PEL, the employer must provide and require protective clothing such as coveralls or similar full-body clothing, head coverings, gloves, and foot coverings. Wherever the possibility of eye irritation exists, face shields, vented goggles, or other appropriate protective equipment must be provided and worn.

Training

Training must be provided for employees involved in each identified work classification. The specific training requirements depend upon the particular class of work being performed. In general industry, training must be provided to all employees exposed above a PEL. Asbestos awareness training must also be provided to employees who perform housekeeping operations covered by the standard. When feasible, warning labels must be placed on all asbestos products, containers, and installed construction materials.

Recordkeeping

The employer must keep an accurate record of all measurements to monitor employee exposure to asbestos. This record must include the date of measurement, the operation involving exposure, the sampling and analytical methods used, and evidence of their accuracy; the number, duration, and results of samples taken; the type of respiratory protective devices worn; and the name, social security number, and the results of all employee exposure measurements. This record must be kept for 30 years.

Hygiene Facilities and Practices

Employers must furnish clean change rooms for employees working in areas with exposure above the TWA and excursion limits. Two lockers or storage facilities must be furnished and separated to prevent the employee's street clothes from being contaminated by protective clothing and equipment. Showers must be equipped so employees can shower at the end of the work shift. Employees must enter and exit the regulated area through the decontamination area.

The equipment room must have impermeable, labeled bags and containers to dispose of contaminated protective clothing and equipment.

Lunchroom facilities for those employees must have positive pressure, a filtered air supply, and be readily accessible. Employees must wash their hands and faces before eating, drinking, or smoking. The employer must ensure that employees do not enter lunchroom facilities with protective work clothing or equipment unless surface fibers have been removed from the clothing or equipment.

Employees may not smoke in work areas where they are occupationally exposed to asbestos.

Medical Exams

In general industry, exposed employees must have a preplacement physical examination before being assigned to an occupation exposed to airborne concentrations of asbestos at or above the action or excursion levels. The physical examination must include a chest X-ray, medical and work history, and pulmonary function tests. Subsequent exams must be given annually and upon termination of employment. However, chest X-rays are required annually only for older workers whose first asbestos exposure occurred more than 10 years ago.

In construction, examinations must be available annually for workers exposed above the action level or excursion limit for 30 or more days per year or who are required to wear negative pressure respirators; chest X-rays are at the physician's discretion.

Silica Exposure

Crystalline silica (SiO_2) is a significant component of the Earth's crust. SiO_2 crystals are minute, very hard, translucent, and colorless in their pure, natural form. Most mined minerals contain some SiO_2. "Crystalline" refers to the orientation of SiO_2 molecules in a fixed pattern instead of a nonperiodic, random molecular arrangement, which os defined as amorphous (e.g., diatomaceous earth). Therefore, silica exposure occurs in various settings, such as mining, quarrying, and stone-cutting operations; ceramics and vitreous enameling; and in the use of filters for paints and rubber. The wide use and multiple applications of silica in industrial applications combine to make silica a major occupational health hazard (silicosis), which can lead to death.

Silicosis is a disabling, nonreversible, and sometimes fatal lung disease caused by overexposure to respirable crystalline silica. More than 1 million U.S. workers are exposed to crystalline silica; yearly, more than 250 die from silicosis (see Tables 8.27 and 8.28). The disease is not curable but is 100% preventable if employers, workers, and health professionals work together to reduce exposure.

The first column is the occupation title. The second column (PMR) is the observed number of deaths from silicosis per occupation divided by the expected number of deaths. Therefore, a value of one indicates no additional risk. A value of ten would indicate a risk ten times greater than the average risk of silicosis. The first table provides risk by occupation, and the second by industry.

Guidelines for Control of Occupational Exposure to Silica

Under the OSHA's standard for air contaminants (29 CFR 1910.1000), employee exposure to airborne crystalline silica shall not exceed an 8-hour time-weighted average limit (variable) as stated in 29 CFR 1910.1000, Table Z-3, or a limit set by a state-administered Occupational Safety and Health Plan is in effect.

TABLE 8.27
Deaths from Silica in the Workplace

Occupation	PMR
Miscellaneous metal and plastic machine operators	168.44
Hand molders and shapers, except jewelers	64.12
Crushing and grinding machine operators	50.97
Hand molding, casting, and forming occupations	35.70
Molding and casting machine operators	30.60
Mining machine operators	19.61
Mining occupations (not elsewhere classified)	15.33
Construction trades (not elsewhere classified)	14.77
Grinding, abrading, buffing, and polishing machine operators	8.47
Heavy equipment mechanics	7.72
Miscellaneous material moving equipment operators	6.92
Millwrights	6.56
Crane and tower operators	6.02
Brick masons and stonemasons	4.71
Painters, construction, and maintenance	4.50
Furnace, kiln, and oven operators, except food	4.10
Laborers, except construction	3.79
Operating engineers	3.56
Welders and cutters	3.01
Machine operators, not specified	2.86
Not specified mechanics and repairers	2.84
Supervisors, production occupations	2.73
Construction laborers	2.14
Machinists	1.79
Janitors and cleaners	1.78

Source: Work-Related Lung Disease Surveillance Report (2002). National Institute for Occupational Safety and Health, U.S. Department of Health and Human Services, Public Health Service, Centers for Disease Control and Prevention, Table 3–8; DHHA (NIOSH) Publication No. 96–134.Publications Dissemination, EID, National Institute for Occupational Safety and Health, 4676 Columbia Parkway Cincinnati, OH.

As mandated by OSHA, the first requirement is to eliminate employee exposure by implementing feasible engineering controls (e.g., dust suppression and ventilation). After all such controls are implemented and they do not control permissible exposure, each employer must rotate its employees to the extent possible to reduce exposure. Only when all engineering or administrative controls have been implemented and the level of respirable silica still exceeds permissible exposure limits, the employer relies on a respirator program under the mandatory requirements of 29 CFR1910.134. Generally, where working conditions or other practices constitute recognized hazards likely to cause death or serious physical harm, they must be corrected.

TABLE 8.28
Deaths from Silica in the Workplace

Industry	PMR
Metal mining	69.51
Miscellaneous nonmetallic mineral and stone products	55.31
Nonmetallic mining and quarrying, except fuel	49.77
Iron and steel foundries	31.15
Pottery and related products	30.73
Structural clay products	27.82
Coal mining	9.26
Blast furnaces, steelworks, rolling and finishing mills	6.49
Miscellaneous fabricated metal products	5.87
Miscellaneous retail stores	4.63
Machinery, except electrical (not elsewhere classified)	3.96
Other primary metal industries	3.63
Industrial and miscellaneous chemicals	2.72
Not specified manufacturing industries	2.67
Construction	1.82

Source: *Work-Related Lung Disease Surveillance Report* (2002). National Institute for Occupational Safety and Health, U.S. Department of Health and Human Services, Public Health Service, Centers for Disease Control and Prevention, Table 3–8; DHHA (NIOSH) Publication No. 96–134.Publications Dissemination, EID, National Institute for Occupational Safety and Health, 4676 Columbia Parkway Cincinnati, OH.

FORMALDEHYDE EXPOSURE

Formaldehyde (HCHO) is a colorless, flammable gas with a pungent, suffocating odor. It is common in the chemical industry. Formaldehyde is the most critical aldehyde produced commercially and is used to prepare urea-formaldehyde and phenol-formaldehyde resins. It is also produced during the combustion of organic materials and is a component of smoke.

The primary sources in workplace settings are manufacturing processes (used in the paper, photographic, and clothing industries) and building materials. Building materials may contain phenol, urea, thiourea, or melamine resins containing HCHO. HCHO resins can be degraded when these materials become damp from exposure to high relative humidity, if the HCHO materials are saturated with water during flooding, or when leaks occur. The release of HCHO occurs when the acid catalysts involved in the resin formulation are reactivated. Outgassing increases when temperatures and relative humidity increase (DOH, Wash., 2003).

Formaldehyde exposure is most common through gas-phase inhalation. However, it can also occur through liquid-phase skin absorption. Workers can be exposed during direct production, material treatment, and resin production. Healthcare professionals, pathology and histology technicians, and teachers and students who handle preserved specimens are potentially at high risk.

Studies indicate that formaldehyde is a potential human carcinogen. Airborne concentrations above 0.1 ppm can irritate the eyes, nose, and throat. The severity of irritation increases as concentrations increase; at 100 ppm, it is immediately dangerous to life and health. Dermal contact causes various skin reactions, including sensitization, which might force persons thus sensitized to find other work.

OSHA requires the employer to conduct initial monitoring to identify all employees exposed to formaldehyde at or above the action level or short-term exposure limit (STEL) and to determine the exposure of each so identified accurately. If the exposure level is maintained below the STEL and the action level, employers may discontinue exposure monitoring until a change could affect exposure levels. The employer must also monitor employee exposure promptly upon receiving reports of formaldehyde-related signs and symptoms.

Source control (if possible) provides the best prevention for exposure control. The best starting point is to select HCHO-free or low-emitting products, such as exterior-grade plywood that uses phenol HCHO resins for indoor use.

Secondary controls include filtration, sealants, and fumigation treatments. Select adsorbents can be used to achieve filtration. Sealants involve coating the materials in question with two or three coats of nitrocellulose varnish or water-based polyurethane. Three coats of these materials can reduce outgassing by as much as 90%.

Training is required at least annually for all employees exposed to formaldehyde concentrations of 0.1 ppm or more significant. The training will increase employees' awareness of specific workplace hazards and the control measures employed. The training also will assist in successful medical surveillance and medical removal programs. These provisions will only be adequate if employees know what signs or symptoms are related to the health effects of formaldehyde and are periodically encouraged to do so.

LEAD EXPOSURE

Lead has been poisoning workers for thousands of years. Most occupational overexposures to lead have been found in the construction trades, such as plumbing, welding, and painting. In plumbing, soft solder (banned for many uses in the U.S.), used chiefly for soldering tinplate and copper pipe joints, is an alloy of lead and tin. Although lead-based paint in residential applications has been banned since it inhibits the rusting and corrosion of iron and steel, it is still used on construction projects. Significant lead exposures can also arise from removing paint from surfaces previously coated with lead-based paint. According to OSHA 93-47 (2002), the operations that generate lead dust and fumes include the following:

Flame-torch cutting, welding, the use of heat guns, sanding, scraping, and grinding of lead-painted surfaces in repair, reconstruction, dismantling, and demolition work

- Abrasive blasting of structures containing lead-based paints
- Use of torches and heat guns, and sanding, scraping, and grinding lead-based paint surfaces during remodeling or abating lead-based paint
- Maintaining process equipment or exhaust ductwork

HEALTH EFFECTS OF LEAD

There are several routes through which lead enters the body. Lead is a toxic substance absorbed into the body in specific doses. It can be absorbed by inhalation and ingestion. Except for certain organic lead compounds not covered by OSHA's Lead Standard (29 CFR 1926.62), such as tetraethyl lead, lead, when scattered in the air as a dust, fume, or mist, can be absorbed into the body by inhalation.

A significant portion of the lead that can be inhaled or ingested enters the bloodstream. Once in the bloodstream, lead is circulated throughout the body and stored in various organs and tissues. Some of this lead is quickly filtered out of the body and excreted, but some remains in the blood and other tissues. As exposure to lead continues, the amount stored in the body will increase if more lead is absorbed than excreted. Cumulative exposure to lead, typical in construction settings, may damage the blood, nervous system, kidneys, bones, heart, and reproductive system and contribute to high blood pressure. Some of the symptoms of lead poisoning include the following:

- Poor appetite
- Dizziness
- Pallor
- Headache
- Irritability/anxiety
- Constipation
- Sleeplessness
- Weakness
- Insomnia
- "Lead line" in gums
- Fine tremors
- Hyperactivity
- "Wrist drop" (weakness of extensor muscles)
- Excessive tiredness
- Numbness
- Muscle and joint pain or soreness
- Nausea
- Reproductive difficulties

LEAD STANDARD DEFINITIONS

According to OSHA's Lead Standard, the terms listed below have the following meanings:

- **Action Level**: Means employee exposure, without regard to respirators, to an airborne concentration of lead of 30 micrograms per cubic meter of air (30 μg/m^3), averaged over 8 hours.
- **Permissible Exposure Limit (PEL)**: Means the concentration of airborne lead to which an average person may be exposed without harmful effects.

OSHA has established a PEL of 50 micrograms per cubic meter of air (50 $\mu g/m^3$) averaged over 8 hours. Suppose an employee is exposed to lead for more than 8 hours in any work day. In that case, the permissible exposure limit, a time-weighted average (TWA) for that day, shall be reduced according to the following formula:

$$\text{Maximum permissible limit } \left(\text{in } \mu g/m^3\right) = 400 \text{ x hours worked in the day.}$$

When respirators are used to supplement engineering and administrative controls to comply with the PEL and all the requirements of the lead standard's respiratory protection rules have been met, employee exposure, to determine whether the employer has complied with the PEL, may be considered to be at the level provided by the respirator's protection factor for those periods the respirator is worn. Those periods may be averaged with exposure levels when respirators are not worn to determine the employee's daily TWA exposure.

- **$\mu g/m^3$**: Means micrograms per cubic meter of air. A microgram is one millionth of a gram. There are 454 g in a pound.

Worker Lead Protection Program

The employer is responsible for developing and implementing a worker-led protection program. This program is essential for minimizing worker risk of lead exposure.

The most effective way to protect workers is to minimize exposure through engineering controls and good work practices.

At a minimum, the following elements should be included in the employer's worker protection program for employees exposed to lead:

- Hazard determination, including exposure assessment
- Engineering and work practice controls
- Respiratory protection
- PPE (protective work clothing and equipment)
- Housekeeping
- Hygiene facilities and practices
- Medical surveillance and provisions for medical removal
- Employee information and training
- Signs
- Recordkeeping

Mold Control

Molds can be found almost anywhere and grow on virtually any organic substance as long as moisture and oxygen are present. The earliest known writings discussing mold infestation and remediation (removal, cleaning up) are in Leviticus, Chapter 14, Old Testament.

Where is mold "typically" found?

As mentioned, name the spot or place; mold has grown in office buildings, schools, automobiles, private homes, and other locations where water and organic matter are left unattended. Mold is not a new issue—just one that has recently received little attention from regulators in the United States. There are no state or federal statutes or regulations regarding molds and IAQ.

Molds reproduce by making spores that cannot usually be seen without magnification. Mold spores waft through the indoor and outdoor air continually. When mold spores land on a damp spot indoors, they may begin growing and digesting whatever they are growing on to survive. Molds generally destroy the things they grow on (USEPA, 2001).

The key to limiting mold exposure is to prevent the germination and growth of mold. Since mold requires water to grow, it is essential to avoid building moisture problems. Moisture problems can have many causes, including uncontrolled humidity. Some moisture problems in workplace buildings have been linked to changes in building construction practices during the 1970s, 80s, and 90s. Some of these changes have resulted in tightly sealed buildings that may lack adequate ventilation, potentially leading to moisture buildup. Building materials like drywall may not allow moisture to escape easily. Moisture problems may include roof leaks, landscaping, gutters that direct water into or under the building, and unvented combustion appliances. Delayed maintenance or insufficient maintenance is also associated with moisture problems in buildings. Moisture problems in temporary structures have frequently been associated with mold problems.

Building maintenance personnel, architects, and builders must know effective ways to avoid mold growth that might arise from maintenance and construction practices. Locating and cleaning existing growths are also paramount to decreasing the health effects of mold contamination. Proper cleaning techniques are essential because molds are incredibly resilient and adaptable (Davis, 2001).

Molds can elicit a variety of health responses in humans. The extent to which an individual may be affected depends upon his or her state of health, susceptibility to disease, the organisms with which he or she came in contact, and the duration and severity of exposure (Ammann, 2000). Some people experience temporary effects that disappear when they vacate infested areas (Burge, 1997). In others, the effects of exposure may be long-term or permanent (Yang, 2001).

It should be noted that systemic infections caused by molds are not expected. Normal, healthy individuals can resist systemic infection from airborne molds.

Those at risk for systemic fungal infection are severely immunocompromised individuals, such as those with HIV/AIDS, individuals who have had organ or bone marrow transplants, and persons undergoing chemotherapy.

In 1994, an outbreak of *Stachybotrys chartarum* in Cleveland, Ohio, was believed by some to have caused pulmonary hemorrhage in infants. Sixteen of the infants died. The CDC sponsored a review of the cases and concluded that the scientific evidence provided did not warrant the conclusion that inhaled mold was the cause of the illnesses in infants. However, the panel also stated that further research was justified, as the original study design appeared flawed (CDC, 1999).

Below is a list of mold components that elicit a response in humans.

- **Volatile Organic Compounds (VOCs)**: "Molds produce a large number of volatile organic compounds. These chemicals are responsible for the musty odors produced by growing molds" (McNeel & Kreutzer, 1996). VOCs also provide the odor in cheeses and the "off" taste of mold-infested foods. Exposure to high levels of volatile organic compounds affects the CNS, producing such symptoms as headaches, attention deficit, inability to concentrate, and dizziness (Ammann, 2000). According to McNeel, mold volatile organic compounds' specific contribution to building-related health problems has not been studied. Also, mold-volatile organic compounds are likely responsible for only a tiny fraction of total VOCs indoors (Davis, 2001).
- **Allergens**: All molds, because of the presence of allergen son spores, have the potential to cause an allergic reaction in susceptible humans (Rose, 1999). Allergic reactions are believed to be the most common exposure reaction to molds. These reactions can range from mild, transitory responses, like runny eyes, nose, throat irritation, coughing, and sneezing, to severe, chronic illnesses such as sinusitis and asthma (Ammann, 2000).
- **Mycotoxins**: Natural organic compounds capable of initiating a toxic response in vertebrates (McNeel & Kreutzer, 1996). Some molds are capable of producing mycotoxins. Molds known to potentially produce mycotoxins, and which have been isolated in infestations causing adverse health effects include certain species of *Acremonium, Alternaria, Aspergillus,* Chaetomium, *Cladosporium, Fusarium, Paecilomyces, Penicillium, Stachybotrys,* and Trichoderma (Yang, 2001).

While a specific mold or mold strain may have the genetic potential to produce mycotoxins, specific environmental conditions are believed to be needed for them to be created. In other words, although a given mold might potentially produce mycotoxins, it will not make them if the appropriate environmental conditions are not present (USEPA, 2001).

Currently, the specific conditions that cause mycotoxin production are not fully understood. The USEPA recognizes that mycotoxins tend to concentrate in fungal spores and that limited information is available regarding the process involved in the release. As a result, the USEPA is currently conducting research to determine "the environmental conditions required for sporulation, emission, aerosolization, dissemination, and transport of [Stachybotrys] into the air (USEPA, 2001).

Mold Prevention

As mentioned, the key to mold control is moisture control. Solve moisture problems before they become mold problems. Several mold prevention tips are listed below:

- Fix leaky plumbing and leaks in the building envelope as soon as possible.
- Watch for condensation and wet spots. Fix the source(s) of moisture problem(s) as soon as possible.

- Prevent moisture due to condensation by increasing surface temperature or reducing the moisture level in the air (humidity). To increase surface temperature, insulate or increase air circulation. To reduce the moisture level in the air, repair leaks, increase ventilation (if outside air is cold and dry), or dehumidify (if outdoor air is warm and humid).
- Keep heating, ventilation, and air conditioning (HVAC) drip pans clean, flowing properly, and unobstructed.
- Vent moisture-generating appliances, such as dryers, to the outside where possible.
- Perform regular building/HVAC inspections and maintenance as scheduled.
- Maintain low indoor humidity, below 60% relative humidity (RH), ideally 30%–50%, if possible.
- Clean and dry wet spots within 48 hours.
- Don't let foundations stay wet. Provide drainage and slope the ground away from the foundation.

Mold Remediation

There are no standardized recommendations for mold remediation at present; however, the USEPA is working on guidelines. Certain aspects of the mold cleanup, however, are agreed upon by many practitioners in the field:

- A common-sense approach should be taken when assessing mold growth. For example, it is generally believed that small amounts of growth, like those commonly found on shower walls, pose no immediate health risk to most individuals.
- Persons with respiratory problems, a compromised immune system, or fragile health, should not participate in cleanup operations.
- Cleanup crews should be appropriately attired. Mold should not be allowed to touch bare skin, and eyes and lungs should be protected from aerosol exposure.
- Adequate ventilation should be provided while containing the infestation to prevent the spread of mold to other areas.
- The source of moisture must be stopped, and all areas infested with mold must be thoroughly cleaned. If thorough cleaning is not possible due to the nature of the material (porous versus semi- and non-porous), all contaminated areas should be removed.

Safety tips that should be followed when remediating moisture and mold problems include:

- Do not touch mold or moldy items with bare hands.
- Do not get mold or mold spores in your eyes.
- Do not breathe in mold or mold spores.
- Consider using PPE when disturbing mold. The minimum PPE is a respirator, gloves, and eye protection.

Mold Cleanup Methods

Various mold cleanup methods are available for remediating damage to building materials and furnishings caused by moisture control problems and mold growth. These include a wet vacuum, damp wipe, HEPA vacuum, and removing damaged materials and sealing them in plastic bags. The specific method or group of techniques used will depend on the type of material affected.

AI APPLICATIONS

Artificial Intelligence (AI) can be leveraged in several innovative ways to monitor and manage air pollution. Typically, models such as random forest models are used to produce exposure pathways for chemicals. These models help estimate the potential risk posed to human health by various chemicals, even with limited information (USEPA, 2022).

Subsequent Generation Emissions Measurement (SGEM) is a new initiative to develop advanced, lower-cost, portable air measurement technologies. Considerable studies have been conducted on emissions from smokestacks and mobile sources, and much is now known about these point sources of air pollution. These sources typically have detailed emissions data—data where AI is used to sift through it all, enabling decisions on abatement or other actions can be taken. However, this is much more difficult to implement with fugitive emissions from stationary air pollution sources because they are challenging to model.

AI and Machine Learning (ML) can also be used to build pollution models that predict pollutant concentrations and assess environmental and health risks. These models can estimate different exposure levels within various communities, helping to address environmental concerns.

Experience has demonstrated that AI-driven initiatives enhance air quality monitoring, improve public health, and ensure environmental compliance.

THE BOTTOM LINE

With the rapid advancement of contemporary industry and transportation, the exponential growth of the human population has become a universal problem. AI-driven technologies and techniques are presently considered the most transformative for forecasting air pollution due to their specific features, such as organic learning (natural learning), high precision, superior generalization, robust fault tolerance, and better specificity. Their use has provided new insights into the environmental factors that significantly impact air pollution levels.

Diverse data sources, including satellite imagery, weather data, and traffic patterns, are all part of the AI techniques employed in air pollution science.

Note that despite being promising in tracking and mitigating air pollution, significant challenges remain.

NOTE

1 Adapted from F.R. Spellman (2008). *The Science of Air*, 2nd ed. Boca Raton, FL: CRC Press.

REFERENCES

Ammann, H. (2000). "Is Indoor Mold Contamination a Threat?" Washington State Department of Health. Accessed August 9, 2003. Posted at: https://www.doh.wa.gov/ehp/ocha/mold.html.

Boubel, R.W., Fox, D.L., Turner, D.B. & Stern, A.C. (1994). *Fundamentals of Air Pollution*. New York: Academic Press.

Buonicore, A.J. & Davis, W.T. (eds) (1992). *Air Pollution Engineering Manual*. New York: Van Nostrand Reinhold.

Burge, H.A. & Hoyer, M.E. (1997). "Indoor Air Quality." In *The Occupational Environment—Its Evaluation and Control*, 51–54. DiNardi, S.R. (Ed.). Fairfax, VA: American Industrial Hygiene Association.

Byrd, R.R. (2003). *IAQ FAG Part 1*. Glendale, CA: Machado Environmental Corporation.

Carson, J.E. & Moses, H. (1969). The validity of several plume rise formulas, *J. Air Pol. Cont. Assoc*. 19(11):862.

CDC. (1999). *Reports of Members of the CDC External Expert Panel on Acute Idiopathic Pulmonary Hemorrhage in Infants: A Synthesis*. Washington, DC: Centers for Disease Control.

Compressed Gas Association, Inc. (1990). *Handbook of Compressed Gases*, 3rd ed. New York: Van Nostrand Reinhold.

Cooper, C.D. & Alley, F.C. (1990). *Air Pollution Control: A Design Approach*. Prospect Heights, IL: Waveland Press, Inc.

Davis, P.J. (2001). *Molds, Toxic Molds, and Indoor Air Quality*. Sacramento, CA: California Research Bureau, California State Library.

Davis, M.L. & Cornwell, D.A. (1991). *Introduction to Environmental Engineering*. New York: McGraw-Hill, Inc.

Di Tommaso, R.M., Nino, E. & Fracastro, G.V. (1999). Influence of the boundary thermal conditions on the air change efficiency indexes. *Indoor Air*. 9:63–69.

DOH. (2003). *Wash. Formaldehyde*. Washington State Department of Health, Office of Environmental Health and Safety. Accessed August 2003 at https://www.doh.wa.gov/ehp/ts/IAQ/Formaldehyde.HTM.

EPA. (2005). *Basic Air Pollution Meteorology*. Accessed 1/12/08 @ www.epa.gov/apti.

EPA. (2006). *Air Pollution Control Orientation Course*. Accessed 1/26/08 @ www//www.epa.gov/air/oaqps/eop/course422/ce6c.html.

EPA. (2007). *Basic Concepts in Environmental Sciences*. Accessed 1/2/2008 @ www.epa.gov/eogapti1/module3/collect/collect.htm.

Freedman, B. (1989). *Environmental Ecology*. New York: Academic Press.

Godish, T. (1997). *Air Quality*, 3rd ed. Boca Raton, FL: Lewis Publishers.

Graedel, T.E. & Crutzen, P.J. (1995). *Atmosphere, Climate, and Climate*. New York: Scientific American Library.

Hesketh, H.E. (1991). *Air Pollution Control: Traditional and Hazardous Pollutants*. Lancaster, PA: Technomic Publishing Company.

Heumann, W.L. (1997). *Industrial Air Pollution Control Systems*. New York: McGraw-Hill, Inc.

Hollowell, C.D. et al. (1979a). Impact of infiltration and ventilation on indoor air quality. *ASHRAE J*. XIV:49–53.

Hollowell, C.D. et al. (1979b). "Impact of Energy Conservation in Buildings on Health." In: *Changing Energy Use Futures.* Razzolare, R.A. & Smith, C.B. (Eds.) New York: Pergamon.

Holmes, G., Singh, B.R. & Theodore, L. (1993). *Handbook of Environmental Management & Technology.* New York: John Wiley & Sons.

Leviticus, Chapter 14.

MacKenzie, J.J. & El-Ashry, T. (1988). *Ill Winds: Airborne Pollutant's Toll on Trees and Crops.* Washington, DC: World Resource Institute.

Masters, G.M. (1991). *Introduction to Environmental Engineering and Science.* Englewood Cliffs, NJ: Prentice-Hall.

McNeel, S. & Kreutzer, R. (1996). Fungi & indoor air quality. *Health Environ. Dig.* 10(2):22–24.

Molhave, L. (1986). Indoor air quality about sensory irritation due to volatile organic compounds. *ASHRAE Trans.* 92(1): Publication #2954, 306–316.

OHSA 93-47. *Lead Exposure in Construction.* Washington, DC: U.S. Department of Labor, Occupational Safety & Health Administration. Accessed August 2022 at https://www.osha-slc.gov/pls/oshaweb/owadisp.show_document?p_table=FACT_Sheet.

Peavy, H.S., Rowe, D.R. & Tchobanoglous, G. (1985). *Environmental Engineering.* New York: McGraw-Hill, Inc.

Postel, S. (1987). "Stabilizing Chemical Cycles." In *State of the World.* Brown, L.R. (Ed.) New York: Norton. 117–127.

Rose, C.F. (1999). *Antigens.* Cincinnati: American Conference of Governmental Industrial Hygienists 1999. ACGIH Bioaerosols Assessment and Control, Ch. 25, 25-1 to 25-11.

Schaefer, V.J. & Day, J.A. (1981). *Atmosphere: Clouds, Rain, Snow, Storms.* Boston, MA: Houghton Mifflin Company.

Urone, P. (1976). "The Primary Air Pollutants-Are Gaseous. Their Occurrence, Sources, and Effects." In *Air Pollution.* Stern, A.C. (Ed.), Vol. 1. 2-3 New York: Academic Press.

USEPA. (1987). Indoor Air Facts No. 1, *EPA, and Indoor Air Quality.*

USEPA. (1988). *National Air Pollutant Emission Estimates 1940–1986.* Washington, DC: Environmental Protection Agency.

USEPA. (2001). *Indoor Air Facts No. 4 (revised) Sick Building Syndrome.* Accessed 9/8/2022. Posted at https://www.epa.gov/iaq/pubs/sbs.html.

USEPA. (2022). *Artificial Intelligence Inventory.* Washington, DC: USEPA.

Wadden, R.A. & Scheff, P.A. (1983). *Indoor Air Pollution: Characteristics, Predications, and Control.* New York: John Wiley & Sons.

Woods, J.E. (1980). Environmental implications of conservation and solar space heating. Engineering Research Institute, Iowa State University, Ames, Iowa, BEUL 80-3. Meeting of the New York Academy of Sciences, New York, January 16.

World Health Organization (WHO). (1982). *Indoor Air Pollutants, Exposure, and Health Effects Assessment. Euro-Reports and Studies No.78: Working Group Report.* Copenhagen: WHO Regional Office.

Yang, C.S. (2001). *Toxic Effects of Some Common Indoor Fungi.* Accessed 6/8/2022. Posted at https://www.envirovillage.com/Newsletters/Enviros/No4_09.htm.

Zhang, Y., Wang, X., Riskowski, G.L. & Christianson, L.L. (2001). Quantifying ventilation effectiveness for air quality control. *Trans. Am. Soc. Agric. Biol. Eng.* 44(2):385–390.

Zurer, P.S. (1988). Studies on ozone destruction expand beyond Antarctic. *C & E News.* May:18–25.

9 Environmental Impact Assessment and Mitigation

WHAT IS ENVIRONMENT IMPACT ASSESSMENT (EIA)?

First, we can say that the Environmental Impact Assessment (EIA) is an early warning system and an ongoing review process. Second, we can say that EIA is a problem-solving approach to decision-making. Third, the EIA process is designed to be systematic, reproducible, and interdisciplinary. Fourth, EIA is both the process and the document. Fifth, the initial EIA is a preliminary evaluation. Sixth, and the bottom line, EIA is designed to improve planning and decision-making and plays a crucial role in evaluating the potential ecological consequences of development projects.

Okay, after this introduction, let's take a more detailed look at EIA.

Again, EIA is a decision-making process, but it is also a document that provides a systematic, reproducible, and interdisciplinary evaluation of the potential effects of a proposed action and its practical alternatives on the physical, biological, cultural, and socioeconomic attributes of a picture geographic area.

Often a key component in national, regional, or local facilities planning and land use planning, EIA is becoming an even more critical component. The idea behind EIA is to ensure that significant environmental resources are recognized early in the planning process and protected through proper planning and decision-making. The EIA analysis should be *systematic* to ensure that all feasible alternatives that meet the proposal's needs are considered and compared. Moreover, the relevant environmental resources are described and evaluated, and all measures that could protect these resources are fully considered in the planning process. Note that the EIA should be *reproducible* to permit independent verification of the findings and conclusions presented in the EIA document. Also, the EIA should be *interdisciplinary* to ensure that experts in the relevant physical, biological, cultural, and socioeconomic disciplines contribute their expertise to the overall assessment so that the evaluation of resources and impacts is accurate and comprehensive.

Regarding the decision-making process, EIA is designed to provide a means for all stakeholders in an action to be heard and to participate in the selection of alternatives and mitigation of adverse impacts. An essential aspect of the EIA is that it brings before decision-makers more alternative courses of action so that they may better achieve several, instead of just one, sets of goals.

AI algorithms excel at examining and evaluating substantial and complex datasets, making them vital tools for environmental impact assessments.

 DOI: 10.1201/9781003601562-11

Finally, EIA helps to identify actions needed to prevent future environmental damage from anticipated impacts.

Steps or Stages of the EIA Process

The steps or stages of the EIA process vary from project to project. However, as a matter of good engineering practice, the basic steps must be applied to each project. The requirements determine the steps involved in the EIA. These steps, common to all EIA projects, are:

- **Project Identification and Definition**: Based on personal experience, I have determined that although this step seems relatively minor, it is not and can be complex, especially for multiple projects. The proposed project is explicitly defined to determine the possible impacts of the proposal so that the entire scope of the environmental effects is evaluated.
- **Screening**: This step is vital because the question must be answered: Does this project require an EIA? Screening enables informed decision-making.
- **Scoping**: This process informs the public about the proposed project and allows them to provide input, ask questions, and so forth. Scoping also includes identifying the key stakeholders, introducing them to the project, and highlighting the value of the EIA in determining whether the project is a go or no-go.
- **Baseline Study**: This step includes a comprehensive study of the project site, using literature, field surveys, measurements, and the collection of representative samples, etc. The components studied include climate, geology, soil type, flora and fauna, and wildlife characteristics (there is a large amount of data—AI to the rescue).
- **Impact Analysis**: After the proposed project's environmental, social, and economic impacts are recognized and forecasted, a detailed explanation of alternatives to the design process is included.
- **Impact Mitigation**: After environmental impacts are foreseen and identified, actions to reduce the level of damage and avoid the potentially harmful consequences of the planned project are suggested.
- **EIA Report**: The report summarizes the Environmental Impact Assessment process and details the Environmental Management Plan; it is the guide used to accomplish the project.
- **Review of Draft EIA Report**: This step provides decision-makers with the necessary information to assess the report's adequacy and effectiveness.
- **Decision-Making**: The project can be approved, rejected, or subjected to further changes. Once approved, the regulatory body issues the proponent an Environmental Impact Statement.

AI APPLICATIONS IN EIA

We must be far thinking about protecting the environment, workers' safety and health, the safety and health of all living organisms, and those yet to arrive on Earth. AI is forward-thinking, and we have only scratched the surface of the potential for

AI-assisted technological advances. AI has been around for some time; however, its use in EIA is recent, with interest growing each day alongside extensive data collection. The complexity of this environmental data is a challenging obstacle for researchers and others to overcome. The good news is that innovative computer programs provide us with solutions for achieving precise predictions, risk evaluations, constant monitoring, and other valuable advantages (Haupt et al., 2022). As pointed out earlier, AI techniques can reason, make decisions, and learn from input data. These techniques transform real-world information into machine-understandable and usable knowledge, making decisions through planned pathways of optimization and solution searching (European Commission, 2020). Environmental data is intricate, and this is where ML and computer vision (neural networks) teach computers and systems to derive meaningful information from digital images, videos, and other visual inputs. These systems then act when spotting issues or defects. Both methods are essential in managing complex data, such as environmental engineering data.

DID YOU KNOW?

The EIA process must be systematic, reproducible, and interdisciplinary. Note that EIA is the process and the document.

THE BOTTOM LINE

AI is revolutionizing EIA by improving data analysis, ecological modeling, biodiversity monitoring, risk assessment, stakeholder engagement, and permitting processes. EIA effectively expresses the national, regional, or local intent to establish a sound and sustainable environmental policy for governmental and private decisions. EIA is both an early warning system and an ongoing review process.

REFERENCES

European Commission. (2020). *White Paper on Artificial Intelligence—A European Approach to Excellence and Trust.* Accessed 9/26/24 @ https://digital-strategy.ec.europa.europa.eu/Consultation/white-paper-artificial-intelligence-European-approach-excellence-and-trust.

Haupt, S.E., et al. (2022). The history and practice of AI in environmental sciences. *Bull. Am. Meteor. Soc.* 103:E1351–E1370.

10 Waste Management Engineering

INTRODUCTION

We are a throwaway society; when we don't want something anymore, we dump it into the trash, throw it outside, burn it, or curb it for garbage pickup service. But here is the problem: It is not gone when we "throw away" waste. Dealing with the waste has only been postponed. Sometimes, this postponement means that when we return, the wastes are helpful and harmless (as with some biodegradable wastes). Still, more often, it means that the problems we must face will be worse—increased by chemistry and entropy—that 55-gallon drum of toxic waste was easier to handle before it rusted out.

Amid the cries of "Not in My Backyard" and "Pick up the Trash, but Don't Put It Down," we need to hear a more realistic and environmentally kinder truth: There is no such thing as a free lunch.

We somehow pay for what we get or use, whether we see the charges or not. The price for our solid waste habits will soon be charged to us. In some places (big cities, for example), the awareness of the size of the bill is sinking in.

Environmentally, what does that mean? In short, if we, as a society, will consume as we do, build as we do, and grow as we do, we have to pay the price for our increase. And that, sometimes, will mean that our waste will be "in our backyard." We will have to increase the amount of solid waste we reuse and recycle; we will have to spend tax dollars to solve the problems with landfills and trash incineration; we will have to seriously look at how we live, how the goods we buy are packaged, how our industries deal with their wastes—because if we don't, the bill will be more than we can afford to pay.

We will swim in it without a life jacket if we do not handle our waste.

Advancements in technology have made our lives more comfortable, safer, healthier, and, in many cases, more enjoyable. Some would say that progress is not without cost. This statement is correct—however, what costs do they refer to? Can we afford the consequences if these costs include more Bhopals, Times Beaches, Love Canals, or another Exxon Valdez? If such disasters are "to be included as a cost of progress," then we must say that the cost outweighs the gain.

To balance technological progress and its environmental results, we must use technological advances, including Artificial Intelligence, to ensure that "progress" is not too costly—or life-threatening—to our environment and ourselves.

RCRA's Waste Management Hierarchy sums up what could/should/would happen with waste—any waste—in a "best of all possible worlds." However, though it is romantic and too simple to say we "should" follow these standards, in practical terms, we benefit long-term by striving to achieve them.

DOI: 10.1201/9781003601562-12

Regulating problem wastes, developing safe and environmentally friendly ways to dispose of them, and using the technologies we develop to control the future of such wastes is in everyone's best interests.

AMERICAN SOCIETY AND WASTE

Concerning teaching college-level courses dealing with solid and hazardous wastes, the portrait and portrayal of American society that I often use in my lectures presented to environmental science/health/engineering students is not flattering. I describe America as a throwaway society. I explain that the American Throwaway Society displays and underscores a characteristic that might be described as habit, trend, custom, or practice—the tendency to discard those objects we no longer want. When we don't want it anymore, we discard it. When it offends us, we dispose of it. Throughout history, a common practice was, "I don't want it any longer. Take it down to the river and dump it." When mile-long caravans of wagon trains crossed the tremendous American prairies, the pioneers discarded anything and everything that overburdened the wagon; this was a common practice for survival—homemaking materials, furnishings that made the trip more complex—some of them necessary for successful homesteading, but too heavy to carry were left along the trail west. Lifelong treasures once so important that they couldn't be left behind now became burdens (and waste) to the American pioneering wagoner's . . . things to be disposed of to make their lives easier. This pioneering throwaway tendency has been called the Frontier Mentality. That is, "we can throw away anything we no longer want out here or in the frontier (the prairie) . . . there is so much endless space, trash placed in the vast, endless prairie won't hurt a thing . . . it's like urinating in the ocean . . . won't hurt a thing!" This ignores the effect of accumulation, which eventually clogs up all that 'empty' space.

Again, when something is no longer of value because it is broken, worn out, out of style, or no longer needed for whatever reason, we feel discarding it is not a big issue. But it is—mainly when the item we throw away is a hazardous substance that is persistent, nonbiodegradable, and poisonous.

What is the magnitude of the problem with hazardous substance/waste disposal? Let's take a look at a few facts.

- Hazardous substances—including industrial chemicals, toxic waste, pesticides, and nuclear waste—are entering the marketplace, the workplace, and the environment in unprecedented quantities.
- The United States produces almost 300 million metric tons of hazardous waste each year—with a population of 320,000,000+, which amounts to more than one ton for every person in the country.
- Through air, soil, and water pollution, hazardous wastes pose short- and long-term threats to human health and environmental quality (Spellman, 2007).

This chapter discusses a growing and significant problem facing environmental practitioners and humanity: anthropogenically produced waste. Specifically, we face daunting questions: What will we do with all the waste we generate? What are the alternatives? What technologies are available to mitigate the "waste problem?"—a problem that grows daily.

Before beginning our discussion, we focus on an important question: When we throw waste away, is it gone? Remember, though we are faced today and in the immediate future with growing mountains of waste we produce (and we are running out of places on earth to dispose of them), an even more pressing two-fold problem is approaching: the waste's toxicity and persistence.

We will discuss waste and the toxicity problem later, but now, think about the persistence of the waste we dispose of. For example, when we excavate a deep trench and place several 55-gallon drums of liquid waste within it, then bury the entire sordid mess, are we disposing of the waste in an earth-friendly way? Are we disposing of it permanently? What happens a few years later when the 55-gallon drums corrode and leak? Where does the waste go? Where does the waste end up? What are the consequences of such practices? Are they insignificant to us today because they are tomorrow's problems? Does anyone care?

We need to ask ourselves these questions and determine the answers now. If we are uncomfortable with the answers we come up with, shouldn't we feel the same about the answers someone else (our grandchildren) will have to come up with later when it is far too late?

Waste is not easily disposed of. We can hide it. We can mask it. We can move it from place to place. We can take it to the remotest corners of the earth. But because of its persistence, waste is not permanently gone when we think it is. It has a way of returning and reminding us—a way of persisting. How persistent is waste? It is very persistent, as thousands of documented cases make clear.

In this section, we define and discuss solid wastes and, later, AI applications. We focus on a significant portion of solid wastes, Municipal Solid Wastes (MSW), because people living in urban areas, where many of the problems associated with solid waste occur, generate these wastes. We also discuss another significant waste problem: Hazardous Waste. We also discuss waste control technologies for minimization, treatment, and disposal.

SOLID WASTE REGULATORY HISTORY (UNITED STATES)

For most of the nation's history, municipal ordinances (rather than federal regulatory control) were the only solid waste regulations in effect. These local urban governments controlled solid waste almost from the beginning of each settlement because of street disposal's inherent severe health consequences. Along with prohibiting dumping garbage in the streets, municipal regulations usually stipulated requirements for proper disposal in designated waste dump sites and mandated owners to remove their waste piles from public property.

The federal government did not begin regulating solid waste dumping until the nation's harbors and rivers were either overwhelmed with raw wastes or headed in that direction. The federal government used its constitutional powers under the *Interstate Commerce Clause* to enact the *Rivers and Harbors Act of* 1899. The U.S. Army Corps of Engineers was empowered to regulate and sometimes prohibit private and municipal dumping practices.

Not until 1965 did Congress finally get into the picture (due to strong public opinion) by adopting the *Solid Waste Disposal Act* of 1965, which became the responsibility of the U.S. Public Health Service to enforce. The intent of this act was to:

1. Promote the demonstration, construction, and application of solid waste management and resource recovery systems that preserve and enhance the quality of air, water, and land resources.
2. Provide technical and financial assistance to state, local, and interstate agencies in planning and developing resource recovery and solid waste disposal programs.
3. Promote a national research and development program for improved management techniques, more effective organizational arrangements, new and improved methods of collection, separation, recovery, and recycling of solid wastes, and the environmentally safe disposal of nonrecoverable residues.
4. Provide for the promulgation of guidelines for solid waste collection, transport, separation, recovery, and disposal systems.
5. Provide training grants in occupations involving designing, operating, and maintaining solid waste disposal systems (Tchobanoglous et al., 1993).

After Earth Day 1970, Congress became more sensitive to waste issues. In 1976, Congress passed solid waste controls as part of the *Resource Conservation and Recovery Act* (RCRA). "Solid waste" was defined as any garbage, refuse, sludge from a waste treatment plant, water supply treatment plant, or air pollution control facility and other discarded material.

In 1980, Public Law 96–510, 42U.S.C. Article 9601, the *Comprehensive Environmental Response, Compensation and Liability Act* (CERCLA), was enacted to provide a means of directly responding to and funding the activities of response to problems at uncontrolled hazardous waste disposal sites. Uncontrolled MSW landfills are facilities that have not operated or are not operating under RCRA (USEPA, 1989).

Many other laws governing the control of solid waste management problems are now in effect. Federal legislation and associated regulations have encouraged the implementation of solid waste management programs at the state level of government. Legislation will continue to be an essential part of future solid waste management.

SOLID WASTE CHARACTERISTICS

Solid waste (also called *refuse, litter, rubbish, waste, trash*, and (incorrectly) *garbage*) refers to any of a variety of materials that are rejected or discarded as being spent, useless, worthless, or in excess. Table 10.1 provides a functional waste classification system.

Solid waste is probably more correctly defined as "any material thing no longer wanted." O'Reilly (1992) points out that defining solid waste is tricky because solid waste is a series of paradoxes:

- Personal in the kitchen trash can—but impersonal in a landfill;
- What one individual may deem worthless (an outgrown or out-of-fashion coat, for example) and fit only for the trash can—another individual may find valuable;

TABLE 10.1
Classification of Solid Waste

Type	Principal Components
Trash	Highly combustible waste paper, wood, and cardboard cartons, including up to 10% treated papers, plastic, or rubber scraps; commercial and industrial sources
Rubbish	Combustible waste, paper, cartons, rags, wood scraps, combustible floor sweepings, domestic, commercial, and industrial sources
Refuse	Rubbish and garbage; residential sources
Garbage	Animal and vegetable wastes, restaurants, hotels, markets; institutional, commercial, and club sources

Source: Adapted from Davis & Cornwell (1991). *Introduction to Environmental Engineering*, 2nd ed., p. 585.

- Of little cost concern to many Americans—yet very costly to our society in the long term;
- An issue of severe federal concern—yet a very localized problem from municipality to municipality.

The famous adage is accurate—everyone wants waste picked up, but no one wants it to be put down. It goes almost without saying that the other adage, "Not in My Back Yard" (NIMBY) is also accurate. The critical point is that it becomes solid waste whenever a material object is thrown away, regardless of its actual or potential value.

Garbage (with its tendency to decompose rapidly and create offensive odors) is often used as a synonym for solid waste. It refers strictly to animal or vegetable waste from food handling, storage, preparation, or consumption.

The solid waste stream is the collective and continual production of all refuse (the sum of all solid wastes from all sources). As stated, an estimated 6 billion metric tons of solid waste are produced annually in the United States. The two most significant sources of solid waste are agriculture (animal manure, crop residues, and other agricultural by-products) and mining (dirt, waste rock, sand, and slag, the material separated from metals during the smelting process). About 10% of the total waste stream is generated by industrial activities (plastics, paper, fly ash, slag, scrap metal, and sludge or biosolids from treatment plants).

From Figure 10.1, we see that paper and paperboard account for the most significant percentage (about 29%) of refuse materials by volume of MSW. Yard wastes account for almost 14%. Glass and metals make up nearly 14% of MSW, food wastes are just under 15%, and plastics are about 12.3%.

USEPA (2009) points out that approximately 243 million metric tons of MSW were generated in the U.S. in 2009, equivalent to more than 5 pounds per person daily. By the year 2010, the EPA estimated that waste generation in the U.S. would rise to more than 250 million metric tons annually, almost 5.5 pounds per person per day.

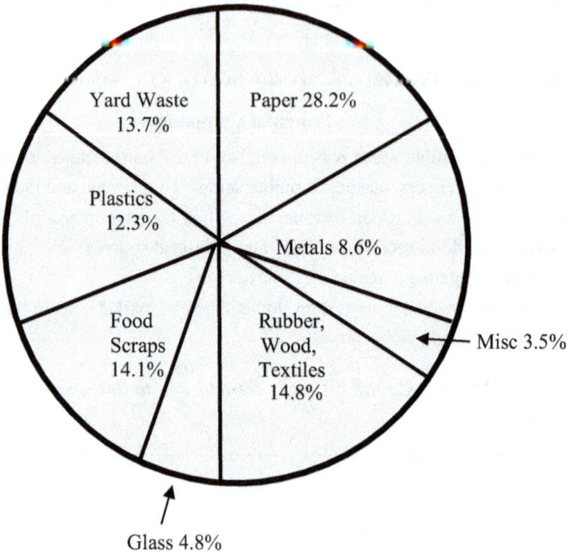

FIGURE 10.1 Composition of municipal solid waste discarded in a typical day by each American. (EPA, Meeting the Environmental Challenge, 2009; Characterization of Municipal Solid Waste in the U.S., 2009 Update, EPA/530-5-019, Washington, DC: US Environmental Protection Agency.)

SOURCES OF MUNICIPAL SOLID WASTES (MSW)

Municipal solid waste (MSW) sources in a community are generally related to land use and zoning. They include residential, commercial, institutional, construction and demolition, municipal services, and treatment plants.

Residential Sources of MSW

Residential sources of MSW are single- and multi-family detached dwellings and apartment buildings. The types of solid waste generated include food waste, textiles, paper, cardboard, glass, wood, ashes, tin cans, aluminum, street leaves, and special bulky items, including yard waste collected separately, white goods (refrigerators, washers, dryers, etc.), batteries, oil, tires, and household hazardous wastes.

Commercial Sources of MSW

Commercial sources of MSW are restaurants, hotels, stores, motels, service stations, repair shops, markets, office buildings, and print shops. The types of solid waste generated include paper, cardboard, wood, plastics, glass, special wastes such as white goods and other bulky items, and hazardous wastes.

Institutional Sources of MSW

*H*ospitals, schools, jails and prisons, and government centers are institutional sources of MSW. The types of solid waste generated by institutional sources are the same as those generated by commercial sources.

Construction and Demolition Sources of MSW

Construction and demolition sources of MSW include new construction sites, the razing of old buildings, road repair/renovation sites, and broken pavement. Standard construction materials such as wood, steel, plaster, concrete, and soil are the types of solid waste generated by these sources.

Municipal Services Sources of MSW

Municipal services (excluding treatment plants) generate MSW through street cleaning, landscaping, parks and beaches, recreational areas, and catch basin maintenance and cleaning activities. Solid wastes generated by municipal services include rubbish, street sweepings, general wastes from parks, beaches, and recreational areas, and basin debris catches.

Treatment Plant Site Sources of MSW

Water, wastewater, and other industrial treatment processes (for example, incineration) are the sources of MSW at treatment plant sites. The principal types of solid waste generated by treatment plant sites are sludges or biosolids, fly ash, and general plant wastes.

WHAT IS A HAZARDOUS SUBSTANCE?

Hazardous wastes can be informally defined as a subset of all solid and liquid wastes that are disposed of on land rather than being shunted directly into the air or water, which can potentially affect human health and the environment adversely. We tend to think of hazardous wastes as resulting mainly from industrial activities. Still, households also play a role in the generation and improper disposal of substances that might be considered hazardous wastes. Hazardous wastes (due to Bhopal and other disastrous episodes) have been given much attention, but surprisingly, little is known about their nature and the actual scope of the problem. In this section, we examine definitions of hazardous materials, substances, wastes, and others and attempt to bring hazardous wastes into perspective as a significant environmental concern.

Unfortunately, defining a *hazardous substance* is largely a matter of "pick and choose," with various regulatory agencies and pieces of environmental legislation defining that term and related terms somewhat differently. Many of the terms are used interchangeably. Even experienced professionals in environmental health and safety fields, like environmental engineers with the Certified Hazardous Materials Manager (CHMM) certification, sometimes interchange these terms. However, the terms are generated by different federal agencies through various pieces of legislation and have somewhat different meanings, depending upon the nature of the problem addressed. To understand the scope of the dilemma we face in defining *hazardous substances*, let's look at the terms commonly used today, used interchangeably, and often thought to mean the same thing.

Hazardous Materials

A *hazardous material* is a substance (gas, liquid, or solid) capable of causing harm to people, property, and the environment. The United States Department of Transportation (DOT) uses the term *hazardous materials* to cover nine categories identified by the *United Nations Hazard Class Number System*, including:

- Explosives
- Gases (compressed, liquefied, dissolved)
- Flammable Liquids
- Flammable Solids
- Oxidizers
- Poisonous Materials
- Radioactive Materials
- Corrosive Materials
- Miscellaneous Materials

Hazardous Substances

The USEPA uses the term hazardous substance for chemicals that, if released into the environment above a certain amount, must be reported, and depending on the threat to the environment, for which federal involvement in handling the incident can be authorized. The USEPA lists hazardous substances in its 40 CFR Part 302, Table 302.4.

The Occupational Safety and Health Administration (OSHA) uses the term *hazardous substance* in 29 CFR 1910.120 (which resulted from Title I of SARA and covers *emergency response)* differently than does the EPA. Hazardous substances (as defined by OSHA) cover every chemical regulated by DOT and the EPA.

Extremely Hazardous Substances

The EPA uses the term "extremely hazardous substance" to refer to chemicals that must be reported to the appropriate authorities if released above the *threshold reporting quantity* (RQ). The list of extremely hazardous substances is identified in Title III of the *Superfund Amendments and Reauthorization Act* (SARA) of 1986 (40 CFR Part 355). Each substance has a threshold reporting quantity.

Toxic Chemicals

The EPA uses the term *toxic chemical* for chemicals whose total emissions or releases must be reported annually by owners and operators of certain facilities that manufacture, process, or otherwise use listed toxic chemicals. The list of toxic chemicals is identified in Title III of SARA.

HAZARDOUS WASTES

> The most alarming of all man's assaults upon the environment is the contamination of air, earth, rivers, and sea with dangerous and even lethal materials. This pollution is, for the most part, irrecoverable; the chain of evil it initiates not only in the world that must support life but also in living tissues is, for the most part, irreversible. In this now universal contamination of the environment, chemicals are the sinister and little-recognized partners of radiation in changing the very nature of the world—the very nature of life.
>
> —*Rachel Carson (1962)*

Rachel Carson combined the insight and sensitivity of a poet with the realism and observations of science more adeptly than anyone before her. Famous for her classic and highly influential book *Silent Spring*, it seems strange to us today that such a visionary as Rachel Carson was (after the publication of her magnum opus) ostracized, vilified, laughed at, criticized, and disregarded. To those guilty of the sins she revealed, Rachel Carson was an enemy to be discredited—and silenced. She was not, however, disregarded by those who understood. To these concerned folks with conscience, her message was clear: waste, if not properly treated and handled, not only threatens human life in the short term but the environment as a whole in the long term. Her plea was also straightforward: stop poisoning the earth.

Examined with a clear vision of retrospect, the environmental missionary Rachel Carson was well ahead of her time. Her fears in 1962 were based on limited data, but they have since been confirmed. Rachel Carson was right.

The EPA uses hazardous waste for chemicals regulated under the Resource, Conservation, and Recovery Act (RCRA-40 CFR Part 261.33). The DOT regulates hazardous waste in transportation (49 CFR Parts 170–179).

For this text, we define hazardous waste as any hazardous substance that has been spilled or released into the environment. For example, chlorine gas is a hazardous material. When chlorine is released into the environment, it becomes a hazardous waste. Similarly, asbestos is a hazardous material when it is in place and undisturbed. It becomes a hazardous waste when broken, breached, or thrown away.

HAZARDOUS CHEMICALS

OSHA uses the term hazardous chemical to denote any chemical that poses a risk to employees if exposed to it in the workplace. Hazardous chemicals cover a broader group of chemicals than the other chemical lists.

AGAIN, WHAT IS A HAZARDOUS SUBSTANCE?

To form the most robust foundation for understanding the main topic of this chapter (hazardous waste), we use RCRA's definition of a hazardous substance, which can also be used to describe *hazardous waste*. RCRA defines something as a *hazardous substance* if it possesses the following four characteristics: *reactivity, ignitability, corrosiveness,* or *toxicity*. Briefly:

- **Ignitability** refers to the characteristic of being able to sustain combustion and includes the category of flammability (ability to start fires when heated to temperatures less than 140°F or 60°C).
- **Corrosive** substances (or wastes) may destroy containers, contaminate soils and groundwater, or react with other materials to cause toxic gas emissions. Corrosive materials provide a specific hazard to human tissue and aquatic life with extreme pH levels.
- **Reactive** substances may be unstable or tend to react, explode, or generate pressure during handling. Pressure-sensitive or water-reactive materials are included in this category.
- **Toxicity** is a function of the effect of hazardous materials (or wastes) that may come into contact with water or air and be leached into the groundwater or dispersed in the environment.

The principal concerns are toxic effects that may occur to humans, fish, or wildlife. Toxicity, until 1990, was tested using a standardized laboratory test called the *extraction procedure* (EP Toxicity Test). The EP Toxicity test was replaced in 1990 by the *Toxicity Characteristics Leaching Procedure (TCLP)* **because the EP test failed to adequately simulate the flow of toxic contaminants into drinking water**. The TCLP test is designed to identify wastes likely to leach hazardous concentrations of particular toxic constituents into the surrounding soils or groundwater due to improper management.

TCLP extracts constituents from the tested waste in a manner designed to simulate leaching actions that occur in landfills. The extract is then analyzed to determine if it possesses any toxic constituents listed in Table 10.2. If the concentrations of the toxic constituents exceed the levels listed in the table, the waste is classified as hazardous.

Recall our general rule of thumb that states that any hazardous substance spilled or released to the environment is no longer classified as hazardous but as hazardous waste. The EPA uses the same definition for hazardous waste as it does for hazardous substances. The four characteristics described in the previous section (reactivity, ignitability, corrosivity, or toxicity) can be used to identify hazardous substances and wastes.

Note that the EPA lists substances that it considers hazardous wastes. These lists take precedence over any other method used to identify and classify substances as hazardous (i.e., legally, if a substance is listed in one of the EPA's lists described below, it is a hazardous substance, no matter what).

EPA Lists of Hazardous Wastes

EPA-listed hazardous wastes are organized into three categories: nonspecific source wastes, specific source wastes, and commercial chemical products. All listed wastes are presumed hazardous regardless of their concentrations. EPA developed these lists by examining different types of wastes and chemical products to determine whether they met any of the following criteria:

TABLE 10.2
Maximum Concentration of Contaminants for TCLP Toxicity Test

Contaminant	Regulatory Level (mg/L)
Arsenic	5.0
Barium	100.0
Benzene	0.5
Cadmium	1.0
Carbon tetrachloride	0.5
Chlordane	0.03
Chlorobenzene	100.0
Chloroform	6.0
Chromium	5.0
Cresol	200.0
2,4-D	10.0
1,4-Dichlorobenzene	7.5
1,5-Dichloroethane	0.5
2.4-Dinitrololuene	0.13
Endrin	0.02
Heptachlor	0.008
Hexachlorobenzene	0.13
Hexachloroethane	3.0
Lead	5.0
Lindane	0.4
Mercury	0.2
Methoxychlor	10.0
Methyl ethyl ketone	200.0
Nitrobenzene	2.0
Pentachlorophenol	100.0
Pyridine	5.0
Selenium	1.0
Silver	5.0
Tetrachloroethylene	0.7
Toxaphene	0.5
Trichloroethylene	0.5
2,4,5-Trchlorophenol	400.0
2,4,6-Trchlorophenol	2.0
2.4,5-TP (Silvex)	1.0
Vinyl chloride	0.2

Source: USEPA (1990b), 40 CFR 261.24.

- Exhibit one or more of the four characteristics of hazardous waste.
- Meet the statutory definition of hazardous waste.
- Are acutely toxic or acutely hazardous.
- Are otherwise toxic.

These lists are described briefly, as:

- **Nonspecific source wastes**: Generic wastes commonly produced by manufacturing and industrial processes. Examples from this list include spent halogenated solvents used in degreasing, wastewater treatment sludge from electroplating processes and dioxin wastes, most of which are "acutely hazardous" wastes because of the danger they present to human health and the environment.
- **Specific source wastes**: Wastes from specially identified industries such as wood preserving, petroleum refining, and organic chemical manufacturing. These wastes typically include sludge, still bottoms, wastewaters, spent catalysts, and residues, such as wastewater treatment sludge from pigment production.
- **Commercial chemical products** (also called "P" or "U" list wastes because their code numbers begin with these letters): Wastes from specific commercial chemical products or manufacturing chemical intermediates. This list includes chemicals such as chloroform and creosote, acids such as sulfuric and hydrochloric, and pesticides such as DDT and kepone (40 CFR 261.31, 32 & 33).

Note that the EPA ruled that any waste mixture containing a listed hazardous waste is also considered hazardous and must be managed accordingly. This applies regardless of the percentage of the waste mixture composed of listed hazardous wastes. Wastes derived from hazardous wastes (residues from the treatment, storage, and disposal of a listed hazardous waste) are also considered hazardous waste (EPA, 1990a).

WHERE DO HAZARDOUS WASTES COME FROM?

Hazardous wastes are derived from several waste generators. Most of these generators are in the manufacturing and industrial sectors, including chemical manufacturers, the printing industry, vehicle maintenance shops, leather products manufacturers, the construction industry, and metal manufacturing. These industrial waste generators produce various wastes, including strong acids and bases, spent solvents, heavy metal solutions, ignitable wastes, and cyanide wastes.

WHY ARE WE CONCERNED ABOUT HAZARDOUS WASTES?

From the environmental scientist's perspective, any hazardous waste release that could alter the environment in any way is of significant concern. The specifics of their concern lie in acute and chronic toxicity to organisms, bioconcentration, biomagnification, genetic change potential, etiology, pathways, change in climate and habitat, extinction, persistence, and aesthetics (visual impact).

Remember that when a hazardous substance or hazardous material is spilled or released into the environment, it becomes a hazardous waste. Since specific regulatory legislation is in place regarding hazardous wastes, responding to hazardous waste leak/spill contingencies, and proper handling, storage, transportation, and treatment of hazardous wastes, this distinction is important—the goal being, of course, protecting the environment—and ultimately, protecting ourselves.

Why is there so much concern about hazardous substances and hazardous wastes? This question is relatively easy to answer because of the hard lessons we have learned in the past. Our answers are based on experience—actual hazardous materials incidents that we know of and have witnessed—that have resulted in tragic consequences, not only to the environment but also to human life. Consider an example (from *Surviving an OSHA Audit* (1998)). Maybe it will provide a better explanation of why the control of hazardous substances and wastes is critically important to us all.

HAZARDOUS WASTE LEGISLATION

A few people (Rachel Carson, for one) could have predicted that a disaster on the scale of Bhopal would occur, but humans are strange in many ways. We may know that a disaster is possible, is likely, could happen, and is predictable. We predict it—but do we act? Do we act before someone dies? No. Not often. Not often enough. We don't think about the human element. We forget the victims of hazardous materials spills—until they suffer or after they die, when we can no longer help them.

Is it fair? Don't mouth that platitude about "life not being fair"—we all know it isn't, but ideally, we strive to make it more so. Is it right? So what do we do about it? We legislate, of course.

Because of Bhopal and other similar (but less catastrophic) chemical spill events, the United States Congress (pushed by public concern) developed and passed specific environmental laws and regulations to regulate hazardous substances/wastes in the U.S. This section focuses on the two regulatory acts most crucial to the current management programs for hazardous wastes. The first (mentioned several times throughout the text) is the Resource Conservation and Recovery Act (RCRA). Specifically, RCRA provides guidelines for prudent management of new and future hazardous substances/wastes. The second act (more briefly mentioned) is the Comprehensive Environmental Response, Compensation, and Liability Act (CERCLA), or Superfund, which deals primarily with past mistakes: inactive and abandoned hazardous waste sites.

RESOURCE CONSERVATION AND RECOVERY ACT (RCRA)

The Resource Conservation and Recovery Act (RCRA) is the most important law for hazardous waste management in the United States. RCRA and its amendment, the *Hazardous and Solid Waste Act* (HSWA-1984), deal with the ongoing management of solid waste throughout the country, emphasizing hazardous waste. Keyed to the waste side of hazardous materials, rather than broader issues dealt with in other acts, RCRA is primarily concerned with land disposal of hazardous wastes. The goal is to protect groundwater supplies by creating a "cradle-to-grave" management system with three key elements: *a tracking system, a permitting system, and control of disposal.*

1. A **tracking system**—a **manifest** document accompanies any waste transported from one location to another.
2. A **permitting** system—helps ensure facilities that treat, store, or dispose of hazardous wastes operate safely.
3. A **disposal control** system—controls and restricts the **disposal** of hazardous wastes onto or into the land (Masters, 1991).

The RCRA regulates five specific areas for managing hazardous waste (with a focus on *treatment, storage,* and *disposal*). These are:

1. Identify what constitutes hazardous waste and provide a classification of each.
2. Publishing requirements for generators to identify themselves, which includes notification of hazardous waste activities and standards of operation for generators.
3. Adopting standards for transporters of hazardous wastes.
4. Adopting standards for treatment, storage, and disposal facilities.
5. Enforcing standards through a permitting program and legal penalties for noncompliance (Griffin, 1989).

Arguably, the RCRA is our single most important law regarding the management of hazardous waste—the most comprehensive piece of legislation that the EPA has promulgated to date.

CERLCA

The mission of the Comprehensive Environmental Response, Compensation, and Liability Act of 1980 (Superfund or SARA) is to clean up hazardous waste disposal mistakes of the past and to cope with emergencies of the present. More often referred to as the *Superfund Law,* a large trust fund (about $1.6 billion) was created due to its key provisions. Later, in 1986, when the law was revised, this fund was increased to almost $9 billion. The revised law is designated as the *Superfund Amendments and Reauthorization Act of 1986 (SARA).* The critical requirements under CERCLA include:

1. CERCLA authorizes the EPA to deal with short-term problems (emergencies triggered by a spill or release of hazardous substances) and long-term issues involving abandoned or uncontrolled hazardous waste sites for which more permanent solutions are required.
2. CERCLA has set up a remedial scheme for analyzing the impact of contamination on sites under a hazard ranking system. This hazard ranking system compiles a list of prioritized disposal and contaminated sites. When promulgated, this list becomes the National Priorities List (NPL). The NPL identifies the worst sites in the nation based on such factors as the quantities and toxicity of wastes involved, the exposure pathways, the number of people potentially exposed, and the importance and vulnerability of the underlying groundwater.
3. CERCLA also requires parties responsible for hazardous waste problems to pay the entire cost of cleanup.
4. Title III of SARA requires federal, state, and local governments and industry to collaborate in developing emergency response plans and reporting on hazardous chemicals. This requirement is commonly known as the Community Right-To-Know Act, which allows the public to obtain information about the presence of hazardous substances in their communities and their releases into the environment.

WASTE CONTROL TECHNOLOGY

> How to handle society's toxic chemical waste now ranks among the top environmental issues in most industrial countries. Without concerted efforts to reduce, recycle, and reuse more industrial waste, the quantities produced will overwhelm even the best treatment and disposal systems.
>
> —*Sandra Postel, p. 37, 1987*

One of the most challenging and pressing current environmental concerns (dilemmas) confronting environmental scientists (and many others) is what to do with all the solid and hazardous wastes our throwaway society produces. In simple (and simplistic) terms, we could say that we should shift from a throwaway society to a recycling one, which would help restore a gain in our living standards. We could also say that since disposing of hazardous waste is so expensive and risky, to make the situation better, we should follow RCRA's Waste Management Hierarchy (in descending order of desirability) to: (1) stop producing waste in the first place; (2) if we cannot avoid producing it, then produce only minimum quantities; (3) recycle it; (4) if it must be produced but cannot be recycled, then treat it; (5) if it cannot be rendered non-hazardous, dispose of it safely; and (6) once it is disposed, continuously monitor it to ensure no adverse effects on the environment.

All of these statements have merit. The question is, are they realistic? To a point, yes. We have developed several strategies to curb the spread of hazardous substances/wastes. One approach is treating hazardous wastes to neutralize or make them less toxic. However, again, a better strategy would be to reduce or eliminate the use of toxic substances and the generation of hazardous waste. To a degree, we can accomplish this, but to think that we can do away with all our hazardous materials, processes that use hazardous materials, and processes that produce hazardous materials is, at present, wishful thinking.

We must refine our waste reduction programs as much as possible and develop technologies to treat waste products we cannot replace, do away with, or reduce. Such technologies and practices are available to us today. Environmental science and technology can be used to develop measures and practices by which hazardous chemical wastes can be minimized, recycled, treated, and disposed of. In this section, we review these measures, practices, and technologies.

WASTE MINIMIZATION

Waste minimization (or source reduction measures) can be accomplished in various ways, including feedstock or input substitution, process modifications, and sound operating practices. However, considerable information must be gathered before any source reduction measure can be implemented.

One of the first steps in the information-gathering process is determining the exact nature of the waste produced. The waste must initially be characterized and categorized by type, composition, and quantity, a task accomplished by performing a **chemical process audit** or **survey** of the chemical process. Remember that during this information-gathering survey, it is essential to look closely for any off-specification input materials that might produce defective outputs, accidental contamination of inputs, process chemicals and outputs, and any obsolete chemicals (which should be appropriately disposed of).

During the survey, particular attention should be given to problem areas—excessive waste per production unit, excessive process upsets or bad batches, or frequent off-specification inputs. The effect of process variables on the waste stream created and the relationship of waste stream composition to the input chemicals and process methods used should be examined. For example, determine precisely how much process water is used. Can the amount of water used be reduced? Can process water be reused? Questions like these should be addressed during the chemical process survey (Lindgren, 1989).

The exact nature of the waste must also be determined to assess the feasibility of reuse, recycling, materials recovery, waste transfer, or proper waste disposal methods. Usually accomplished through sampling the waste stream and then analyzing the sample in the laboratory, the nature of the waste can yield valuable information about the industrial process and the condition of process equipment.

Substitution of Inputs

After completing the chemical process survey, the information gathered may suggest or justify the substitution of certain chemicals, process materials, or feedstock to enable the process of hazardous wastes to be reduced in volume or no longer produced. Note that input substitutions are often inseparable from process modifications. A few specific examples of possible input substitutions include:

- Use of synthetic coolants in place of emulsified oil coolants;
- Use of water-based paints instead of solvent-based paints;
- Use of non-cyanide-based electroplating solutions;
- Use of cartridge filters instead of earth filters.

Process Modifications

One key benefit of performing a chemical process audit or survey is that audits often suggest modifications to production systems that minimize hazardous waste stream production. Whenever a chemical process can be made more efficient, the volume and toxicity of the residuals usually reduce.

Good Operating Practices

Good operating practices are directly related to reducing wastage, preventing accidental chemical releases, and increasing the useful lifetime of process chemicals. Ensuring good operating practices by workers can only be accomplished through practical worker training. This training should include proper process operations and practical spill response training.

RECYCLING

> Use it up, wear it out, make it do, or do without.
>
> —*New England Proverb*

If waste generation is unavoidable, strategies such as recycling should be pursued to minimize the waste to the greatest extent possible. Recycling is collecting and processing materials that would otherwise be thrown away as trash and turning them into new products. Recycling can benefit communities and the environment. Benefits of recycling include:

- Reduces the amount of waste sent to landfills and incinerators
- Conserves natural resources such as timber, water, and minerals
- Prevents pollution by reducing the need to collect new raw materials
- Saves energy
- Reduces greenhouse gas emissions that contribute to global climate change
- Helps sustain the environment for future generations
- Helps create new well-paying jobs in the recycling and manufacturing industries in the United States

Steps to Recycling Materials

Recycling includes three steps: collection and processing, manufacturing, and purchasing new products made from recycled materials, which create a continuous loop.

Collection and Processing

Recyclable methods include curbside collection, drop-off centers, and deposit or refund programs. After collection, recyclables are sent to a recovery facility to be sorted, cleaned, and processed into materials that can be used in manufacturing. Recyclables are bought and sold just like raw materials, and prices fluctuate depending on supply and demand in the United States and the world.

Manufacturing

More and more of today's products are being manufactured with recycled content. Common household items that contain recycled materials include:

- Newspapers and paper towels
- Aluminum, plastic, and glass soft drink containers
- Steel cans
- Plastic laundry detergent bottles

Recycled materials are also used in new ways, such as recovered glass in asphalt to pave roads or recovered plastic in carpeting and park benches.

Purchasing New Products Made from Recycled Materials

By buying new products made from recycled materials, you help close the recycling loop. There are thousands of products that contain recycled content. When you go shopping, look for:

- Products that can be easily recycled
- Products that contain recycled content

Here are some of the terms used:

- **Recycled-Content Product**: This means the product was manufactured with recycled materials, either collected from a recycling program or waste recovered during the normal manufacturing process. Sometimes, the label will tell you how much of the content was from recycled materials.
- **Postconsumer Content**: This is similar to recycled content, but the material comes only from recyclables collected from consumers or businesses through a recycling program.
- **Recyclable Product**: These products can be collected, processed, and manufactured into new products after use. These products do not necessarily contain recycled materials.

Some of the standard products you can find that can be made with recycled content include:

- Aluminum cans
- Car bumpers
- Cereal boxes
- Comic books
- Egg cartons
- Glass containers
- Laundry detergent bottles
- Motor oil
- Nails
- Newspapers
- Paper towels
- Steel products
- Trash bags

Recycling Hazardous Wastes

Regarding hazardous wastes, various strategies have been developed to *recycle* (and thus minimize) the volume of hazardous wastes to dispose of. These strategies recover or recycle resources, either materials or energy, from the waste stream. The critical point to note in chemical process recycling is that the product must be processed before reuse. Wastes generally recognized as having components of potential value include:

- Flammable and combustible liquids
- Oils
- Slags and sludge
- Precious metal wastes
- Catalysts
- Acids
- Solvents

The list above shows that one such recycling or recovery effort involves reclaiming organic solvents. Usually accomplished using highly effective distillation techniques, solvents contaminated with metals and organics are heated to produce liquid and vapor phases. Lighter components with high volatiles rise to the top of the liquid phase and begin to vaporize. By carefully controlling the waste mixture's temperature, the desired substance can be vaporized and recovered by condensation, leaving the heavier contaminants behind. A concentrated, highly toxic mixture (far reduced in volume) referred to as still bottoms remains. Bottoms may contain usable metals and other solvents. As distillation technology improves, more bottom materials will be recovered and possibly reused.

TREATMENT TECHNOLOGIES

Because of the 1984 and 1991 amendments to RCRA, hazardous wastes must be treated before ultimate disposal in a landfill. Even with process modifications, material substitution, and recycling, some portions of waste streams may still be hazardous—and must be properly contained. These hazardous waste components require additional treatment. Such treatment occurs in vessels (tanks), reactors, incinerators, kilns, boilers, or impoundments.

Several technologies are currently available to treat hazardous waste streams. This section discusses a few examples, including biological treatment, thermal treatment, activated carbon sorption, electrolytic recovery techniques, air stripping, stabilization and solidification, and filtration and separation treatment systems.

***Note*:** Some of these treatment techniques were covered in greater detail earlier, such as in situ and non-in situ soil contamination treatment; some technologies combine two or more of these basic technologies.

Biological Treatment

Several *biological treatment processes* are available for treating "liquid" hazardous waste streams (contaminated soils and solids are more difficult to treat), including activated sludge, aerobic lagoons, anaerobic lagoons, spray irrigation, trickling filters, and waste stabilization ponds. These processes are typically associated with the biological treatment of municipal and industrial wastewater and are generally used to remove organic pollutants from wastewater. Generally effective on wastewaters with low-to-moderate concentrations of simple organic compounds and lower concentrations of complex organics, these are usually ineffective in attacking mineral components and useless against heavy metals.

Biological treatment of toxic organic components requires considerably more sophisticated operational control (including pretreatment) than is necessary with nontoxic wastewater. Rapid increases in food rate can easily destroy microorganisms used in biological treatment processes. Acclimation and development of a functional population of biota may require considerable time, and the system is continuously subject to upset (Blackman, 1993).

The two biological processes used for treating toxic waste are aerobic (treatment in the presence of oxygen—conventional aeration) and anaerobic (treatment in the absence of oxygen—in a simple septic tank).

In aerobic treatment, organisms require both an energy source and a carbon source for growth, and both affect what type of organisms will grow in a particular environment. Many hazardous waste streams satisfy both basic requirements, and if appropriate nutrients are present, a thriving organic population for waste treatment can exist. Under these conditions, if pH and temperature are controlled, substances that are toxic to the active organisms can be eliminated.

The most essential aspect limiting the applicability of aerobic biological treatment of hazardous waste is the waste's biodegradability—its conversion by biological processes to simple inorganic molecules and biological materials. The biodegradability of a particular waste is very system-specific, and the correct conditions for successful treatment (detoxification or biological conversion of a toxic substance to one less toxic) must be maintained to encourage the correct microbe mixture.

Anaerobic treatment of toxic waste streams has been effectively practiced on many different types of toxic waste streams. This treatment is a fermentation process in which organic waste is oxidized and reduced.

Thermal Processes

Thermal treatment processes (incineration) are commonly used to treat both liquids and solids, either to destroy the hazardous components or to allow the disposal of the process residue or treated waste in an EPA-approved hazardous waste landfill.

During *incineration*, carbon-based (organic) materials are burned at high temperatures—typically ranging from 1,500°F to 3,000°F—to break them down, chiefly into hydrogen, carbon, sulfur, nitrogen, and chlorine. These constituent elements combine with oxygen to form inorganic gases like water vapor, carbon dioxide, and nitrogen oxides. After combustion, the gases pass through a pollution control system to remove acidic gases and particulate matter before being released into the atmosphere.

The advantages of hazardous waste incineration are twofold: (1) it permanently reduces or eliminates the hazardous character of the waste, and (2) it substantially reduces the volume of the waste being disposed of.

Waste characteristics and treatment requirements determine the incinerator design to accommodate liquid or solid wastes. Temperature, turbulence, and retention time (commonly known as the 3 T's of incineration) are the prime factors determining incineration treatment design for both solid and liquid wastes.

The EPA regulates hazardous waste incinerators and requires a permit for operation. To receive an operating permit, an incineration facility must demonstrate a

99.99% destruction and removal efficiency (DRE) for each principal organic hazardous constituent in the feed material.

Non-in situ thermal processes (primarily incinerators) include designs such as liquid injection and boilers, rotary kilns, fluidized beds, and catalytic processes. More sophisticated and less common thermal treatment systems include wet oxidation, pyrolytic, and plasma processes. Some of these processes, such as steam injection, radio frequency heating, and vitrification (molten glass treatment), can be conducted in situ.

Activated Carbon Sorption

Organic substances may be removed from aqueous hazardous waste streams with activated carbon by sorption. Sorption is the transfer of a substance from a solution to a solid phase. In adsorption (a final chemical reaction that forms a cementitious precipitated sludge, not to be confused with absorption, defined as a physical process that does not chemically stabilize a waste material), chemical substances are removed from the waste stream onto a carbon matrix. The carbon may be used in granular or powdered form depending on the application.

The effectiveness of activated carbon in removing hazardous constituents from aqueous streams is directly proportional to the amount of surface area of the activated carbon; in some cases, it is adequate for complete treatment. It can also be applied to the pretreatment of industrial hazardous waste streams before follow-up treatment. Activated carbon sorption is most effective for removing dangerous waste materials that are not water-soluble from water.

Electrolytic Recovery Techniques

The *electrolytic recovery technique* (used primarily to recover metals from process streams to clean process waters or to treat wastewater before discharge) is based on the oxidation-reduction reaction, where electrode surfaces collect the metals from the waste stream.

An electrolytic recovery system typically consists of a treatment vessel (tank, etc.) with electrodes, an electrical power supply, and a gas handling and treatment system. Recovered metal must be removed from the electrodes periodically when the design thickness is achieved for the recovered metal.

Air Stripping

The *air stripping* technique for removing hazardous constituents from waste streams, although not particularly effective, has been used for many years. *Stripping* separates volatile components from less volatile ones in a liquid mixture by partitioning the more volatile materials into a gas phase of air or steam. In air stripping, the moving gas is usually ambient air, which removes volatile dissolved organic compounds from liquids, including groundwater and wastewater. Additional treatment must be applied to the exhaust vapors to destroy and capture the separated volatiles. The process is driven by the concentration gradient between air and liquid phase equilibrium for particular molecules according to *Henry's Law*, which states that at a constant temperature, the weight of gas absorbed by a given volume of a liquid is proportional to the pressure at which the gas is supplied.

Stabilization and Solidification

Stabilization and *solidification* are techniques used to convert hazardous waste from its original form to a physically and chemically more stable material. They accomplish this by reducing the mobility of hazardous compounds in the waste before its land disposal. Stabilization and solidification are beneficial when recovery, removal, or conversion of hazardous components (as required by RCRA) from waste before disposal in a landfill is not possible.

Various stabilization and solidification treatment processes use Portland cement as a binding agent. Waste/concrete composites can be formed that have exceptional strength and excellent durability and retain waste very effectively (Blackman, 1993). Stabilization and solidification treatment processes improve handling and physical characteristics and reduce the solubility or limit the leachability of hazardous components in waste.

Filtration and Separation

Filtration and separation are physical processes for hazardous waste treatment. Filtration (the separation of solid particles from a liquid stream through the use of semi-porous media) is driven by a pressure difference across the media. This pressure difference is caused by gravity, centrifugal force, vacuum, or elevated pressure.

Filtration applied to hazardous waste treatment falls into clarification and dewatering. Clarification occurs when liquids less than 200 ppm are placed within a clarifier, and the solids are allowed to settle out, producing a cleaner effluent. Dewatering is performed on slurries and sludge. Dewatering aims to concentrate the solids into a semi-solid form for further treatment or land disposal.

ULTIMATE DISPOSAL

Most of us are familiar with open dumps. However, you might not be familiar with some of the dumping practices that occurred because of environmental legislation of the 1970s, which placed increasingly stringent controls on releases to the atmosphere and the nation's waterways. To protect our atmosphere and our waterways, the 1970s legislative mindset drove us to dump hazardous materials into open dumps. Why not? Land disposal is safer and proper, isn't it?

No, it is not. But we didn't realize this until the tragic consequences of these practices became apparent later—and today, we're still cleaning up the resulting mess.

We are now well aware that the land is not a bottomless sink that can absorb all of our discards. We've learned that we must pretreat our wastes to detoxify them, degrade them, make them less harmful, and make them more earth-friendly before we deposit them on or in the ground, the soil, the land—our Earth.

Regardless of the treatment, destruction, and immobilization techniques used, some residue(s) from hazardous wastes that must be contained somewhere will remain. This "somewhere" is a burial in the land, deep-well injection, surface impoundments, waste piles, or landfills. In this section, we discuss each of these ultimate disposal methods.

Deep-Well Injection

The practice of *deep-well injection* is not new; it was used in the 1880s by the petroleum industry to dispose of salt water produced when drilling for oil. However, disposing of hazardous materials by deep-well injection is relatively recent. The EPA estimates that about 9 billion gallons of all the hazardous waste produced in the United States (about 22% of the total) is injected deep into the ground. Most deep-well injection sites are in the Great Lakes region and along the Gulf Coast.

Deep-well injection involves injecting liquid waste under pressure into underground strata isolated by impermeable rock, where geologists believe it will be contained permanently and isolated from aquifers. This is typically done at a depth of more than 700 m below the surface. A high-pressure pump forces the hazardous liquids into pores in the underground rock, where they displace the water, oil, and gases originally present. Sandstone and other sedimentary rock formations are used because they are porous and allow the movement of liquids.

In theory, when properly constructed, operated, and monitored, deep-well injection systems may be the most environmentally sound disposal method for toxic and hazardous wastes currently available. However, with anything else that, in theory, is "perfect" or affords us the "best available technology," deep-well injection has its problems. For example, though constructed at a depth below the groundwater table, fractures in the underground geology could allow waste to go where it is not wanted, namely, into the groundwater. The biggest problem with deep-well injection concerns the unknown. We are not sure of the exact fate of hazardous substances after injection—another example of the "we don't know what we do not know" syndrome.

Because of our uncertainty about the results of our hazardous waste disposal practices, the 1984 amendments to RCRA banned unsafe, untreated wastes from land disposal. For those land disposal facilities allowed to accept hazardous substances, USEPA (1986) implemented restrictions requiring:

- Banning liquids from landfills;
- Banning underground injection of hazardous waste within 1/4-mile of a drinking water well;
- Requiring more stringent structural and design conditions for landfills and surface impoundments, including two or more liners, leachate collection systems above and between the liners, and groundwater monitoring;
- Requiring cleanup or corrective action if hazardous waste leaks from a facility;
- Requiring information from disposal facilities on pathways of potential human exposure to hazardous substances;

- Requiring location standards that are protective of human health and the environment.

Surface Impoundments

Surface impoundments are diked or excavated areas used to store liquid hazardous wastes. They have been popular for many years because they are usually temporary, relatively cheap to construct, and allow easy access to treatment.

Unfortunately, in the past, surface impoundments were poorly constructed (quickly dug out or diked and put into operation), poorly sited (built on a thin layer of porous soil that allowed leachate to infiltrate to groundwater), located too close to sources of high-quality drinking water (wells or running water sources), and either not monitored at all or poorly monitored. In 1984, the USEPA estimated that of the more than 180,000 surface impoundments surveyed before 1980, only about 25% were lined, and fewer than 10% had monitoring systems.

Because of the problems associated with poor siting, construction, and management of the early surface impoundments, EPA regulations have toughened the requirements for constructing new surface impoundments. Under the *Hazardous and Solid Waste Amendments (HSWA)* of 1984, for example, the EPA now requires new surface impoundments to include

- The installation of two or more liners;
- A leachate collection system between liners;
- Groundwater monitoring.

Provisions must also prevent liquid from escaping due to overfilling or run-on and erosion of dams and dikes. During construction and installation, liners must be inspected for uniformity, damage, and imperfections. These liners must also meet permit specifications for materials and thickness.

Waste Piles

Waste piles are normally associated with industrial sites. For years, it was common practice to pile up industrial waste literally, and later, when the pile became "too large," dispose of it in a landfill. The industrial practice has been to list such piles as "treatment" piles and even 40 CFR 264/265 subpart L refers to such piles as treatment or storage units.

The environmental problem with such piles is similar to the issues we discussed regarding mining waste. Like mining, industrial waste piles are subject to weather exposure, including evaporation of volatile components into the atmosphere and wind and water erosion. The most significant problem related to industrial waste piles is precipitation—leaching of contaminants (producing leachate), which may percolate into the subsurface.

RCRA specifications for waste piles are similar to those for landfills (to be discussed in the next section) and are listed in 40 CFR 264/265 subpart L. Under the RCRA guidelines, the owner or operator of a waste pile used for storing or treating noncontainerized solid hazardous wastes can choose between compliance with the waste pile or landfill requirements. If the waste pile is used for disposal, it must comply with landfill requirements. The waste pile must be placed on an impervious surface, and a control and monitor system must be in place if leachate is produced. Waste piles must also be protected from wind dispersion.

Landfilling

Landfilling wastes has a history of causing environmental problems—including fires, explosions, production of toxic fumes, and storage problems when incompatible wastes are commingled. Landfills also have a history of contaminating surface and groundwater (EPA, 1990a).

Sanitary landfills are designed and constructed to dispose of municipal solid wastes only. They are not designed, constructed, or allowed to be operated to dispose of bulk liquids and/or hazardous wastes. Landfills that can legally receive hazardous wastes are known as *secure landfills.*

Under RCRA, the design and operation of hazardous waste landfills have become much more technically sophisticated. Instead of the past practice of gouging out a massive maw from the subsurface and then dumping countless truckloads of assorted waste materials (including hazardous materials) into it until complete, a hazardous waste landfill is now designed as a modular series of three-dimensional control cells. Design and operating procedures have evolved to include elaborate safeguards against leakage and migration of leachates.

Secure landfills for hazardous waste disposal are equipped with double liners. Leakage detection, leachate collection and monitoring, and groundwater monitoring systems are required (see Figure 10.2). Liners used in secure landfills must meet regulatory specifications. For example, the upper liner must consist of a 10–100-mil thick *flexible-membrane liner* (FML), usually made of rubber or plastic sheets. The lower liner is usually FML, but recompacted clay at 3 ft thick is also acceptable.

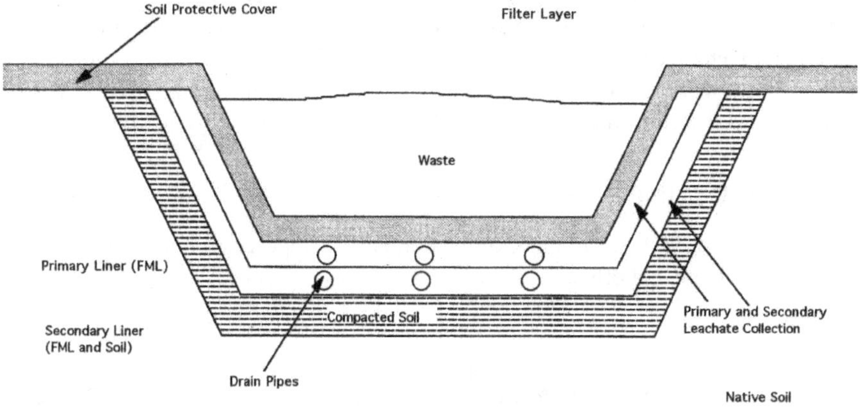

FIGURE 10.2 Cross-section of a secure landfill double liner system.

Secure landfills must be constructed to allow the collection of leachate (usually via perforated drainage pipes with an attached pumping system) that accumulates above each liner. Leachate control is critical. A low permeability cap must be placed over completed cells to aid in this control process (especially from leachate produced by precipitation). When the landfill is finally closed, a cap that will prevent leachate formation via precipitation must be implemented. This cap should be sloped to allow drainage away from the waste.

A landfill cannot be wholly abandoned, ignored, or forgotten when filled and capped. The site must be monitored to ensure leachate does not contaminate the groundwater. This is accomplished by installing a test downgradient to ensure the detection of any leakage from the site.

AI APPLICATIONS

Something must be done about the growing waste problem, whether in management, disposal, or reuse. The significantly rising amount of worldwide waste is bringing the issues of pollution, waste management, and recycling to the forefront of a problem begging for mitigation. We need to find new strategies to improve the waste ecosystem. The question is whether the application of artificial intelligence (AI) will be the White Knight that will solve this growing problem.

Currently, AI is being applied to waste-to-energy, waste-sorting robots, smart bins, waste generation models, waste monitoring and tracking, plastic pyrolysis, distinguishing fossil and modern materials, logistics, disposal, illegal dumping, resource recovery, intelligent cities, process efficiency, cost savings, and improving public health (Fang et al., 2023).

The paradigm shift from the old to the new, AI-based way of managing waste is applied in several areas of waste management to enhance efficiency and effectiveness. The critical applications of AI include predictive analysis for enforcement, which involves enforcing environmental regulations. This predictive analysis helps identify facilities that are more likely to violate rules, improving the efficiency of inspections. Moreover, AI models, such as random forest models, predict chemical exposure pathways. This is a better way to assess the potential risk posed to humans by various chemicals, aiding in better waste management practices. AI is also being leveraged to integrate into intelligent recycling bins and MRFs to improve the sorting and recycling process. These systems are critical because they identify and sort different types of waste more accurately, reducing contamination and increasing recycling rates.

THE BOTTOM LINE

AI applications used in waste management enhance the processes involved, making them more efficient and effective.

REFERENCES

Blackman, W.C. (1993). *Basic Hazardous Waste Management*, Boca Raton, FL: Lewis Publishers.

Carson, R. (1962). *Silent Spring*. Boston: Houghton Mifflin Company.

DOT 49 CFR-170-179, U.S. Department of Transportation

40 CFR 261.24, USEPA (1990).
40 CFR 261.31,32,33.
40 CFR 264.1.
40 CFR 264.52b.
40 CFR 302.4.
40 CFR 355.
Fang, B. et al. (2023). Artificial intelligence for waste management in smart cities: A review. *Environ. Chem. Lett*. 21:1–31.
Griffin, R.D. (1989). *Principles of Hazardous Materials Management*. Chelsea, MI: Lewis Publishers.
Lindgren, G.F. (1989). Managing Industrial Hazardous Waste. Chelsea, MI: Lewis Publishers.
Masters, G.M. (1991). *Introduction to Environmental Engineering and Science*. New York: Prentice-Hall.
O'Reilly, J.T. (1992). *State & Local Government Solid Waste Management*. Deerfield, IL: Clark, Boardman, Callahan.
Postel, S. (1987). *Defusing the Toxics Threat: Controlling Pesticides and Industrial Wastes, Worldwatch Paper 79*, Washington, DC: Worldwatch Institute, pp. 36–37.
Spellman, F.R. (1998). *Surviving an OSHA Audit: A Manager's Guide*. Lancaster, PA: Technomic Publishing Company.
Spellman, F.R. (2007). *The Science of Water: Concepts and Applications*. Lancaster, PA: Technomic Publishing Company.
Tchobanoglous, G., Theisen, H., & Vigil, S. (1993). *Integrated Solid Waste Management: Engineering Principles and Management Issues*. New York: McGraw-Hill.
USEPA. (1986). *Solving the Hazardous Waste Problem: EPA's RCRA Program*. Washington, DC: EPA Office of Solid Waste.
USEPA. (1989). *Decision-Makers Guide to Solid Waste Management*. Washington, DC: United States Environmental Protection Agency, EPA/530-SW89-072.
USEPA. (1990a). *Land Restriction for Scheduled Wastes*. Washington, DC: US Environmental Protection Agency.
USEPA. (1990b). *RCRA Orientation Manual*. Washington, DC: United States Environmental Protection Agency.
USEPA. (1992). *Characterization of Municipal Solid Waste in the U.S.: 1992 Update*. Washington, DC: United States Environmental Protection Agency, EPA/530--5--92-019.
USEPA. (2009). *MSW in the United States*. Washington, DC: United States Environmental Protection Agency.

11 Forestry and Agriculture

ENVIRONMENTAL ENGINEERING APPLICATIONS

Forestry, agriculture, and the environment are interconnected. Note that for our purposes in this discussion, forest stands are analogous to agricultural fields. A healthy environment is necessary for successful crop and livestock production and forest sustainability, including soil, forest lands, and water quality. Environmental engineers in agriculture and forestry play a crucial role in addressing the environmental impact of farming and forestry practices. They develop strategies for water quality, waste management, fire prevention and fire-fighting, and pollution prevention. Their work on projects related to soil conservation, watershed management, and sustainable land use planning is instrumental in maintaining a healthy environment.

The increasing demand for food and forest products in the twenty-first century necessitates the use of advanced technologies. Artificial intelligence (AI) and the Internet of Things (IoT) sensors, for instance, can monitor farmland and forest lands, recognizing changes and increasing demands. This technological advancement is a significant step toward ensuring the sustainability of our food supply and forest resources.

AI IN FOREST FIRE-FIGHTING

In August of 1960, during a hot season with little to no rain, I was one of several young men outfitted in helmets, goggles, gloves, clothing, boots, and masks called upon to fight forest fires around the Tri-Cities of Washington State (Kennewick, Pasco, and Richland), near the confluence of the Snake, Yakima, and Columbia Rivers. Other than the extensive fire damage, I remember my assigned duty during all the fires: swinging a hoedad back and forth or using a Pulaski (a real workhorse, and we all called them hoedads too—combined axe and adze) or a shovel, chainsaw, fire rake (aka McLeod, similar to hoedad), and equipped with a water backpack (quickly drained), safety gear—clearing sagebrush, groundcover, dry branches, twigs, and leaves to make a fire line that served as a fire break.

Did I ever actually see or combat the fire, the burn?

No. Only the results of the fires afterward. As it turned out, all 40 of us smoke-eaters sweltered in the 90-degree (°F) heat (and about half suffered heat stroke or other heat-related issues), but not from the fire; it went in a different direction. It was the high plains end-of-summer heat that was so hot.

Today, I think about the hoedad or Pulaski I used for hours, days, and months to create fire breaks and hopefully cut fires off from continuing along their path. When not fighting forest fires, I used those tools to plant thousands of Douglas Fir seedlings in remote western parts of Washington State.

DOI: 10.1201/9781003601562-13

I also think about forest firefighting today—2024—and compare it with firefighting in 1960. I have already listed the tools used in 1960, but we also had aircraft flying (air tankers) above the fire, watering down hotspots here and there, and D-9 bulldozers. The only other tool we had at our disposal—when they were operable—was walkie-talkies.

Today, forest firefighters have digital technologies to help fight wildfires; none were available in 1960. Current digital technologies play a significant part in both preventing and fighting wildfires. For example, consider satellites and their associated technologies. Satellites and software are used to pinpoint potentially dangerous fires. Drones are employed to monitor the fire with regular and thermal imaging cameras and can fly into areas where manned aircraft can't. Another technological development that keeps humans out of harm's way is the firefighting robot; these can fight fire in the worst or most extreme conditions. One of the most significant technological advancements is provided by virtual reality (VR), which trains firefighters. The IoT connects sensors to gather data from potential wildfire hotspots in remote areas.

Environmental engineers play a vital role in forestry management. They apply AI in forest firefighting activities, wildfire prevention and control programs, and sustainability. Their responsibilities include developing and implementing reforestation plans, conducting environmental impact assessments, designing and maintaining forest roads and infrastructure, and monitoring forest health. These tasks are essential for the sustainable management of our forest resources.

The Bottom Line: Digital technology provides early warning and detection systems, remote technologies, and digital connectivity, helping to make wildfire fighting a more practical exercise and driving down prevention costs.

AI IN AGRICULTURE

The interface between environmental engineering, agriculture, and forestry is a fascinating area of study, mainly because of environmental concerns and the ability to use current digital technology to make our food supply system more productive, sustainable, and efficient.

Making the food supply more productive, sustainable, and efficient becomes even more critical when considering the projected global population growth of almost 10 billion by 2050. Population growth will put more significant pressure on agriculture to produce food.

So, how do we address the looming food shortages? One argument I often hear about this topic is that some think the answer is simple: expand land use, plant more food, and harvest more.

Land, Land, and More Land! This is an accurate statement in one sense of the word: there is a lot of unused land throughout the globe. However, there is a problem for those who argue that we can solve all future food supply issues simply by farming more land.

These promoters of using more available and untilled, uncultivated land to grow more food forget that land is just one of the crop needs. If there is no water, there are no crops. But water is one of two other limiting factors. Land plus water without fertile soil is like a chocolate cake without the chocolate. We could also add the environmental conditions where land has water and might be fertile, but growing on a mountaintop, a glacier, or in many frozen soils is not a starter.

So, we have one alternative to growing more food: embracing innovative practices and technology. Innovative practices include introducing AI to solve many challenges and help diminish old, outdated, and inefficient farming practices.

Altering past practices and applying AI in farming does not precisely mean robots cultivating endless rows of soybeans (this practice may not be that far-fetched—technology is triggering changes quickly). What it does mean is automating farm machinery like driverless tractors. These autonomous tractors are set to revolutionize the agriculture industry by removing the human element and easing the labor supply.

AI in agriculture can help evaluate soil health, collect insights, monitor weather conditions, and recommend the application of fertilizer and pesticides.

The goal is to improve farm yields. Combined with AI, precision agriculture (giving soil and plants precisely what they need to ensure growth) can help farmers grow more crops with fewer resources. AI algorithms enable autonomous crop management, and when combined with IoT sensors that monitor soil moisture levels and weather conditions, they can aid in irrigation by deciding in real time how much water to provide to crops. AI-derived data can maximize yields while minimizing costs.

The labor problem in farming is not a new issue. However, AI offers a solution that doesn't involve hiring more workers. Traditional farming involves a range of manual processes, but AI models have several advantages in this regard. An intelligent agriculture system can handle many tasks, providing a sense of relief from the labor problem in farming.

AI plays a pivotal role in agriculture by detecting leaks in irrigation systems, thereby preventing water wastage and reducing operational costs. It also ensures the continuity of soil nutrients, aids in pest and disease control, and monitors livestock. AI's use of machine learning (ML) algorithms in yield management, analyzing large datasets in real-time, assists farmers in understanding their crops' characteristics and planning more effectively.

IOT SENSORS USED IN AGRICULTURE AND FORESTRY

As mentioned, AI applied in forestry and agriculture depends on various types of sensors to optimize operations. One type of sensor currently used is *location sensors*, which use GPS to triangulate a position. These devices allow for mapping fields, delineating boundaries, identifying variations in topography, and optimizing planting patterns. Because water is vital to agriculture and forestry, *soil moisture sensors* provide real-time. They operate on the principle of dielectric permittivity (which describes charge displacement when a voltage potential is applied), measuring the ability of soil to hold an electrical charge, which is affected by the amount of water present. This monitoring is essential for soil health, preventing waterlogging and soil compaction.

Optical sensors provide information about a crop's physiological status using light waves to assess critical indicators (such as chlorophyll in leaves) of food crops and forestry products. This helps farmers and foresters detect nutrient deficiencies and stress conditions early, allowing preemptive intervention to mitigate potential losses.

Engineers can use biosensors to gain comprehensive insights into the physiological status of their crops, forest stands, and the surrounding environment. These sensors, equipped with biological compounds like enzymes or antibodies, can detect specific biological molecules or pathogens in food, forest plants, or soil. Their ability to monitor plant stress responses and physiological changes induced by temperature fluctuations or drought is particularly noteworthy, making them valuable in maintaining plant health.

Another indispensable tool for real-time monitoring of critical environmental factors is *electrochemical sensors*. These sensors, often mounted on land-based robots convert chemical reactions into measurable electrical signals, allowing engineers to assess soil quality, monitor nutrient levels, and detect environmental contaminants. Moreover, these robots determine pH levels, nutrient concentrations, salinity levels, and the presence of any harmful pollutants.

Temperature and humidity are crucial in determining crop and forest plant health and productivity in the delicate balance of agricultural and forest land ecosystems. *Temperature and humidity sensors*, ranging from simple thermometers to sophisticated digital probes, are critical tools for optimizing growth conditions and mitigating the impact of climatic fluctuations.

Another type of sensor, *the mechanical sensor*, monitors soil quality, compaction, and resistance. These sensors also track repetitive stress and load cycles in tractor axles, bearings, and drive shafts (Sharp, 2024).

THE BOTTOM LINE

Some examples of where and how AI-type sensors are used in agriculture and forestry include automated farm machinery, driverless tractors, innovative irrigation fertilization systems, intelligent spraying, vertical farming software, and AI-based harvesting robots. One significant advantage of AI-assisted agriculture and forestry is the use of technology to sort produce and trees and grade them based on condition, color, shape, and size. Implementing AI and IoT in agriculture and forestry is not without challenges. One significant challenge is the integration of legacy systems with new technologies.

REFERENCE

Sharp, N. (2024). *7 Types of Agricultural Sensors Driving the Smart Farming Revolution*. Accessed 9/28/24 @www.escatec.com/blog/7-steps-of-agricultural-sensors-driving-the-smart-farming-recolution.

12 Green Energy

Conventional Energy: Coal, natural gas, oil, nuclear
Carbon Neutral Energy: Wind, solar, hydro, geothermal, biomass
Renewable Energy: Wind, solar, all hydro, geothermal, biomass
Green Energy: Wind, solar, small hydro, geothermal, biomass

WHAT IS GREEN ENERGY?

Studying green energy through science allows us to grasp the pressing need for green (a subset of renewable) energy and the complicated task of developing viable energy sources for the future.

Regarding future energy needs, there is no question or doubt that we are facing an international crisis of unprecedented proportions—we are headed for a train wreck—the end of affordable and accessible petroleum products and the resulting economic and political turmoil. Keep in mind that we not only need to replace liquid hydrocarbon fuels because the wells are almost dry but also that we need to clean up the environment—we need to produce reliable green/renewable energy that will not destroy or pollute our fragile environment. Even though peak oil is here, hydrocarbon energy resources are soon to be strained to the limit, and the train wreck of world economies collapsing will then move beyond a pending disaster to an actual catastrophe itself; we need to follow Adrienne Rich's perfect words and "dive into the wreck." A substantial and reliable supply of green energy—a subset of renewable energy—is the wreck-preventer we need. The train wreck can be avoided by pursuing emerging technologies. These include making oil from garbage, algae, plastics, sewage, agricultural and forestry biomass/waste to power fuel cells and full-sized plug-in diesel hybrid SUVs. Moreover, the development of these emerging technologies must be accompanied by advancements made in solar, wind, and ocean technologies.

Many of us have realized that we pay a price for what we call "the good life." Consumption of the world's resources makes us all at least partially responsible for the shortages and the resulting environmental pollution. Simply, it is a matter of record, proven that pollution (and its ramifications) is one of the inevitable products of "the good life" when maintained by fossil fuels. However, fuel shortages and pollution from fossil fuel usage are not something any single individual causes, can prevent, or correct. To reduce fuel shortages and their harmful effects, we must band together as an informed, knowledgeable electorate and pressure our elected decision-makers to manage the problem now for the future. We need to replace fossil fuels with clean and efficient renewable energy. Keep in mind, however, that we need to do this slowly. Change takes time, and we must ensure not only energy availability in the future but also that we have enough energy to support our needs. We need to quiet that vocal minority, the doomsayers and naysayers who shout that transitioning from our current fossil fuel-based economy to an economy based

DOI: 10.1201/9781003601562-14

entirely on renewable energy would be too "costly" and "impractical." This text was written not to silence these doubting Thomases but to make their arguments hollow and, at best, moot. We have an infinite source of green energy but only a limited amount of fossil fuel.

Because this is a science text, we have adhered to scientific principles, models, and observations. However, you need not be a scientist to understand the principles and concepts we present—we go easy on the complex math and science and present the material user-friendly—there is no failure to communicate. However, you need an open mind, a love for the challenge of wading through the muck, an ability to decipher problems, and the patience to answer questions relevant to each topic presented, maybe only for yourself. Real-life situations are woven throughout the fabric of this text and presented in straightforward, plain English to equip you with the facts, knowledge, and information you need to understand complex issues and make informed decisions.

The move toward green energy is driven by a relatively new awareness of what we are doing to our environment: our impact on our surroundings. Moreover, because of this awareness of our impact on the environment, the importance of transitioning from conventional energy resources to renewable sources—green energy sources—has become apparent to the average person, the consumer. The movement toward a sustainable environment is possible when we switch from conventional energy sources (hydrocarbon sources) to green energy sources (aka green power). The problem with the green energy movement is that many commonly use it to refer only to renewables, and others are confused by the term.

So, to clear up confusion and misunderstandings about the term green energy, we can say that it most commonly refers to the United States market, which provides a range of services and products with green energy. Remember that the term 'green energy' should not be confused with renewable energy. Still, unfortunately and incorrectly, the two terms are often confused and used interchangeably.

Green energy is generated from natural resources like water, wind, or sunlight. The confusion between green energy and renewable energy is that green energy comes from renewable energy sources, but there are some differences between renewable and green energy.

Remember that these two energy resources do not harm the environment regarding releasing greenhouse gas emissions into the atmosphere.

Again, green energy often comes from renewable technologies such as solar, wind, biomass, geothermal, and hydroelectric power. Note that these technologies work in different ways, whether taking power from the sun, as with solar panels, using water flow, or using wind turbines to generate energy.

The critical point here is that to be regarded as green energy, a resource can't produce pollution. Of course, this is not the case with fossil fuels. What this means is that not all sources of renewable energy are green. A good example is power generation, which burns organic material from sustainable forests. This may be renewable, but it is not necessarily green due to the carbon dioxide (CO_2) produced by the burning process itself.

With green energy sources, it is all about many of them being able to be replenished naturally. In contrast, fossil fuel sources like natural gas or coal can't be replenished quickly—millions of years are required to renew these sources. A considerable advantage of green energy sources is that they do not disturb ecosystems by drilling or mining.

Now, let's shift gears regarding future energy needs. Undoubtedly, we are facing an international crisis of unprecedented proportions—we are headed for a train wreck—the end of affordable and accessible petroleum products and the resulting economic and political turmoil. Keep in mind that we not only need to replace liquid hydrocarbon fuels because the wells are almost dry but also that we need to clean up the environment—we need to produce reliable renewable energy that will not destroy or pollute our fragile environment. Even though peak oil is here, hydrocarbon energy resources are soon to be strained to the limit, and the train wreck of world economies collapsing will then move beyond a pending disaster to the actual disaster itself; we need to follow Adrienne Rich's perfect words and "dive into the wreck." A substantial and reliable supply of renewable energy is the wreck-preventer we need. The train wreck can be avoided by pursuing emerging technologies. These include making oil from garbage, algae, plastics, sewage, agricultural and forestry biomass/waste to power fuel cells and full-sized plug-in diesel hybrid SUVs. Moreover, the development of these emerging technologies must be accompanied by advancements made in solar, wind, and ocean technologies. Remember that if we shift to producing battery-powered vehicles versus internal combustion engines, we need to develop the technology involved further; we need to lessen range anxiety and the total end of operation of battery-powered cars because the batteries are integral and irreplaceable in the present configuration of the battery-powered cars—that is, when the vehicle's battery dies, so does the vehicle. We need to develop batteries with a more fantastic range and accessible, easy gas-station battery change-out capability to enable the continued operation of the car whenever battery changeout is required.

These statements about EV range anxiety demonstrate how complicated, perplexing, confusing, befuddling, and confounding it is when/while we attempt to transition from hydrocarbon-fueled to electric-powered vehicles. Umar Shakir (2023) points out that at the "present time, EV owners are overall not satisfied with the reliability of the charging infrastructure available in the U.S. … things are looking worse than last year."

Hypothetical Case Study 12.1

The race is on to do whatever it takes to develop a super battery or fuel cell to power the cars of the future. Accompanying specifications and directions state that this battery or fuel cell needs to have a large capacity (large ampere-hour capacity) and be continuously recharged via light-striking super-sensitive solar sensors. A certain amount of current or electron flow is necessary to charge any rechargeable battery. The solar super sensors we invented will produce substantial current flow. These super sensors are wired from the battery to at least four positions on the roof of our experimental automobile, where they can constantly be exposed to daylight. In addition, we know what you're thinking: it does get dark at night, so how will you continuously recharge the car's battery, fuel cell, or whatever else we call it? Good question.

The answer is simple: two additional wires and ultra-super light sensors (we also invented) will be run from the battery to the car's headlamps, one to each. Thus, during nighttime driving, the light from the headlamps will be sensed by the ultra-super sensors and sent back to the battery. We know electrical engineers are scratching their heads and wondering where we got such an unworkable idea. We are not finished with our explanation—hold onto your brain cells, digital or otherwise. An amplification or booster device (like a transistor that amplifies a signal but is much more powerful) is placed within the wiring between the battery and the external and headlamp cell connections. We are not talking about perpetual motion here but about perpetual energy.

Is this scenario possible? Absolutely. As Americans, we believe in the power of innovation, and history has shown that we can achieve the impossible. This project has the potential to revolutionize the future of renewable energy, and we are excited and hopeful about its possibilities.

The time for this project is now. We are at a critical juncture in the fight against climate change and renewable energy development. We must act with urgency and determination to make a difference.

Let's say we can create a power plant arrangement consisting of a powerful battery or fuel cell with a built-in recharge system like the one described. What's next? Another good question. What's next is the final ingredient in any innovative process: education, training, and on-the-job training (OJT). To produce innovation and research and development grants, we must have personnel who are highly educated and trained. Unfortunately, in the U.S., we are losing (or have already lost in some areas) our edge in higher education. The U.S. will not restore itself to preeminence in science, technology, engineering, and mathematics until we shift our focus, money, and talent to educate our youth in these critical subjects.

THE 411 ON ENERGY[1]

Defining energy can be accomplished by providing a technical definition or a characterization in layman's terms. Because this book aims to reach technical readers and a wide range of general readers, the definitions provided herein and hereafter are best described as technical-nontechnical-based. Energy can be defined in several ways. In the broad sense, energy means the capacity of something—a person, an animal, or a physical system (machine)—to do work and produce change. In layman's terms, energy is the amount of force or power that, when applied, can move one object from one position to another. It can also describe someone doing energetic things such as running, talking, and acting lively and vigorous. Science uses it to describe how much potential a physical system must change. It is also used in economics to describe the part of the market where energy is harnessed and sold to consumers. In this text, we define energy in technical-nontechnical-based terms as something that can do work or the capacity of a system to do work.

There are two primary forms of energy: kinetic and potential energy. *Kinetic energy* is energy at work or in motion; that is, moving energy—a car in motion or a rotating shaft has kinetic energy. In billiards, a player gives the cue kinetic energy when she strikes the ball with the cue. As the ball rolls, it exerts kinetic energy. When the ball meets another ball, it transmits its kinetic energy, accelerating the next ball. *Potential energy*, on the other hand, is stored energy, like the energy stored in a coiled or stretched spring or an object stationed above a table. A roller coaster has the most potential energy when stopped at the top of a long drop. Another source of potential energy is when a carbonated soda can remains unopened. The can is pressurized with gas that is not in motion but has potential energy. Once the can is opened, the gas is released, and the potential energy is converted to kinetic energy.

According to the Law of Conservation of Energy, energy cannot be created or destroyed but can be made to change forms. Moreover, energy stays the same when it changes from one form to another. Let's consider an example of the Law of Conservation of Energy: The energy of something is measured at the start; the energy changes from potential (stored) energy to kinetic (moving) energy and back again. At the end, the energy is measured again. The energy measured at the start is the same as that measured at the end; it will always be the same. One caveat to this explanation is that we now know matter can be converted into energy through processes that modify or amplify the Law of Conservation of Matter and Energy.

TYPES OF ENERGY

The primary sources of green energy are wind, solar, and hydroelectric power (including pumped hydropower storage systems and tidal energy, which uses ocean energy from the tides in the sea). Both solar and wind power can be produced on a small scale for private use or business, or they can be designed and constructed on a larger, industrial scale.

There are many types of energy, including:

- Kinetic (motion) energy
- Water energy
- Potential (at rest) energy
- Elastic energy
- Nuclear energy
- Chemical energy
- Sound energy
- Internal energy
- Heat/Thermal energy
- Light (radiant) energy
- Electric energy

Energy sources can also be categorized as renewable or nonrenewable. The Energy Information Administration (EIA) (2009) points out that when we use electricity in our homes, the electrical power is probably generated by burning coal, a nuclear reaction, or a hydroelectric plant at a dam. Therefore, coal, nuclear, and hydropower are called energy sources. When we fill up a gas tank, the source might be petroleum or ethanol made by growing and processing corn.

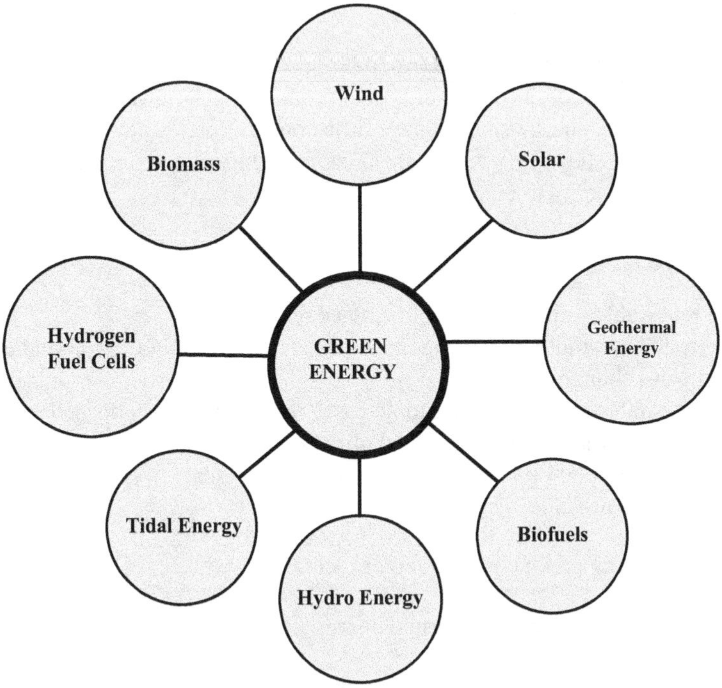

FIGURE 12.1 Renewable-green energy sources.

As mentioned, energy sources are divided into two groups*:* *Renewable*: An energy source that can be quickly replenished, including green energy and *Nonrenewable*: An energy source that we are using up and cannot recreate. For example, petroleum was formed millions of years ago from the remains of ancient sea plants and animals. In the United States, most of our energy comes from nonrenewable sources. Coal, petroleum, natural gas, propane, and uranium are nonrenewable energy sources. They are used to make electricity, heat our homes, move our cars, and manufacture all kinds of products. Both green (renewable) and nonrenewable energy sources can produce secondary energy sources, including electricity and hydrogen. Renewable/green energy sources include (see Figure 12.1).

NONRENEWABLE ENERGY VS. GREEN ENERGY

Green energy (an energy source that can be quickly replenished) is the focus of this text. Unfortunately (depending on your point of view), nonrenewable energy sources on Earth are available in limited quantity and may vanish within the next 100 years. Moreover, remember that nonrenewable sources are not environmentally friendly and can seriously affect our health. Notwithstanding the environmental and health impacts of using nonrenewable energy sources, it is essential to point out both sides of the argument, i.e., the benefits derived and non-benefits from using these sources. For example, nonrenewable energy sources are beneficial in that:

Nonrenewable Energy Benefits

- Nonrenewable sources are easy to use.
- A small amount of nuclear energy will produce much power.
- Nonrenewable energy sources have little competition.
- Nonrenewable energy sources are relatively cheap when converting from one type of energy to another.

Non-Benefits

- Nonrenewable sources will expire someday.
- The speed at which such resources are used can bring about profound environmental changes.
- Nonrenewable sources release toxic gases into the air when burned and can further exacerbate ongoing cyclical climate change.
- Because nonrenewable sources are becoming scarcer, prices of these sources will begin to soar.

The benefits of using green energy sources include:

- Wind, sun, ocean, and geothermal energy are available in abundant quantities and free to use.
- Green energy sources have low carbon emissions and are considered environmentally friendly.
- Green energy helps stimulate the economy and creates job opportunities.
- Green energy sources enable the country to become independent, not relying on foreign (often hostile) sources.

Non-benefits of green energy sources include:

- Initial setup costs of renewable energy sources are pretty high.
- Solar energy is limited to daytime use and not during the night or rainy season.
- Geothermal energy can bring toxic chemicals beneath the earth's surface onto the top, creating environmental damage.
- Hydroelectric dams are expensive and can affect natural flow and wildlife.
- Wind energy production requires high winds and must be appropriately sited. Also, they are tall structures that can affect bird populations. There have also been incidents where wind turbines tip over or are poorly constructed internally. Note this finding is based on the author's personal experience and studies.

DID YOU KNOW?

The largest consumer of fossil fuels in modern agriculture is ammonia production (for fertilizer) via the Haber process, which is essential to high-yielding intensive agriculture. The specific fossil fuel input to fertilizer production is primarily natural gas, to provide hydrogen via steam reforming.

Four major sectors of the economy share the use of energy in the United States. Each end-use sector consumes electricity produced by the electric power sector (EIA, 2023):

- **Commercial**: 18%—includes buildings such as offices, malls, stores, schools, hospitals, hotels, warehouses, restaurants, places of worship, and more.
- **Industrial**: 33%—includes facilities and equipment used for manufacturing, agriculture, mining, and construction.
- **Residential**: 21%—consists of homes and apartments.
- **Transportation**: 28%—comprises vehicles that transport people or goods, such as cars, trucks, buses, motorcycles, trains, subways, aircraft, boats, barges, and even hot air balloons.

Primary energy consumption in the United States was almost three times greater in 2012 than in 1949. Between 1949 and 2012, it increased over the previous year in all but 18 years.

The year 2009 sharply contrasted the historical trend, partly due to the economic recession. Actual gross domestic product (GDP) fell 2% compared to 2008, and energy consumption declined by nearly 5%, the most significant single-year decline since 1949. Decreases occurred in all four major end-use sectors: residential—3%, commercial—3%, industrial—9%, and transportation—3% (EIA, 2023).

GREEN ENERGY FORMS

The forms of green energy include, as shown in Figure 12.1, and are listed as follows:

- Solar
- Hydro
- Wind
- Geothermal
- Biomass
- Ocean/Marine Power (tidal power)
- Biofuels
- Hydrogen fuel cells

AI IN ENERGY

Artificial intelligence (AI) has the potential to revolutionize several vital industries, and the renewable energy sector is no exception. AI has applications in every facet of green energy production. However, for this discussion, wind power generation is singled out to provide an excellent example of where AI can help renewable energy in wind-produced energy. AI is already employed effectively in several aspects of wind energy production. Wind turbines rely on the wind to produce energy, and AI improves planning and operational efficiency. The idea, of course, is to keep the wind turbines turning to produce energy. Unnecessary shutdowns due to environmental hazards and weather are reduced, and AI can aid in preventing equipment malfunctions and damage caused by atmospheric conditions. AI also predicts maintenance needs to optimize wind turbine performance. AI tools can also be used to monitor the performance of turbines in real-time and automate turbine inspection. The benefit is that humans may miss or overlook defects, while AI tools can identify potential problems, helping to optimize system operations.

THE BOTTOM LINE

Green energy production and use are increasing because it's cheaper, much better for our health, and great for the environment. AI accurately predicts demand for green energy generation, production, and power usage, enabling green energy sources to transmit power more efficiently and safely. In this author's opinion, we have only scratched the surface of using AI to make energy generation sustainable—it is the future because green energy coupled with AI holds great promise for the future.

NOTE

1 Much of the information in this section is from F.R. Spellman (2014). *The Environmental Impacts of Renewable Energy.* Boca Raton: CRC Press.

REFERENCES

EIA. (2023). *Use of Energy in the United States Explained.* U.S. Energy Information Administration. Accessed 02/10/23 @ https://www.eia.gov/energyexplianed/index.cfm?

Shakir, U. (2023). *EV Charging in the US Is Still a Bad Time—And Somehow Getting Worse.* Accessed 8/16/23 @ The Verge. https://www.theverge.com/about-the-verge.

13 Land Surveying

INTRODUCTION

Integrating artificial intelligence (AI) and machine learning (ML) into land surveying increases accuracy, efficiency, and safety. Sixty-five years ago, I was part of a State of Alaska surveying party working on what would become the Fairbanks to Anchorage (or Anchorage to Fairbanks) highway. I was a rod and chainman and then a level operator. We were surveying the wild, close to Denali Park, 75 miles from Fairbanks and roughly 200 miles north of Talkeetna, Alaska. During my six months of hiking the terrain following survey flags that had been hung during the previous summer—surveying in the wild stretches of Alaska in the winter was a non-starter.

Anyway, my crew of three surveyed miles of the flagged terrain, all armed to the teeth. We were warned that Grizzly Bears were common in the area—at least, some environmental engineers told us. Having been an early student of wildlife management and the wild, I knew that Grizzly Bears in that area were unlikely for one main reason: Wolverines.

For the uninformed, a Grizzly Bear-Wolverine fight is unlikely. We were also told once that the Earth is the center of the universe, even though Galileo proved this false. Knowing that the Wolverine and Grizzly Bear are territorial animals and avoid each other unless there is a food shortage in one or the other's territory, I still carried a handgun. I also kept a loaded 30.06 rifle within reach at all times. For defense against bears? No. I feared the Wolverine most; they are fearless and will take on both the bear and the human. They are tenacious animals that are not afraid of anything, no matter what. During my time on the flagged-to-be highway stretch, I saw one Wolverine in 6 months—it was way off in the distance, and my crew and I waited until it moved off.

So, along with being aware of the surroundings and the potential danger roaming alongside us, when I was promoted to level operator, not only did I go through on-the-job training on how to operate the device properly, but I also learned the leveling terms then and the ones used today in level operation. First, I learned about the *dumpy level* (builder's level or leveling instrument) versus the automatic level. I found the dumpy level—using optics to establish or check points in the same horizontal plane with a vertical staff to measure higher distances and transfer, measure, and set heights—is much easier to use or manipulate than the automatic level. The *automatic level* is a variant of the dumpy level that makes use of a compensator that ensures that the line of sight remains horizontal once the operator has roughly leveled the instrument; it uses adjustment screws to level the instrument. I was already familiar with the term *benchmark*—a fixed point of known elevation for which other points' *reduced level* (RL; it equates to elevations of survey points regarding a commonly assumed datum) is to be determined. The *datum surface* is the reference plane to which the RL of the other survey points is determined. Another term I learned was

DOI: 10.1201/9781003601562-15

backsight (BS). This is the first staff reading taken by the surveyor after the instrument is set up and leveled. Then, *foresight* (FS) is the last reading taken before changing the instrument to the other position. The *intermediate sights* are all readings taken between backsight and foresight. The *spirit level* indicates whether a surface is horizontal (level) or vertical (plumb). Later, I used a *laser level* consisting of a laser beam projector that can be affixed to a tripod and projects a fixed red or green beam along the horizontal and vertical axis. It turned out that the laser level became one of my best friends. This was the case even though, by then, I was an accomplished leveler; I measured borrow pits to ensure the contractor was accounting for the correct amount of gravel removed from the pit so that the State got an honest reading on the material within the barrow pit and was not overcharged. Even with my experience, I found that, on occasion, I had to remeasure a couple of barrow pits because my measurements did not make any sense to me. So, when I used the laser level, my accuracy in measurements improved dramatically.

THE BENEFIT OF AI IN SURVEYING

Technology is changing, advancing, and improving, which is true of surveying. Today, with the incorporation of AI, it would be almost impossible for me to make an inaccurate measurement. AI-powered drone software can perform repetitive tasks and complete time-consuming tasks like outlining standard features, including buildings, roads, and vegetation. In addition to allowing surveyors to convert aerial datasets into CAD site plans with fewer people, the AI-integrated drones provide high-quality images, videos, and analytics to complete report and analysis tasks—this reduces human error. AI technology also enables surveyors to analyze data and predict outcomes at a fraction of the time required to complete manual tasks. As AI technology advances, surveying will become more accurate and accessible.

THE BOTTOM LINE

The integration of AI brings several benefits to land surveying. However, it also brings challenges and ethical considerations. There are also ethical considerations related to data privacy, the responsible use of AI, and the potential for some of the surveyors I worked with to lose employment. As with all technological advancements, a balance must be struck between technological advancement and considerations essential for the sustainable growth of the profession.

14 Biotechnology

BY ANY OTHER NAME

Environmental biotechnology is used to study the natural environment and is applied to this study. It harnesses biological processes for their exploitation and use.

How is environmental biotechnology used and exploited?

Well, think of water and wastewater operations. Environmental technologists are often called water and wastewater operators.

Why?

Well, consider wastewater treatment. This green technology (this is what I call it) is nothing more than a stream in a box.

A stream in a box?

Yes, wastewater treatment uses the same microbiological technologies as a natural stream, which, given enough time and travel distance downstream, performs similarly to purify contaminated water.

Environmental engineers use their tools. Tools? Yes, absolutely. Environmental engineers have a toolbox complete with tools. What are these tools? Good question. The short list of these tools includes: knowledge of fluid mechanics, ecology, principles of toxicology, risk assessment, management principles, hydrogeology, modeling contaminant migration through multimedia systems, aquatic chemistry, environmental microbiology, applied statistical analyses, open channel hydraulics, field methods in habitat analysis and wetland delineation, principles of estuarine environments, hydrology, resources modeling, environmental sampling, sediment transport and river mechanics, geomorphic and ecological foundations of stream restoration, atmospheric chemistry, environmental chemicals, economic foundations for public decision-making, business law for engineers, environmental impact assessment, geographic information systems (GIS), global environmental sustainability, water resources management, sustainable development, green engineering, energy planning, renewable energy, innovative growth strategies for sustainable urban development and revitalization, environmental safety and health, and many other specialties.

Wastewater operations mimic Nature's Way (Biomimicry) in purifying contaminated water. Environmental biotechnology is often described as optimally using nature's microorganisms in an integrated process to clean up contaminated water. And this is a good thing. The environmental impact of wastewater treatment cannot be ignored. Biomimicry provides a fresh perspective by offering Nature's Way, which is efficient, environmentally friendly, and cost-effective.

A critical aspect of their profession is that environmental engineering and engineers should never forget Nature's Way, its models, systems, and elements because solutions to pollution are available if mimicked.

DOI: 10.1201/9781003601562-16

AI TO THE RESCUE

Have you ever wondered what today would be like if the great geniuses of the past had access to AI, ML, deep learning (DL), and the Internet of Things (IoT)?

Think about it. If we had the incomparable genius of Einstein, Marie Curie, Isaac Newton, Charles Darwin, Nikola Tesla, Galileo Galilei, Ada Lovelace (the world's first computer programmer), Pythagoras, Carl Linnaeus (Father of Taxonomy), Rosalind Franklin (whose groundbreaking work in molecular biology/DNA), and then Jonas Salk, Louis Pasteur and a few others hooked up with AI technology, what would this world look like today? Would we have a cure for cancer? Would we have the solution to climate change? Would we have renewable energy in full swing? Would we have a solution to capture carbon? Would we be better prepared for pandemics? Would we have a colony of humans on Mars? Would we have ready access to quantum computing? Would we have an environmentally sustainable economy? And there are so many other problems or issues that we have not solved to date—would genius coupled with AI be the solution to the issues facing humanity today?

Could the brilliance of history's past geniuses enhance AI and propel it to performance levels we have yet to conceive? Could they expedite scientific breakthroughs, leading us to a future of unprecedented discovery?

We need to consider and find the answers to the questions mentioned herein. Will AI and its subsets lead us to the promised land of solving problems—life-saving solutions?

I am optimistic that AI will assist us in ways we can't imagine. In biotechnology, AI already plays a crucial role in problem-solving, mainly because of its ability to sift through mountains of data in record time. AI is a groundbreaking force with unlimited potential and is already important in its ability to process substantial complex biological data and recognize patterns—AI might be able to identify those technical principles beyond human skills. AI is enabling medical professionals to perform precision medicine and personalized healthcare. Hopefully, healthcare professionals can use AI to administer predictive and precision healthcare. AI application in research to discover new drugs is significant because it can accelerate the discovery process and save lives. AI-driven robots can perform many repetitive tasks without getting bored or hurt. AI is used in image analysis and medical imaging to help find tumors and other medical disorders.

THE BOTTOM LINE

AI-driven technologies are being used in some applications, but we have not yet reached the edge of the possibilities of AI and ML for public health. Consider, for example, that AI technology can forecast trends in opioid overdose mortality—again, AI technologies and uses are a work in progress. Used ethically, AI-driven biotechnology is promising—it is a shame that the geniuses of the past did not have access to computers and AI technology. Environmental biotechnology is genuinely irreplaceable as a professional discipline.

15 Environmental System Modeling

WHAT IS A MODEL

According to the Environmental Protection Agency (EPA, 2009a), a model is defined as:

> A simplification of reality constructed to gain insights into select attributes of a physical, biological, economic, or social system. A formal representation of the behavior of system processes, often in mathematical or statistical terms. The basis can also be physical or conceptual.

According to NRC (2007), a model is defined as:

> A simplification of reality that is constructed to gain insights into select attributes of a particular physical, biological, economic, or special system.

For all readers who have read my written works, you know that I always say, at the beginning of any of my books, as Voltaire advised, "If you wish to converse with me, please define your terms."

Note: Models should be and typically are developed for a well-defined system and a set of conditions under which the use of the model is scientifically defensible—**the application niche** (app niche or niche app). An app is a unique program created with a specific purpose and targets a particular industry. Examples of app niches include artificial intelligence (AI), machine learning (ML), the Internet of Things (IoT), augmented reality (AR), and virtual reality (VR).

During my tenure as an assistant professor of environmental engineering and health, more than one of my students mentioned that they wanted vital terms and definitions right up front in any presentation and not stuck away at the end as a glossary. So, key terms and definitions are presented here at the beginning of this presentation.

KEY TERMS AND DEFINITIONS (USEPA, 2009A)

Algorithm: A precise rule (or set of rules) for solving some problem. Linking environmental models with evolutionary and other metaheuristic optimization algorithms ("guided search") seeks to find the global optimal solution to optimization problems. However, finding a genuinely global solution is not guaranteed. Machine learning methods are used when modeling the objectives or constraints.

Analogous Models: When nonhuman species are used to demonstrate the potential health effects of chemicals on humans.

Calibration: Adjusting model parameters within physically defensible ranges until the resulting predictions best fit the observed data. In some disciplines, calibration is also called "parameter estimation."

DOI: 10.1201/9781003601562-17

Computational Models: Express the relationships among system components using mathematical representations (Van Waveren et al., 2000).

Conceptual Models: A hypothesis regarding the critical factors that govern the behavior of an object or process of interest. This can be an interpretation or working description of the characteristics and dynamics of a physical system.

Ecological Effects Models: Provide a statistical relationship between a level of pollutant exposure and a particular environmental indicator.

Economic Impact Models: Used in rulemaking, priority setting, and enforcement model output as a monetary value.

Emissions Models: Estimate the rate or amount of pollutant emissions to water bodies and the atmosphere.

Exposure Models: Estimate the dose of pollutant to which humans or animals are exposed.

Fate and Transport Models: Calculate the movement of pollutants in the environment. They are further classified into Subsurface Water Quality Models, Surface Water Quality Models, and Air Quality Models.

Human Activity Models: Simulate human activities and the behaviors that result in the emission of pollutants.

Human Health Effects Models: Provide a statistical relationship between the chemical and an adverse human health effect.

Model: A simplification of reality constructed to gain insights into select attributes of a physical, biological, economic, or social system. A formal representation of the behavior of system processes, often in mathematical or statistical terms. Note that the term model can be ambiguous when describing an "abstraction (or parameterization) of reality." Models can take on many forms; computational and conceptual models are the most common and relevant. There are many kinds of models (EPA, 2009a), including:

- Computational models
  - **Analytical Models**: Unique computational models that can be solved mathematically in terms of analytical functions.
- Conceptual models
- **Physical Models**: Not a conventional model
- **Analogous Models**: Not a conventional model

Environmental models can be used to inform a variety of activities, including:

- Research
- Policy analysis
- Toxicity screening
- National regulatory decision making
- Implementation applications

Natural Systems Process: Simulate dynamics of ecosystems that give rise to fluxes of nutrients and emissions.

Noneconomic Impact Models: Evaluate the effects of contaminants on various noneconomic parameters (e.g., crop yields).

Mode (of a model): How a model operates. Prognostic (or predictive) models are designed to forecast outcomes and future events, while diagnostic models work "backward" to assess causes and precursor conditions (EPA, 2009a).

Model Framework: The system of governing equations, parameterization, and data strategies that represent the formal mathematical specification of a conceptual model (EPA, 2009a).

Parameter: Terms in the model that are fixed during a model run or simulation but can be changed in different runs to conduct sensitivity analysis or achieve calibration goals.

Peer Review: Performed by independent and objective experts, a review of and judgment on a mode's underlying science, its developed process, and its overall "trustworthiness" and "reliability" for prediction.

Post-auditing: Assesses a model's ability to provide valuable predictions of future conditions for management decisions.

State Variable: The dependent variable calculated within the model. These are also often the performance indicators of the models that change over the simulation.

System: A collection of objects or variables and the relations among them.

Transparency: The clarity and completeness with which data, assumptions, and analysis methods are documented. Experimental replication is possible when information about modeling processes is appropriately and adequately communicated.

Uncertainty: Describes a lack of knowledge about models, parameters, constants, data, and beliefs.

Variability: Refers to observed differences attributable to heterogeneity or diversity. Variability results from natural random processes and is usually not reducible by further measurement or study (although it can be better characterized).

Note: The definition of environmental models was not included in the terms and definitions critical to this discussion. Environmental models are a catchall phrase for various models used to protect human health and safeguard the natural environment upon which life depends. The main objectives of the material presented in this chapter are to provide a basic introduction to environmental modeling, provide definitions and types of environmental models, explain how and why models are used in environmental sciences, and discuss the model "life cycle."

ENVIRONMENTAL MODEL USE

Environmental models have been helping to explain scientific phenomena and predict outcomes and behaviors in settings where empirical observations have been limited or unavailable for several years (USEPA, 2009a). These models help guide researchers and their policies and are based on simplifying assumptions of environmental processes. Note, however, that models cannot wholly replicate the integral complexity of the entire environmental system.

ENVIRONMENTAL MODEL STRUCTURE

The structure of environmental models begins with a foundation: their system. The system should be defined as step one in the environmental modeling process.

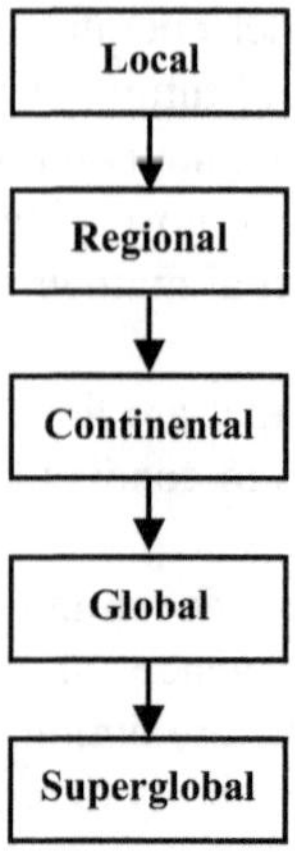

FIGURE 15.1 Examples of increasing scales.

This is important because the system determines the model's boundaries and describes how the model can be applied and to which systems or situations. Before defining the system to be used, it is essential to determine what processes the model attempts to reproduce and include. Additionally, the spatial and temporal scale (i.e., the specific application of the term scale, the size of space, and the spatial scale at which the included processes occur; see Figure 15.1) must be determined. This means that the model structure includes processes (chemical, physical, or biological) and scope or scale (time or space).

COMPUTATIONAL MODELS

In this text, we focus on the types of computational models. The available data determine the types, the intended use, and the interpretation of model-generated results. However, note that the types of models are not mutually exclusive. The types of computational models are:

- **Empirical Models (Aka a Statistical Model)**: These contain very little information on the underlying mechanisms and depend upon observed relationships—rather than theory—among experimental data. What the deal is that you observe a particular outcome following some circumstance, which allows you to make a reliable prediction of the future outcome. These are commonly thought of as "best-fit"' models whose parameters may or may not have real-world interpretation.
- **Mechanistic Models (Parametric Modeling):** Unlike empirical models, these include the mechanisms or processes between the state variables (EPA, 2009b). Note that when data quality is otherwise equivalent, extrapolation from mechanistic models often carries higher confidence than empirical models.

- **Deterministic Models:** These provide a solution for the state variable(s) rather than a set of probabilistic outcomes. Their operation is based on predefined logic and rules. Its generic equation:

$$N = e^{K(a-\beta 2)}$$

  N is a function of K, a, and β
- **Probabilistic Models**: These are the opposite of deterministic models. They are based on probability theory and utilize the entire range of input data to develop a probability distribution of model output (i.e., risk or exposure) rather than a single point value. Its generic equation:

$$P(N \mid a, \beta)$$

  The probability of N given a and β
- **Dynamic Models:** These predict how a system changes with time or space. They are concerned with time and the sequencing of operations. Its generic equation:

$$N_{t+1} = N_0 + a \cdot N_t$$

- **Static Models**: These make predictions about how a system changes as the value of an independent variable changes. Its generic equation:

$$N = \sigma^2 + \mu/\rho$$

THE ROLE OF ENVIRONMENTAL MODELING

Environmental modeling, or environmental niche modeling, uses ecological models to predict species distribution across geographic space. In simplest terms, it is the creation and use of mathematical models of the environment. The use of models has increased significantly. Although models do not generate "truth," they can provide analyses and information to inform decision-making. Even though research is confronted with many constraints when obtaining data (e.g., time, access, funding, equipment, and staff), policy decisions should be informed by the best information and data.

Models have become handy tools that provide insight where more data and information are needed. They can help users study the behavior of ecological systems, design field studies, interpret data, and generalize results (EPA, 2009a). Models are used to make long- and short-term forecasts to extrapolate from the past and answer "what-if" questions. One of the significant positives of these models is that they can also produce concise data summaries in diagnostic and regulatory contexts (NRC, 2007).

Note that the uncertainties and assumptions associated with any model must be considered—just as with observational data—before model-generated results are applied in any context.

ENVIRONMENTAL MODELS USED BY REGULATORS

Environmental models are categorized into groups representing a continuum of processes that translate the interactions between human activities and natural processes into human health and environmental impacts (see Figure 15.2). Regulators use classes of ecological models, including:

- Human Activity Models
- Natural Systems Processes
- Emissions Models
- Fate and Transport Models
- Exposure Models
- Human Health Effects Models
- Ecological Effects Models
- Economic Impact Models
- Noneconomic Impact Models

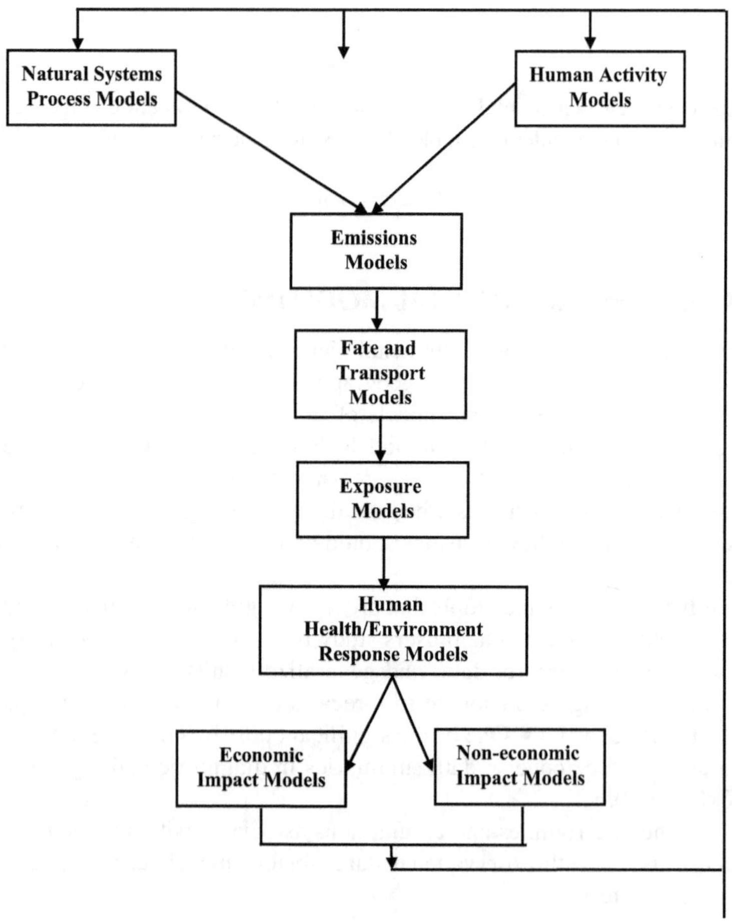

FIGURE 15.2 Classes of environmental models. Modified from NRC (2007).

MODEL LIFE CYCLE

The model's life cycle is constant, and there are many instances when earlier stages are revisited to refine the model. The life cycle follows a general iterative progression described below (USEPA, 2009a):

- **Identification**
  - Determine correct decision-related questions and establish objectives
  - Define the purpose of the modeling activity
  - Specify the model application context
- **Development**
  - Develop the conceptual model that reflects the underlying science of included processes
  - Derive the mathematical representation of that science and then encode it into a computer program
- **Evaluation**
  - Peer review
  - Conduct formal testing to ensure model expressions have been encoded correctly
  - Test model outputs by comparison with empirical (independent) data
- **Application**
  - Run the model and analyze outputs to inform a decision

THE ALTERNATIVE MODEL LIFE CYCLE

You probably thought fashioning a model to complete a complex project was a good idea, and many agree. However, remember that not all projects require the full development of a new model; often, existing models can be applied to a specific situation. In these instances, we can use an alternative model life cycle involving model evaluation, application, and post-auditing (if needed).

The problem specifications dictate which model to use as long as the chosen model meets the specific requirements. Once selected, a model may require site-specific values or calibration (see Sidebar 15.1). Similarly, other qualitative evaluations of the model may further corroborate its application.

SIDEBAR 15.1: SITE-SPECIFIC CALIBRATION (EPA, 2009A)

When data for quantifying one or more parameter values are limited, calibration exercises can be used to find solutions that result in the "best-fit" model. These solutions are based on measured physically defensible ranges to provide meaningful information. The point is that this type of calibration should be undertaken with caution.

The use of calibration to improve model performance varies because of many concerns associated with it. The appropriateness of calibration may often be a function of the modeling activities undertaken.

An example is the EPA's Office of Water's standard practice of calibrating well-established model frameworks such as CE-QUAL-W2 (a model for predicting river temperature fluctuations), to specific systems (e.g., the Snake River). The Snake River, about 1,080 miles (1,740 km) long, is significant in the interior Pacific Northwest region of the United States. The EPA evaluated water temperature regimes of the mainstem Snake River using transect measurements and the RBM 10 one-dimensional heat budget model (Cope, 2001). This work relied on detailed monitoring information collected by the Columbia River Inter-Tribal Fish Commission and Fisheries and Aquatic Sciences in the early 1990s at 18 locations in the Snake River (Karr et al., 1998). This transect data can also examine and simulate vertical temperature structures in the mainstream river. In the report, the two-dimensional CE-QUAL-W2 model framework, developed by the U.S. Army Corps of Engineers (Cole & Buchak, 1995), is used to simulate temperature regimes in Lower Granite Pool from the confluence of the Clearwater and Snake Rivers to Lower Granite Dam. The study period is from July through October of 1992.

Based on the measurements and simulation outputs using CE-QUAL-W2, the following observations are offered:

1. The CE-QUAL-W2 model framework captures most of the observed patterns of stratification in the pool in 1992. The model also predicts the arrival time of cold water overflows at the dam after the flow augmentation from Dworshak Dam to within approximately 1 day of the observed arrival time.
2. The predictive outlet temperature was generally lower than the measured temperature, even with the wind-sheltering coefficient set to zero.
3. Flow augmentation appears to have little effect on temperatures within this surface layer; augmentation may cause temperature increases at the surface.
4. After passing a cold air mass, the pool's temperature regime resembled the pattern observed during flow augmentation, with cooler input waters at the upstream boundary plugging beneath a warmer surface layer within the pool.
5. An exploratory simulation assuming the release of all water over the spillway (instead of the powerhouse) resulted in slightly lower surface temperatures at the downstream end of the pool during the first augmentation period in July 1992 when compared to the simulation of actual conditions (released through the powerhouse).

The bottom line: differences between the two simulations were more challenging to discern during the other augmentation episodes.

DID YOU KNOW?

A modified version of the life cycle is appropriate when model development is not required. If an existing model will work for the specified problem, model development (and design) is circumvented, leaving three stages in the life cycle—modeling needs/requirements/model design and model application (EPA, 2009a).

DATA QUALITY

Data quality is fundamental to environmental modeling and is pertinent during model application and the modeling life cycle. The indicators of data quality include (EPA, 2009a):

- **Precision**: The quality of being reproducible in amount or performance
- **Bias**: Systematic deviation between a measured (i.e., observed) or computed value and its "true" value
- **Representativeness**: The measure of the degree to which data accurately and precisely represent a characteristic of a population, parameter variations at a sampling point, a process condition, or an environmental condition
- **Comparability**: A measure of the confidence with which one data set or method can be compared to another
- **Completeness**: A measure of the amount of valid data obtained from a measurement system
- **Sensitivity**: The degree to which the model outputs are affected by changes in selected input parameters.

The predicted model outcome(s) were observed during the post-audit. After implementing a remedial or management action, the model post-audit process involved monitoring the modeled system to determine whether the actual system response concurs with that predicted by the model. Post-audits can also evaluate how well stakeholder and decision-making roles were integrated during development (Manno et al., 2006; EPA, 2009a).

The importance of data quality cannot be overstated; it is fundamental to environmental modeling. A model's quality is also governed by its structure, scientific understanding, evaluation, etc. Quality assurance is, therefore, necessary throughout the stages of the modeling lifecycle. Data quality indicators include the quantitative and qualitative measures of principal quality attributes (EPA, 2009a). In modeling, transparency is paramount. The way to achieve transparency is through quality assurance and control, along with peer review, playing essential roles in modeling efforts. The environmental model is fashioned in the way shown here:

$$\text{Qualitative \& Quantitative Data} + \text{Understanding} + \text{Conceptual Model} = \text{Model}$$

Environmental modeling is not new, but thanks to digitization and AI, it is much more accessible and easier to fashion into operable and accurate models; this is the case because AI can handle mountains of data quickly, accurately, and efficiently. Case Study 15.1 provides a condensed example, for illustrative purposes, of a USEPA Tier I model used for estimating pesticide concentrations in rice paddies.

Case Study 15.1: EPA Tier I Rice Model

The EPA's Tier I Rice Model estimates surface water exposure from pesticide use in rice paddies. The single screening-level concentration calculated with this model represents both short- and long-term surface water exposure. It can be used for aquatic ecological and drinking water exposure assessments for human health risk assessments. The formula of the Tier I Rice Model v1.0 is as follows:

$$C_w = \left(m_{ai}{}^{1}\right) / \left(0.00105 + 0.00013K_d\right)$$

And if appropriate:

$K_d = 0.01_{Koc}$

where

C_w = Water concentration (μg/L)

$M_{ai}{}^{1}$ = Mass applied per unit area (kg/ha)

K_d = Water-sediment partitioning coefficient (L/kg)

K_{oc} = Organic carbon partitioning coefficient (L/kg)

The Conceptual Model

The Tier 1 Rice Model v1.0 relies on an equilibrium partitioning concept to provide conservative estimates of environmental concentrations from applying pesticides to rice paddies. When a pesticide is applied to a rice paddy, the model assumes it will instantaneously partition between the water and sediment phases. The aqueous concentration that results from such partitioning is described as follows:

$$C_w = \left(m_{ai}\right) / \left(V_w + m_{sed}K_d\right) \tag{15.1}$$

where

C_m = Water concentration (mass/volume)

m_{ai} = Mass of active ingredient applied to paddy (mass)

V_w = Volume of water column plus pore water (volume)

m_{sed} = Mass of sediment at equilibrium with water column (mass)

K_d = Water-sediment partitioning coefficient (volume/mass)

Note that neither the degradation of the pesticide nor the mass transfer from the aqueous phase to the sediment is considered in this conceptualization, which significantly simplifies the model. The absence of degradation adds conservatism to the model (i.e., estimated concentrations should be higher than those usually found in rice paddies). The absence of a mass transfer process can either add or reduce conservatism depending on numerous conditions, such as whether the pesticide is soil or water applied and whether actual degradation occurs preferentially in the sediment or the water compartment; however, consideration of this parameter is a refinement beyond a Tier 1 conceptualization.

$$msed = d_{sed}A\rho_b \tag{15.2}$$

$$V_w = d_wA + d_{sed}\theta_{sed}A \tag{15.3}$$

where:

d_{sed} = Sediment depth (length)

d_w = Water column depth (length)

A = Area of the rice paddy (area)

θ_{sed} = Porosity of sediment (-)

p_b = Bulk density of sediment (mass/volume)

Further, the mass per unit area is defined as:

$$m_{ai}' = m_{ai}/A \quad (15.4)$$

where

m_{ai} = Mass applied per unit area (mass/area)

Substitution of Equations 15.2, 15.3, and (15.4) into Equation 15.1 produces the conceptual model with commonly understood parameters:

$$C_w = (m_{ai'})/(d_w + d_{sed}(\theta_{sed} p_b k_d) \quad (15.5)$$

Parameterization

The physical parameter values that describe the rice paddy for the Tier 1 Rice Model v1.0 represent a reasonably conservative system and include water column depth at 0.10 m, sediment depth at 0.01 m, organic content to sediment at 0.01, sediment bulk density at 1,300 kg/m^3, grain density at 2,650 kg/m^3, and sediment porosity (Brady, 1984; Hillel, 1982; USDA, 1998, 2002).

- **Water Depth, Bulk Density, Porosity:** The water column depth (d_w) of 10 cm is the typical average rice paddy water depth at flood release (USDA, 1998, 2002). Bulk density is assumed to be roughly the same as agricultural soils (Brady, 1984; Hillel, 1982). Grain density is considered to be that of quartz (2.65 g/mL); thus, the porosity of the sediment is 0.509.
- **Organic Carbon Content**: The organic carbon content (f_{oc}) represents the mean of sediments or in sediments taken from 15 rice fields. These sediments are described in 17 studies submitted supporting the registration of rice pesticides and are assumed to be typical for rice paddies. Thirteen rice fields are located in the southeastern U.S., and two are in California. The mean organic carbon content of the sediment in these 15 rice fields is 0.01.

  The Interim Rice Model assumes an organic carbon content of 0.02. To show how the different modeled organic carbon contents affect the exposure estimates, the Interim Rice Model exposure estimates are compared with those of the Tier 1 Rice Model v1.0 for applications of 1 kg/ha of compounds of differing K_{oc}. The Tier 1 Rice Model v1.0 exposure estimates are approximately twice those of the Interim Rice Model for compounds with a K_{oc} 100,000. As compounds of decreasing K_{oc} are modeled, partitioning to sediment becomes a less critical dissipation process, and Tier 1 Rice Model v1.0 exposure estimates approach those of the Interim Rice Model.

- **Water Calibration of Sediment Depth**: The sediment interaction depth (d_{sed}) was calibrated using direct paddy measurement from 15 sites in 17 studies. The results of this study showed that the maximum observed concentration was that of the overlying water at any time after application. The application rate was the total amount of pesticide applied during the study. In these studies, multiple pesticide applications were typically made within one season. During multiple-application research, no attempt was made to associate concentration measurements with any particular application. Thus, there are cases where the highest concentration was observed before the final application. This situation is consistent with the Tier 1 model concept in which the measurement timing is not considered. Also consistent with the Tier 1 concept is the lack of differentiation among various application methods, such as applications before flooding.

  The model sediment depth was calibrated by comparing maximum observed concentrations to model outputs based on a range of depths (0.5–3 cm).

The Formula of the Tier 1 Rice Model v1.0

Substituting the values found into Equation 15.5 and using standard units for K_d (L/kg), $M_{ai'}$ (kg/ha), and C_w (μg/L), the Tier 1 rice conceptual model is simplified to the Tier 1 Rice Model v1.0.

$$C_w = m_{ai'}/(0.00105 + 0.00013K_d) \tag{15.6}$$

where (in this case):

C_w = Water concentration (μ/L)
$M_{ai'}$ = Mass applied per unit area (kg/ha)
K_d = Water-sediment partitioning coefficient (L/kg)

When collecting data, model input values for the water-sediment partitioning coefficient (K_d) should represent the mean K_d of relevant soil (or sediment). A mean K_{oc} should be used in cases where sorption K_d values correlate with soil organic matter content. In these cases, K_d model input values should be calculated from the mean K_{oc} using a fraction of organic carbon (f_{oc}) of 0.01 (for example). The following equations can thus estimate K_d:

$$K_d = f_{oc}k_{oc} \tag{15.7}$$

$$K_d = 0.01k_{oc} \tag{15.8}$$

Alternatively, mean K_{oc} values
can be directly entered into Equation 15.9:

$$C_w = m_{ai'}/(0.00105 + 0.0000013K_{oc}) \tag{15.9}$$

Assumptions

Most of the assumptions in this model help ensure that the outputs protect most environments associated with rice agriculture. The model assumptions include the following:

- Movement of pesticide on suspended sediment is not considered.
- Degradation does not occur.
- Volatilization and other dissipation processes are not considered.
- Partitioning to sediment is instantaneous.
- Water is available for human or wildlife exposure instantaneously.

Assuming the following:

- Water column depth is 10 cm.
- Sediment depth is 1 cm.
- All pore space is saturated with water.
- Organic fraction is 0.01.
- Bulk density is 1,300 kg/m^3.
- Grain density is 2,650 kg/m^3.

Model Evaluation for Drinking Water Exposure

Note that Tier 1 model estimates are screening estimates; as such, they are expected to exceed peak values found in the environment in most cases. This expectation is due to the model's protective assumptions, excluding degradation and uncontaminated water outside the paddy. To evaluate the Tier 1 Model v1.0 for drinking water exposure estimation, model estimates calibrated to indicate exposure within rice paddies were compared to measured concentrations of pesticides in rice-growing environments well downstream of the actual rice paddies. The results of these comparisons are summarized below.

- **Louisiana Downstream Measurement 1:** The Louisiana Department of Agriculture and Forestry collected data from 23 monitoring sites in Acadia, Calcasieu, Cameron, Jefferson-Davis, and Vermillion parishes where surface water bodies received rice paddy discharge water. The maximum detected fipronil concentration (a broad-spectrum insecticide) from March 6, 2000, to May 15, 2000, was 8.41 μg/L. Although the history of fipronil use is unknown in this area, the labeled maximum application rate of 0.05 lbs/acre was assumed to compare rice model results. The rice model gives an estimate that is 3.5 times greater than the maximum detected value.
- **Louisiana Downstream Measurements 2:** Fipronil surface water monitoring dates were submitted for the Mermentau River and Lake Arthur (MRID 43349901). The Mermentau River drains a large portion of the rice acreage in southern Louisiana from the mouths of Bayou Plaquemine and Bayou Nezpique. The area has 250,000 acres of rice, 70% of which was treated

with fipronil in 1999. The monitoring program was designed to provide a snapshot of concentrations on May 11, 1999, just after a 0.45-inch rainfall event, at 0–1 and 4–6 ft depths. The maximum detected concentrations were sampled from the mouth of the Bayou Plaquemine, where fipronil was detected at a maximum concentration of 12.118 μg/L, and total residues of fipronil and three of its degradates were detected at a maximum concentration of 3.509 μg/L. The rice model estimates approximately one order of magnitude greater than the maximum detected value for fipronil.

- **California Indirect Measurements:** The California Department of Pesticide Regulation collected data from six monitoring sites in the Sacramento Valley that were sampled from March 31, 1997, to May 25, 1997, as part of a cooperative water quality monitoring program (MRID 44632501). Residues of molinate, thiobencarb, and methyl parathion (all herbicides) were monitored in water bodies near rice agriculture. They may not have necessarily contained concentrations representative of those found in rice paddies or released paddy discharge water. Maximum concentrations of 0.11 μg/L of methyl parathion, 12.3 μg/L of thiobencarb, and 25.7 μg/L of molinate were detected in the Colusa Basin Drain at the monitoring site near State Highway 20 on the 13th, 20th, and 22nd of May 1997, respectively. The rice model estimates two to three orders of magnitude more significant than these pesticides' maximum detected values.
- **Summary:** Evaluation of the Tier 1 Rice Model v1.0 indicates that modeled within-paddy estimates are conservative, exceeding peak measurements of pesticides in water bodies downstream of rice paddies by less than one to multiple orders of magnitude. The range of modeled estimate exceedances over measured concentrations is most likely in response to chemical, environmental, and study-specific factors.

THE BOTTOM LINE

Note that estimates from the Tier 1 model generally do not represent typical concentrations in human drinking water, as they represent paddy discharge water. However, these concentrations may reasonably estimate acute concentrations for use in environmental/ecological assessments where exposure occurs at or near the rice paddy. In both cases—human drinking water and environmental exposure—the chronic concentrations are expected to be conservative. A higher-tier rice model should be used to estimate chronic exposure to compounds that degrade rapidly into degradates that are not of risk concern.

Also, note that if Tier 1 estimates calculated by the screening method do not exceed the level of concern in a risk assessment, there is high confidence that there will be little or no risk above the level of concern for exposure through water resources. However, because of the uncertainties associated with a screening method, when a level of concern is exceeded, it cannot be determined whether the exceedance will occur or whether this method has overestimated the exposure.

REFERENCES

Brady, N.C. (1984). *The Nature and Properties of Soil*, 9th ed. New York: Macmillan Publishing Company.

Cole, T. & Buchak, E. (1995). CE-QUAL-W2: A Two-dimensional, Laterally-Averaged, Hydrodynamic and Water Quality Model, Version 2.0. User Manual. U.S. Army Corps of Engineers.

Cope, B. (2001). Evaluating Water Temperature Regimes in the Snake River Using Transect Measurements and the RMB Model. EPA 910-R-01-008. Washington, DC.

EPA (US Environmental Protection Agency). (2009a). Guidance on the Development, Evaluation, and Application of Environmental Models (PDF). EPA/100/K-09/003. Washington, DC.

EPA (US Environmental Protection Agency). (2009b). Using Probabilistic Methods to Enhance the Role of Risk Analysis in Decision-Making with Case Study Examples DRAFT (PDF). EPA/100/R-09/001. Washington, DC.

Hillel, D. (1982). *Introduction to Soil Physics*. Orlando, FL: Academic Press.

Karr, M., Fryer, J., & Mundy, P. (1998). Snake River Water Temperature Control Project. Phase II., Methods for managing and monitoring water temperatures of salmon in the lower Snake River. Boise, Idaho.

LDAF (2000). Summary of Actions Taken in the Monitoring of ICON® (Fipronil) Uses in Rice Culture in South Louisiana Department of Agriculture and Forestry, AES Pesticide and Environmental Programs, June 15, 2000.

Manno, J., Smardon, R., DePinto, J.V., Cloyd, E.T. & Del Granado, S. (2006). The Use of Models in Great Lakes Decision Making: An Interdisciplinary Syntheses Randolf G. Pack Environmental Institute, College of Environmental Science and Forestry. Occasional Paper 16. New York.

MRID 42560501. Kruger, H., Homer, A. & McNinch, S (1992). Dissipation of Tilt® in two Arkansas Rice Fields. An unpublished study performed by Wildlife International Ltd. Easton, MD and EN-CAS Analytical Laboratories, Winston-Salem, NC, sponsored and submitted by CIBA-GEIGY Corporation, Greensboro, NC. Study Number: 108261. November 9, 1992, 251 pp.

MRID 43431901. Evans, J. (1994). Basagran® Herbicide Dissipation in a Rice Paddy, 1992 Study: Final Report. An unpublished study performed by JensenAg Consultants Inc., Washington, LA. Stoneville Associates, Inc., Greenville, MS, BASF Corporation, RTP, NC, and ALTA Laboratories, El Dorado Hills, CA; sponsored and submitted by BASF Corporation, RTP, ND; and South Texas Ag Research, Inc., Pattison, T. Study numberEC-92187: June 20, 1995, 374 pp.

NRC (National Research Council) (2007). *Models in Environmental Regulatory Decision Making*. Washington, DC: National Academies Press.

USDA (1998). *Crop Profile for Rice in California*. Washington, DC: United States Department of Agriculture, Pest Management Centers. October 1988.

USDA (2002). *Crop Profile for Rice in Louisiana*. Washington, DC: Department of Agriculture, Pest Management Centers. September 2002.

Van Waveren, R.H., Groot, S., Scholten, H., Van Geer, F., Wosten, H., Koeze, R., & Noort, J. (1999). Good Modeling Practice Handbook PDF. STOWA report 99-05. Leystad, The Netherlands.

16 Industrial Hygiene

WHAT IS INDUSTRIAL HYGIENE?

According to the Occupational Safety and Health Administration (OSHA), industrial hygiene is the science of anticipating, recognizing, evaluating, and controlling workplace conditions that may cause workers' injury or illness. Industrial hygienists use environmental monitoring and analytical methods to detect the extent of worker exposure and apply engineering controls, administrative controls, and other processes, such as personal protective equipment (PPE), to control potential health hazards (OSHA, 1998).

THE SURPRISE

In 60 years of teaching safety, health, industrial hygiene, environmental health, and environmental engineer (EE)—along with some nursing topics in the military and college courses—I have often noticed the surprised looks on my trainee's and students' faces whenever I finish a semester course in environmental engineering with lectures on industrial hygiene.

The trainees and students look puzzled, saying to themselves (and sometimes even out loud to me), "Industrial hygiene ... what does it have to do with environmental engineering?"

In response to their questions about the interface between environmental engineering and industrial hygiene, I developed an often-used reply:

"Okay, consider that environmental engineering and industrial hygiene are related in several ways. To begin with, let's talk about environmental engineering and where it touches base, so to speak, with industrial hygiene."

Pause: By this time, I have my students' full attention. There is nothing new about this. I have a way of mixing excellent, entertaining, crazy, and goofy humor with pointed humor in my lectures to communicate the importance of the topic—whatever the topic. If you are trying to teach anyone anything, you need to communicate. A failure to communicate is not a resident of my ditty bag.

Continuing: "You have all been taught, to this point, that environmental engineering is the professional engineering discipline related to environmental science. In my classes, you have also been taught or told that to be an environmental engineer, you must be a generalist. Being a generalist in environmental engineering means you have been exposed to several broad scientific topics, including physics and chemistry. But those two important disciplines are just scratching the surface of what you need to be familiar with. EE students and practitioners also have been trained in biology, ecology, geology, hydraulics, hydrology, microbiology, mathematics, and artificial intelligence applications. Unless you are blessed with innate knowledge and a boatload of common sense, you cannot create solutions to protect and improve living

 DOI: 10.1201/9781003601562-18

organisms' health and environmental quality unless you have been appropriately trained. The bottom line is that you must have an understanding of mass and energy transfer, environmental chemistry, growth modeling, water pollution, air pollution, global changes such as the greenhouse effect and global temperature change, water, carbon, nitrogen, and oxygen cycles, solid waste management, resource recovery, and finally, you must be equipped with the tools provided by artificial intelligence and its subfields such as deep learning, machine learning, the Internet of Things (IoT), and others."

At this point in my canned presentation, there is usually a moment of silence in the classroom before I continue.

Continuing: "Okay, some of you have asked, and others, the rest of you, are probably wondering where industrial hygiene (IH) practiced in the workplace is an integral part of EE. Good question. The answer lies in risk assessment and mitigation. Environmental engineers who are active in the workplace or have helped design workplaces must be concerned with risk. For example, if an EE is part of an original design team putting together a workplace, the engineers work to engineer out a problem, a risk, and a hazard. Consider, for example, a new or modified workplace with a new machine installation and associated piping secured to the floor, running to its power source. This is a classic trip hazard; the EE and other design team members will engineer the proposed floor pipe run out, eliminating the risk. The team has identified the danger and opted for a 10-ft-high overhead run of the piping so no one can trip over it. The EE identifies and mitigates the hazard, conducts a dose-response assessment (if necessary), looks at exposure assessment, characterizes all risks, and performs a comprehensive risk analysis."

This simple presentation seems to satisfy those who asked or wondered about the EE nexus and the working world. However, after satisfying their curiosity about the EE and IH connections, the students often focus on artificial intelligence (AI) and its growing connection to workplace safety and health.

This always happens with my presentation on AI and IH in the workplace. I often get that "Katie bar the door—no AI, please" look from many students, especially those already working in real-world workplaces, not work-at-home places. These students typically work in businesses that have incorporated some aspect(s) of AI into their workplaces, where robots have replaced many human workers. Instead of realizing the business owners' thinking—that AI is used to increase productivity and efficiency and reduce on-the-job injuries and all the follow-up needs (time off, medical expenses, lawsuits, and so on).

These students see AI as a replacement for human jobs. However, the truth is that someone has to control automated work tasks and perform maintenance and repairs on computerized robotic equipment. The irony is that after more than 40 years of full-time safety and health management experience in various workplaces, I have found that most worker accidents occur during maintenance, repair, or replacement of workplace equipment.

The bottom line: The EE working to make workplaces hazard-free and safe must look at all angles of the work and ensure that even if robots are employed instead of humans, the robot equipment is designed to be worked on safely. Note that preventing humans from bypassing safety guidelines (I have never figured out the foolproof fix to this problem) is a manager's job and is not generally relegated to EE supervision.

My final classroom comments on this topic: "I have noticed that almost all of you have laptop computers, tablets, or other digital recording devices … and this is good. We are on the cusp of a digital revolution, the Fourth Industrial Revolution (4IR); thus, capable digital EEs are the thing of the present and definitely of the future."

"The bottom line is that you must often adjust and think differently from how you would for a more narrowly defined environmental engineering problem. This is not a practice of thinking outside the box … but instead the practice of thinking inside the box—inside the environment—inside the workplace."

REFERENCE

OSHA. (1998). *Industrial Hygiene.* Accessed 10/24/24 @ https://www.360traINING.COM/OSHA/iND-HYGIENE...

Part III

Conclusion

17 Data Center Limiting Factors

THE OTHER SIDE OF A LIMIT

It would be difficult for anyone, especially those involved with the environment and ecology in any way or fashion, not to know or easily be reminded of certain important things or natural laws. One of the natural laws familiar to the environmentally inclined and those included has to do with limits. This law is about limiting factors and is known as *Liebig's Law of the Minimum.* Simply, this law is best explained through examples. But first, recall that limit can mean many different things to many different people, such as a maximum, a minimum, a check, a constraint, a cutoff point, a restriction, and others. However, the bottom line is that some might say we need to keep a tight rein on things—helping to create a limit or a maximum limit.

Example 17.1

Think of a garden plot. This plot is well-fertilized, well-watered, planted in the desired growth zone, and is to be well-monitored and maintained. You introduce healthy, properly spaced tomato plants. You wait for the plants to grow so you can harvest the tomatoes. The problem is there is no growth.

The problem?

The planter did not take into account the trees surrounding the plot. The trees, hardwoods, are in full spring bloom. The tree canopy blocks sunlight.

No sunlight = no growth

The point is that without sunlight, the planted tomatoes will not grow. The lack of sunlight is the limiting factor.

Example 17.2

Let's get back to Liebig's Law of the Minimum. In this example consider or imagine Liebig's Barrel. This is a staved barrel made up of wooden staves of unequal length.

You use your garden hose to slowly fill the barrel. You fill the barrel until the water level reaches the overflow level. The barrel's overflow level is its shortest stave—the barrel's limiting factor.

Earlier in this discussion, limits were described, and for humans, we are usually looking for a maximum limit or an operation with no limits—limitlessness, infinity.

DOI: 10.1201/9781003601562-20

For AI data centers, there are two limiting factors—electricity and water. Before discussing these two limiting factors, an actual data center operation is presented in Case Study 17.1.

Case Study 17.1: Microsoft and the City of Quincy, Washington (USEPA, 2023)[1]

In a nutshell, the Quincy Water Reuse Utility (QWRU) was constructed in a partnership between Microsoft and the city of Quincy, Washington. The Reuse Utility treats cooling water from a Microsoft data center and recirculates it to the data center for the same purpose, reducing reliance on local potable groundwater. The groundwater also has high levels of total dissolved solids, which can cause problems when used in the data center's cooling equipment. The water supplied from the QWRU is of more suitable quality for the data center's cooling equipment than the conventionally treated, mineral-rich groundwater.

In context, the City of Quincy is in the arid eastern part of the State of Washington and receives only 7.8 inches (199 mm) of annual rainfall. The potable water required to service residential, commercial, and industrial needs within the city is provided by five groundwater wells. These wells in basaltic rocks reach depths up to 1,000 ft (305 m) into the Columbia Plateau regional aquifer system. Although the residential part of Quincy is relatively small, with about 8,100 people, its large industrial sector means its water demand is akin to that of 30,000 inhabitants. Yahoo, Sabey, Intuit, Vantage, and Microsoft have data centers along with a large food processing industry within the Quincy region.

So, why Quincy, Washington? Why are major companies building their data centers in Quincy? The attraction, in one word, is affordability, affordability, affordability. Simply, Quincy is an attractive area to build these data centers because of the area's affordable land (as compared to land in the Silicon Valley region to the south of Washington State in California), cool temperatures, and the availability of renewable, abundant, and cheap hydropower. The surrounding area is also a hub of agricultural activity.

Let's look at the city's water needs.

Typically, on a yearly basis, the city uses approximately 2.1 billion gallons (8 million liters) of groundwater. Until recently, food processors used 57% of the pumped groundwater; 33% went to residents and public and commercial services such as schools, hotels, or restaurants; and the remaining 10% was used by data centers for cooling purposes.

Note that since 2007, Microsoft has operated one of the biggest data centers in Quincy on a 270-acre (110-hectare) campus. The large volume of water needed to cool high-density cloud computing server farms is making water management a growing priority for data center operators, including Microsoft. In the past, Microsoft received potable groundwater from Quincy for cooling; they would discharge the water to the sewer system for treatment at the city's municipal wastewater reclamation facility. However, after four to five cooling cycles, this water was rich in minerals, including calcium, magnesium, and silica. This was not a good situation because the high mineral content meant that long cycles in the data centers could damage heat transfer equipment or generally impair heat transfer, as water evaporated during cooling and the concentration of total dissolved solids (TDS) increased. To mitigate this mineral buildup, water had to be "blown down" or "bled" (i.e., drained) more often from the cooling equipment to remove mineral build-up.

Complicating the TDS problem in the cooling water was the original wastewater treatment approach. The city possessed and operated a municipal reclamation facility (MRF) that uses reclaimed water to irrigate public areas and to recharge shallow aquifers in the area through groundwater recharge basins. However, the treatment plant could not handle the high TDS concentrations in the discharges from the regional data centers to the MRF. Thus, the data centers were charged an extra fee to accommodate adjustments to the facility's process and increased maintenance needs. Another fly in the ointment or icing on the "cake," so to speak, is the high levels of TDS, calcium, magnesium, and silica that together formed a visible cake on the surface of the groundwater recharge basins in the summer. This was causing it to exceed discharge and indirect potable reuse requirements, e.g., for aquifer recharge, set by the Washington State Department of Ecology.

The City of Quincy and Microsoft both recognized the technical difficulty of using mineral-rich municipal groundwater for cooling data center equipment; moreover, they also recognized the subsequent challenge of treating the data center's industrial wastewater with high levels of TDS at the city's municipal wastewater reclamation facility. So what the City and Microsoft did was to form a partnership to create the QWRU, basically an entirely new facility specifically designed to treat the cooling water from Microsoft's data center. Ten years of planning and construction were finally completed when QWRU officially became operational on June 30, 2001. Pipes more than 30 miles (48 km) long run across Quincy's industrial area connected to ten distinct treatment systems. To date, only Microsoft's data center campus is connected to the QWRU.

The QWRU is a closed-loop system and neither discharges industrial wastewater with high mineral content to the environment nor discharges industrial wastewater to the city's MWRF. Note that in order to prevent mineral buildup in the cooling water, salts are removed from the water prior to reuse at the data center and concentrated in lined brine ponds to form a solid, which is disposed of in a manner that prevents entry into state ground or surface waters. What this does is ensure that the cooling water quality does not adversely affect cooling equipment or heat transfer within the data center. Just as important (maybe even more so), the QWRU substantially reduces the data center's reliance on potable groundwater for data center cooling, saving millions of gallons of water per year.

For its treatment system, the QWRU relies on a combination of:

- New online systems, including a lime softening facility (like alum treatment using calcium hydroxide), an ultrafiltration facility, a permeate tank facility [used in reverse osmosis (RO)], a blend tank facility storage, and brine ponds.
- Previously existing infrastructure, which was part of the city's industrial wastewater treatment systems, including a high-efficiency softening (HES) facility and reverse osmosis (RO) plant.

The water is cycled four to five times at the data center and then the resulting blowdown water is transferred to a lime softening facility that adds limewater (calcium hydroxide) to remove hardness (deposits of calcium and magnesium salts). Next in the treatment train is the ultrafiltration facility.

After ultrafiltration, some of the water is transferred to a blend tank, while the rest is further softened in the HES facility and filtered at the RO plant. A pipe transfers the HES calcium- and magnesium- laden brine and the RO brine silica to brine ponds. Nature takes over, and the brine undergoes maximum evaporation, with the residuals disposed of separately. Note that Quincy expects it will need to do this every 2 or 3 years.

The treated wastewater continues to flow from the RO system and is stored in a permeate tank for approximately 8 hours and acts as a buffer whenever the system is down for maintenance of the HES or RO systems. The ultrafiltration and RO water are blended at various ratios throughout the day depending on the conductivity levels (ideally: 300–350 μS). Finally, the treated industrial wastewater is returned to the data center for cooling water.

In operation, the QWRU system must be topped off or topped up often because water evaporates during the cooling process, and a small volume of highly concentrated brines needs disposal. Because the data centers do not need potable water for cooling, the City of Quincy and Microsoft have been able to use make-up water from untreated surface water sources by accessing rights to water drawn from the Columbia Basin Project in the U.S. Bureau of Reclamation and the Quincy Irrigation District canal.

DID YOU KNOW?

According to the U.S. Bureau of Reclamation (USBR, 2023), more than 670,000 acres in east-central Washington serve the Columbia Basin Project. The main facilities of the project include Grand Coulee Dam, Franklin D. Roosevelt Lake, three power plants, four switchyards, and a pump-generating plant. Primary irrigation facilities include the Feeder Canal, Banks Lake, the Main, West, East High, and East Low canals, O'Sullivan Dam, Potholes Reservoir, and Potholes Canal. There are over 300 miles of main canals, 2,000 miles of laterals, and 3,500 miles of drains and wasteways on the project. In addition to supplying water for irrigation, producing electricity, controlling floods, providing recreation, and regulating streamflow, the Columbia Basin Project also provides water for cities, industries, navigation, and endangered species.

During the summer, when cooling demand in data centers is highest, the QWRU uses almost 300 million gallons per year of Columbia Basin Project water from the irrigation canal. This is a positive practice because it substantially lowers the need to use groundwater for cooling and reduces the city's potable water demand and treatment costs since there is no need to soften the water before sending it back to the data center. Note that the irrigation canal, where the water is extracted, runs dry in winter—but this is not a significant issue since cooling demand is comparatively lower in the winter months.

Five percent of the make-up water, on an annual basis, is still potable groundwater, which is mixed and softened before entering the cooling systems. Forging ahead, the City of Quincy aims to purchase more water from the Bureau of Reclamation and get permission from the Quincy Irrigation District to withdraw more water from the canal because it requires minimal treatment for the data centers and could eliminate the data center's reliance on potable groundwater supplies.

THE BOTTOM LINE

AI data centers need and use huge amounts of electricity for operation and water for cooling. The future need for more electricity and greater access to water for powering the exponential growth of AI is a given. The problem is that predictions of future energy and water demand are weighed down with uncertainties due to the lack of visibility into proprietary private sector planning for new model training. This is the case even though there is some visible movement toward increasing the amount of nuclear power as the generator of more electricity. There is also a concern about procuring electrical components (availability of), such as transformers, switching equipment, advanced transmission lines, and generation equipment, in the near and longer term. Access to water supplies for data centers is a considerable issue and is aligned concurrently with the need for more electrical power sources—the current movement toward water reuse may ameliorate the problem of finding adequate supplies of cooling water.

Oh, a tight rein on humans in everything we do or want to do and in some cases not what we want but it happens anyway because we make a mistake or mistakes—in some cases, they are not checked until a limit enters the scene, so to speak. In the case of expanding AI, two limiting factors are present—the need for large amounts of electricity and water—and they need to be addressed and sources found or developed.

Some would say, including this author, that the recent advances in artificial intelligence have been hailed as potentially transformative for a long list of industries and other sectors, including national security.

The question is: What is the limit?

NOTE

1 From F.R. Spellman (2024). *The Science of Water Reuse* (in production). Boca Raton, FL: CRC Press.

REFERENCES

USEPA. (2023). *Case Studies That Demonstrate the Benefits of Water Reuse*. Accessed 12/22/23 @ https://epa.gov/waterreuse/forms/contact-us--about-water-reuse-and-recycling.

USBR. (2023). Columbia Basin Project. Accessed 12/12/23 @https://www.usbr.gov/pn/grandcoulee/cbp.

Glossary[1]

Abiotic: The nonliving part of the physical environment (for example, light, temperature, and soil structure).

Absorption: (1) Movement of a chemical into a plant, animal, or soil. (2) Any process by which one substance penetrates the interior of another substance. In chemical spill cleanup, this process applies to chemical uptake by capillaries within certain sorbent materials.

Absorption units: Devices or units designed to transfer the contaminant from a gas phase to a liquid phase.

Absorber: In a photovoltaic device, the material that readily absorbs photons to generate charge carriers (free electrons or holes).

AC: Alternating current.

Accidental spills: The unintended release of chemicals and hazardous compounds or materials into the environment.

Acid Hydrolysis: A chemical process in which acid converts cellulose or starch to sugar.

Acid Mine Drainage: Refers to water pollution that results when sulfur-bearing minerals associated with coal are exposed to air and water and form sulfuric acid and ferrous sulfate. The ferrous sulfate can further react to form ferric hydroxide, or yellow boy, a yellow-orange iron precipitate found in streams and rivers polluted by acid mine drainage.

Acid Rain: Also called acid precipitation or acid deposition, acid rain contains harmful amounts of nitric and sulfuric acids formed primarily by sulfur dioxide and nitrogen oxides released into the atmosphere when fossil fuels are burned. It can be wet precipitation (rain, snow, or fog) or dry precipitation (absorbed gaseous and particulate matter, aerosol particles, or dust). Acid rain has a pH below 5.6. Average rain has a pH of about 5.6, which is slightly acidic. The term pH measures acidity or alkalinity and ranges from 0 to 14. A pH measurement of 7 is regarded as neutral. Measurements below 7 indicate increased acidity, while those above indicate increased alkalinity.

Acre-foot: The volume of water that will cover an area of 1 acre to a depth of 1 ft.

Acreage: An area, measured in acres, subject to ownership or control by those holding total or fractional shares of working interests. Acreage is considered developed when development has been completed. A distinction may be made between "gross" acreage and "net" acreage:

- **Gross**: All acreage covered by any working interest, regardless of the percentage of ownership in the interest.
- **Net**: Gross acreage adjusted to reflect the percentage of ownership in the working interest in the acreage.

Active Power: The component of electric power that performs work, typically measured in kilowatts (kW) or megawatts (MW). Also known as "real power." The terms "active" and "real" are used to modify the base term "power" to differentiate it from Reactive Power.

Active Solar: As an energy source, energy from the sun is collected and stored using mechanical pumps or fans to circulate heat-laden fluids or air between solar collectors and a building.

Adaptation: Adjustment to environmental conditions.

Adaptive Management: Focuses on learning and adapting through partnerships of managers, scientists, and other stakeholders who learn together how to create and maintain sustainable ecosystems. Adaptive management helps science managers maintain flexibility in their decisions, knowing that uncertainties exist and providing managers the latitude to change direction to improve their understanding of ecological systems and achieve management objectives. It is about taking action to enhance progress towards desired outcomes.

Adverse Water Conditions: Reduced stream flow, lack of rain in the drainage basin, or low water supply behind a pond or reservoir dam result in a reduced gross head that limits hydroelectric power production or forces restrictions on multipurpose reservoirs or other water uses.

Adverse Weather Conditions: Reduced streamflow, lack of rain in the drainage basin, or lower water supply behind a pond or reservoir dam result in a reduced gross head that limits hydroelectric power production or forces restrictions on multipurpose reservoirs or other water uses.

AFUE (Annual Fuel Utilization Efficiency): A measure of space heating equipment efficiency defined as the fraction of energy output/input.

Agriculture: An energy-consuming subsector of the industrial sector that comprises all facilities and equipment for growing crops and raising animals.

Albedo: The ratio of light reflected by a surface to the light falling on it.

Air Cleaner: A device using filters or electrostatic precipitators to remove indoor air pollutants such as tobacco smoke, dust, and pollen. Most portable units are 40 W on low speed and 100 W on high speed.

Air Collector: A medium-temperature collector used predominantly in space heating, utilizing pumped air as the heat-transfer medium.

Air Pollution Abatement Equipment: Equipment used to reduce or eliminate airborne pollutants, including particulate matter (dust, smoke, fly ash, dirt, etc.), sulfur oxides, nitrogen oxides (NO_x), carbon monoxide, hydrocarbons, odors, and other pollutants.

Air Temperature Adjustment: Change in air temperature associated with a change in tree canopy cover (°C per 1% change in tree canopy cover).

Alternating Current (AC): An electric current that reverses its direction at regularly recurring intervals, usually 50 or 60 times per second.

Alternative Fuel: Alternative fuels for transportation applications include the following:

- methanol
- denatured ethanol and other alcohols

- fuel mixtures containing 85% or more by volume of methanol, denatured ethanol, and other alcohols with gasoline or other fuels
- natural gas
- liquefied petroleum gas (propane)
- hydrogen
- coal-derived liquid fuels
- fuels (other than alcohol) derived from biofuels such as soy diesel fuel)
- electricity (including electricity from solar energy)

"... any other fuel the Secretary determines, by rule, is substantially not petroleum and would yield substantial energy security and environmental benefits." The term "alternative fuel" does not include alcohol or other blended portions of primarily petroleum-based fuels used as oxygenates or extenders, i.e., MTBE, ETBE, other ethers, and the 10% ethanol portion of gasohol.

Alternator: A device that turns the rotation of a shaft into alternating current (AC).

Ambient: Natural condition of the environment at any given time.

Anadromous Fish: Fish that spend a large portion of their life cycle in the ocean and return to freshwater to breed; from the Greek, literally "up-running."

Anaerobic Decomposition: In the absence of oxygen, as in an anaerobic digester or lagoon, which produces carbon dioxide and methane.

Anaerobic Digestion: A biochemical process by which organic matter is decomposed by bacteria in the absence of oxygen, producing methane and other byproducts.

Anaerobic Lagoon: A liquid-based organic waste management installation characterized by waste in water at a depth of at least 6 ft for 30–200 days.

Anemometer: Wind speed measurement device used to send data to the controller and also used to conduct wind site surveys.

The Angle of Incidence: The angle that a ray of sun makes with a line perpendicular to the surface. For example, a surface that directly faces the sun has a solar angle of incidence of zero, but if the surface is parallel to the sun (for instance, sunrise striking a horizontal rooftop), the incidence is 90°.

Anion: A negatively charged ion that is attracted to the anode.

Annual Removals: The net volume of growing stock trees removed from the inventory during a specified year by harvesting, cultural operations such as timber land improvement, or land clearing.

Annualized Growth Rate: Calculated as follows: $(x_n/x_1)1/n$ where x is the value under consideration and n is the number of periods.

Anode: The electrode at which oxidation (a loss of electrons) occurs. The anode is the negative terminal for fuel cells and other galvanic cells; for electrolytic cells, the anode is the positive terminal.

Anthropogenic: Made or generated by humans or caused by human activity. The term is used in the context of global climate change to refer to gaseous emissions that are the rules of human activities and other potentially climate-altering activities, such as deforestation.

API Gravity: The American Petroleum Institute measures the specific gravity of crude oil or condensate in degrees. An arbitrary scale expressing the gravity or the density of liquid petroleum products. The measuring scale is calibrated in terms of degrees API; it is calculated as follows:

$$\text{Degrees API} = \left(141.5 / \text{sp. Gr. } 60°\text{F}/60°\text{F}\right) - 131.5$$

Apparent Power: The product of the voltage (in volts) and the current (in amperes). It comprises both active and reactive power. It is measured in "volt-amperes" and often expressed in "kilovolt-amperes" (kVA) or "megavolt-amperes" (MVA).

Appropriate Use: A proposed or existing use on a refuge that meets at least one of the following conditions: (1) the use is a wildlife-dependent one; (2) the use contributes to fulfilling the refuge purpose(s).

Aquatic: Growing in, living in, or dependent upon water.

Aquifer: Water-bearing stratum of permeable sand, rock, or gravel.

Aromatics: The simultaneous purchase and sale of identical or similar assets across two or more markets in order to profit from a temporary price discrepancy.

Attenuator: This device is a long floating structure like the terminator but is oriented parallel to the waves rather than normal to them. It rides the waves like a ship, and movements of the device at its bow and along its length can be restrained to extract energy. The theoretical advantage of the attenuator over the terminator is that its area normal to the waves is small and therefore the forces it experiences are much lower.

Availability Factor: A percentage representing the number of hours a generating unit is available to produce power (regardless of the amount of power) in a given period, compared to the number of hours in the period.

Average Stream Flow: The rate, usually expressed in cubic feet per second, at which water passes a given point in a stream over a set period of time.

Average Water Conditions: The amount and distribution of precipitation within a drainage basin and the runoff conditions present as determined by reviewing the area's water supply records over a long period of time.

Azimuth Angle: The angle between the true south and the point on the horizon directly below the sun.

Bacilli (pl.), bacillus (sing): Members of a group of rodlike bacteria that occur everywhere in soil and air. Some are responsible for diseases such as anthrax or for causing food spoilage.

Bacteria: One-celled microorganisms.

Bacteriophage: A virus that infects bacteria; often called a *phage*.

Bagasse: The fibrous material remaining after the extraction of juice from sugarcane; often burned by sugar mills as a source of energy.

Baghouse filter: A closely woven bag for removing dust from dust-laden gas streams. The fabric allows passage of the gas with retention of the dust.

Bare Rock Succession: An ecological succession process whereby rock or parent material is slowly degraded to the soil by a series of bio-ecological processes.

Barrel: A unit of volume equal to 42 U.S. gallons.

Barrier: Any obstruction to fish passage or aquatic barrier.

Basin: The land surrounding and draining into a water body.

Benzene (C_6H_6): An aromatic hydrocarbon present in small proportion in some crude oils and made commercially from petroleum by the catalytic reforming of naphthenes in petroleum naphtha. Also made from coal in the manufacture of coke. Used as a solvent in the manufacture of detergents, synthetic fibers, petrochemicals, and as a component of high-octane gasoline.

Bilateral Agreement: A written agreement signed by two parties that specifies the terms for exchanging energy.

Binary Cycle: Binary geothermal systems use the extracted hot water or steam to heat a secondary fluid to drive the power turbine.

Biochemical Conversion: The use of fermentation or anaerobic digestion to produce fuels and chemicals from organic sources.

Biodiesel: Fuel derived from vegetable oils or animal fats. It is produced when a vegetable oil or animal fat is chemically reacted with an alcohol.

Bioenergy: Useful, renewable energy produced from organic matter, which may either be used directly as a fuel or processed into liquids and gases.

Biofuels: Liquid fuels and blending components produced from biomass (plant) feedstocks, used primarily for transportation.

Biogas: A combustible gas derived from decomposing biological waste. Biogas normally consists of 50%–60% methane.

Biogenic: Produced by biological processes of living organisms.

Biogenic Emissions: Emissions that are naturally occurring and are not significantly affected by human actions or activities.

Biological Diversity or Biodiversity: The variety of life and its processes, which includes the variety of living organisms, the genetic differences among them, and the communities and ecosystems in which they occur.

Biological Integrity: Biotic composition, structure, and functioning at genetic, organism, and community levels comparable with historical conditions, including the natural biological processes that shape genomes, organisms, and communities.

Biodiversity Conservation: The goal of conservation biology is to retain indefinitely as much of the Earth's biodiversity as possible, with emphasis on biotic elements most vulnerable to human impacts.

Biomass: Any organic nonfossil material of biological origin constituting a renewable energy source. It is produced from organic matter that is available on a renewable or recurring basis, including crops and trees, wood and wood residues, plants (including aquatic plants), grasses, animal manure, municipal residues, and other residue materials. Biomass is generally produced sustainably from water and carbon dioxide by photosynthesis. There are three main categories of biomass—primary, secondary, and tertiary.

Biomass Gas (Biogas): A medium-Btu gas containing methane and carbon dioxide, resulting from the action of microorganisms on organic materials such as a landfill.

Biomass Waste: Organic non-fossil material of biological origin that is a byproduct or a discarded product. Biomass waste includes municipal solid waste from biogenic sources, landfill gas, sludge waste, agricultural crop byproducts, straw, and other biomass solids, liquids, and gases. Still, it excludes wood and wood-derived fuels (including black liquor), biofuel feedstock, biodiesel, and ethanol fuel. Biomass waste also includes energy crops grown specifically for energy production, which would not normally constitute waste.

Biomaterials: Products derived from organic (as opposed to petroleum-based) products.

Biota: The plant and animal life of a region.

Blackbody: An ideal substance that absorbs all radiation falling on it and reflects nothing.

Black Liquor (Pulping Liquor): The alkaline spent liquor removed from the digesters by chemically pulping wood. After evaporation, the liquor is burned as fuel in a recovery furnace that permits the recovery of certain basic chemicals.

Bone Dry: Having 0% moisture content.

Breakwater: A barrier protecting a harbor or shore from the impact of waves.

Breccia: A coarse-grained clastic rock composed of angular broken rock fragments held together by a mineral cement or in a fine-grained matrix.

Breeding Habitat: Habitat used by migratory birds or other animals during the breeding season.

Brine: A geothermal solution containing appreciable amounts of sodium chloride or other salts.

British Thermal Unit (Btu): This is a basic thermal (heat) energy measure. A Btu is defined as the energy required to increase the temperature of 1 lb of water by 1°F at normal atmospheric pressure. 1 Btu = 1,055 J.

Btu Conversion Factor: A factor for converting energy data between one unit of measurement and the British thermal unit (Btu). Btu conversion factors are generally used to convert energy data from physical units of measure (such as barrels, cubic feet, or short tons) into the energy-equivalent measure of Btu.

Btu per Cubic Foot: The total heating value, expressed in Btu, produced by the combustion, at constant pressure, of the amount of gas that would occupy a volume of 1 ft^3 at a temperature of 60°F, if saturated with water vapor and under pressure equivalent to 30 inches of mercury at 32°F and standard gravitational force (980.665 cm/s^2) with air of the same temperature and pressure as the gas, when the products of combustion are cooled to the initial temperature of gas and air when the water formed by combustion is condensed to the liquid state. (Sometimes called gross heating value or total heating value.)

Buffer Zones: Land bordering and protecting critical habitats or water bodies by reducing runoff and nonpoint source pollution loading; areas created or sustained to lessen the negative effects of land development on animals, plants, and their habitats.

Bulk Density: Weight per unit of volume, usually specified in pounds per cubic foot.

Burnup: Amount of thermal energy generated per unit mass of fuel, expressed as Gigawatt-Days Thermal per Metric Ton of Initial Heavy Metal (GWDT/MTIHM), rounded to the nearest gigawatt day.

Butane (C_4H_{10}): A straight-chain or branch-chain hydrocarbon extracted from natural gas or refinery gas streams, gaseous at standard temperature and pressure. It includes isobutene and normal butane and is designated in ASTM Specification D1835 and Gas Processors Association specifications for commercial butane.

Butylene (C_4H_8): An olefinic hydrocarbon recovered from refinery or petrochemical processes, gaseous at standard temperature and pressure. Butylene is used in the production of gasoline and various petrochemical products.

Calcination: A process in which a material is heated to a high temperature without fusing, so that hydrates, carbonates, or other compounds are decomposed and the volatile material is expelled.

Calcium Sulfate: A white crystalline salt, insoluble in water. Used in Keene's cement, in pigments, as a paper filler, and as a drying agent.

Calorie: The amount of heat required to raise the temperature of 1 g of water 1°C.

Canopy: The layer of foliage formed by the crowns of trees in a stand. For stands with trees of different heights, foresters often distinguish among the upper, middle, and lower canopy layers. These represent foliage on tall, medium, and short trees. The uppermost layers are called the overstory.

Cap Rocks: Rocks of low permeability that overlie a geothermal reservoir.

Capacity Factor: The ratio of the electrical energy produced by a generating unit for the period considered to the electrical energy that could have been made at continuous full-power operation during the same period.

Capacity, Gross: The full-load continuous rating of a generator, prime mover, or other electric equipment under specified conditions designated by the manufacturer. It is usually indicated on a nameplate attached to the equipment.

Capitol Cost: The cost of field development, plant construction, and the equipment required to generate electricity.

Carbon: A nonmetallic element (symbol C) found in all organic and inorganic substances, such as coal and natural gas. The atomic weight of C is 12, and CO_2 is 44. To convert emissions reported as mass or weight of C to mass of $CO_{2,}$ multiply CO_2 by 12/44.

Carbon Accounting: Generally, it refers to processes undertaken to *measure* amounts of carbon dioxide equivalents emitted by an entity. It is used, among other things, by nation-states, corporations, and individuals.

Carbon Budget: The balance of the exchanges (incomes and losses) of carbon between carbon sinks (e.g., atmosphere and biosphere) in the carbon cycle.

Carbon Cycle: All carbon sinks and carbon exchanges from one sink to another by various chemical, physical, geological, and biological processes.

Carbon Dioxide (CO_2): A combustion product; the most common greenhouse gas.

Carbon Dioxide Equivalent (CDE): The amount of carbon dioxide by weight emitted into the atmosphere that would produce the same estimated radiative forcing as a given weight of another radiatively active gas. Carbon dioxide equivalents are computed by multiplying the weight of the measured gas (for example, methane) by its estimated global warming potential (21 for methane). "Carbon equivalent units" are defined as carbon dioxide equivalents multiplied by the carbon content of carbon dioxide (i.e., 12/44).

Carbon Intensity: The amount of carbon by weight emitted per unit of energy consumed. A standard measure of carbon intensity is the weight of carbon per British thermal unit (Btu) of energy. The carbon intensity and the emissions coefficient are identical when only one fossil fuel is considered. When there are several fuels, carbon intensity is based on their combined emissions coefficients weighted by their energy consumption levels.

Carbon Monoxide (CO): A colorless, odorless gas produced by incomplete combustion. Carbon monoxide is poisonous when inhaled.

Carbon Output Rate: The amount of carbon by weight per kilowatt-hour of electricity produced.

Carbon Sequestration: The process through which carbon dioxide is removed from the atmosphere, for example, in forests. During this process, carbon dioxide is taken up through plants' leaves and incorporated into the plants' wood biomass.

Carbon Sinks: Carbon reservoirs and conditions that take in and store more carbon (carbon sequestration) than they release. Carbon sinks can partially offset greenhouse gas emissions. Forests and oceans are common carbon sinks.

Casing: Pipe placed in a wellbore as a structural interface between the wellbore and the surrounding formation. It typically extends from the top of the well and is cemented to maintain the diameter of the wellbore and provide stability.

Cast Silicon: Crystalline silicon is obtained by pouring pure molten silicon into a vertical mold and adjusting the temperature gradient along the mold volume during cooling to obtain slow, vertically-advancing crystallization of the silicon. The polycrystalline ingot thus formed is composed of large, relatively parallel, interlocking crystals. The cast ingots are sawed into wafers for further fabrication into photovoltaic cells. Cast-silicon wafers and ribbon-silicon sheets fabricated into cells are usually called polycrystalline photovoltaic cells.

Cathode: The electrode at which reduction (a gain of electrons) occurs. The cathode is the positive terminal for fuel cells and other galvanic cells; for electrolytic cells (where electrolysis occurs), the cathode is the negative terminal.

Cation: A positively charged ion.

Cells: Refers to the unencapsulated semiconductor compounds of the module that convert solar energy to electricity.

Cells to OEM (non-PV): Cells are shipped to non-photovoltaic original equipment manufacturers (OEMs), such as boat and car manufacturers.

Cellulose: The main carbohydrate in living plants. Cellulose forms the skeletal structure of the plant cell wall.

Cellulosic Ethanol: Ethanol derived from biomass's cellulosic and cellulosic parts.

Characterization: Sampling, monitoring, and analysis activities to determine the extent and nature of contamination at a facility or site. Characterization provides technical information to develop, screen, analyze, and select appropriate cleanup techniques.

Chemical Separation: A process for extracting uranium and plutonium from dissolved spent nuclear fuel and irradiated targets. The fission products that are left behind are high-level waste. Chemical separation is also known as reprocessing.

Chips: Woody material cut into short, thin wafers. They are used as a raw material for pulping and fiberboard or as biomass fuel.

Chlorofluorocarbon (CFC): Any of various compounds consisting of carbon, hydrogen, chlorine, and fluorine used as refrigerants. CFCs are now thought to be harmful to the Earth's atmosphere.

Climate: The average weather (usually taken over a 30-year time period) for a particular region and time period. Climate is not the same as weather; instead, it is the average weather pattern for a specific area. Weather describes the short-term state of the atmosphere. Climatic elements include precipitation, temperature, humidity, sunshine, wind velocity, phenomena such as fog, frost, hailstorms, and other weather measures.

Climate Change: The term "climate change" is sometimes used to refer to all forms of climatic inconsistency. However, because the Earth's climate is never static, the term is more appropriately used to imply a significant change from one climatic condition to another. In some cases, climate change has been used synonymously with global warming; however, scientists use the term broadly to include natural climate changes.

Climate Effects: Impact on residential space heating and cooling (kg CO_2/tree/year) from trees located greater than approximately 15 m (50 ft) from a building (far trees) due to associated reductions in wind speeds and summer air temperatures.

Coarse Materials: Wood residues suitable for chipping, such as slabs, edgings, and trimmings.

Co-firing: The practice of introducing biomass into the boilers of coal-fired power plants.

Cogeneration: The production of electrical energy and another form of sound energy (such as heat or steam) through the sequential use of energy.

Cogeneration System: A system using a common energy source to produce electricity and steam for other uses, increasing fuel efficiency.

Commercial Sector: An energy-consuming sector comprising service-providing facilities and equipment for businesses, federal, state, and local governments, as well as other private and public organizations, such as religious, social, or fraternal groups. The commercial sector includes institutional living quarters. It also includes sewage treatment facilities. Common uses of energy associated with this sector include space heating, water heating, air conditioning, lighting, refrigeration, cooking, and running various other equipment. Note: This sector includes generators that produce electricity and/or useful thermal output primarily to support the activities of the above-mentioned commercial establishments.

Commercial Species: Tree species suitable for industrial wood products.

Condensate: Water formed by condensation of steam.

Condenser: Equipment that condenses turbine exhaust steam into condensate.

Conifer: A tree or shrub in the phylum Gymnospermae whose seeds are in woody cones. There are 500–600 species of living conifers.

Conservation: Managing natural resources (including preservation, restoration, and enhancement) to prevent loss or waste.

Conservation Corridor: Connections between suitable habitats that allow passage of plant or animal species.

Conservation Easement: A non-possessory interest in real property owned by another imposing limitation or affirmative obligations with the purpose of returning or protecting the property's conservation value.

Conservation Feature: A feature in the building designed to reduce the usage of energy.

Conservation Program: A utility company furnishes home weatherization services free or at reduced cost or provides free or low-cost devices for saving energy, such as energy-efficient light bulbs, flow restrictors, weather stripping, and water heater insulation.

Conservation Reserve Program (CRP): CRP provides farm owners or operators with an annual per-acre rental payment and half the cost of establishing a permanent land cover in exchange for retiring environmentally sensitive cropland from production for 10–15 years. In 1996, Congress reauthorized CRP for an additional round of contracts, limiting enrollment to 36.4 million acres at any time. The 2002 Farm Act increased the enrollment limit to 39 million acres. Producers can offer land for competitive bidding based on an Environmental Benefits Index (EBI) during periodic signups or can automatically enroll more limited acreages in processes such as riparian buffers, field windbreaks, and grass strips. CRP is funded through the Commodity Credit Corporation (CCC).

Conservation Status: Assessment of the status of ecological processes and the viability of species or populations in an ecoregion.

Conventional Hydroelectric (hydropower) Plant: A plant in which all the power is produced from natural streamflow as regulated by available storage.

Correlation (statistical term): In its most general sense, correlation denotes the interdependence between quantitative and qualitative data. It would include the association of dichotomized attributes and the contingency of multiple classified attributes. The concept is general and may be extended to more than two variables. The word is most frequently used in a somewhat narrower sense to denote the relationship between measurable variates or ranks.

Criteria Pollutant: A pollutant that is determined to be hazardous to human health and is regulated under EPA's National Ambient Air Quality Standards. The 1970 amendments to the Clean Air Act required the EPA to describe a pollutant's health and welfare impacts as the "criteria" for inclusion in the regulatory regime.

Critical Habitat: According to U.S. Federal law, the ecosystems upon which endangered and threatened species depend, specific geographic areas, whether occupied by a listed species or not, that are essential for its conservation and have been formally designated by rule published in the Federal Register.

Crop Failure: This includes mainly the acreage on which crops failed because of weather, insects, and diseases, but it also includes some land not harvested due to lack of labor, lower market prices, or other factors. The acreage planted to cover and improve crops not intended for harvest is excluded from crop failure and is considered idle.

Crop Residue: Organic residue remaining after the harvesting and processing of a crop.

Cropland: Total cropland includes five components: cropland harvested, crop failure, cultivated summer fallow, cropland used only for pasture, and idle cropland.

Cropland Harvested: This includes row crops and closely sown crops, hay and silage crops, tree fruits, small fruits, berries, tree nuts, vegetables and melons, and miscellaneous other minor crops. In recent years, farmers have double-cropped about 4% of the acreage.

Cropland Pasture: Land used for long-term crop rotation. However, some cropland pastures are marginal for crop uses and may remain in the pasture indefinitely. This category also includes land used for pasture before crops reached maturity and some land used for pasture that could have been cropped without additional improvement.

Cropland Used For Crops: Cropland used for crops includes cropland harvested, crop failure, and cultivated summer fallow.

Crust: Earth's outer layer of rock. Also called the lithosphere.

Dam: A physical barrier constructed across a river or waterway to control the flow of or raise the water level. The purpose of construction may be for flood control, irrigation needs, hydroelectric power production, and/or recreational use.

Daylighting: Direct, diffuse, or reflected sunlight to provide supplemental lighting for building interiors.

Daylighting Controls: A system of sensors that assesses the amount of daylight and controls lighting or shading devices to maintain a specified lighting level. The sensors are sometimes referred to as "photocells."

dbh: The diameter measured at approximately breast height (usually 4.5 ft above the ground), commonly used by foresters to describe tree size.

DC: Direct current.

Deadweight Tons: The lifting capacity of a ship expressed in long tons (2.240 lbs), including cargo, commodities, and crew.

Decatherm: Ten therms or 1,000,000 Btu.

Decontamination: Removal of unwanted radioactive or hazardous contamination by a chemical or mechanical process.

Deepest Total Depth: The deepest total depth of a given well is the distance from a surface reference point (usually the Kelly bushing, also called the drive bushing) to the point of deepest penetration measured along the wellbore. If a well is drilled from a platform or barge over water, the depth of the water is included in the total length of the well bore.

Densification: A mechanical process to compress biomass (usually wood waste) into pellets, briquettes, cubes, or densified logs.

Defoliator: An agent that damages trees by destroying leaves or needles.

Deforestation: The net removal of trees from forested land.

Degasification System: This is a method for removing methane from a coal seam that could not otherwise be removed by standard ventilation fans and thus would pose a substantial hazard to coal miners. These systems may be used prior to or during mining activities.

Degradable Organic Carbon: The portion of organic carbon present in such solid waste as paper, food waste, and yard waste susceptible to biochemical decomposition.

Degradation: The loss of native species and processes due to human activities, such that only certain components of the original biodiversity persist, often including significantly altered natural communities.

Demand Indicator: A measure of the number of energy-consuming units, or the amount of service or output, for which energy inputs are required.

Demonstrated Resources: Same qualifications as identified resources but include measured and indicated degrees of geologic assurance and exclude the inferred.

Dependable Capacity: The load-carrying ability of a station or system under adverse conditions for a specified period of time.

Depleted Resources: Resources mined include recovered coal, lost coal in mining, and coal reclassified as sub-economic because of mining.

Depletion Factor: The annual percentage of the depletion of the thermal resource.

Design Head: The achieved river, bondage, or reservoir surface height (forebay elevation) that provides the water level to produce the full flow at the gate of the turbine to attain the manufacturer's installed nameplate rating for generation capacity.

Designated Wilderness Area: An area designated by Congress as part of the National Wilderness Preservation System.

Diesel Fuel: A fuel composed of distillates obtained in petroleum refining or blends of such distillates with residual oil used in motor vehicles. The boiling point and specific gravity are higher for diesel fuels than gasoline.

Digester Gas: Biogas that is produced using a digester, which is an airtight vessel or enclosure in which bacteria decompose biomass in water to produce biogas.

Direct Methanol Fuel Cell (DMFC): A fuel cell in which the fuel is methanol (CH_3OH) in gaseous or liquid form. The methanol is oxidized directly at the anode instead of first being reformed to produce hydrogen. The electrolyte is typically a polymer electrolyte membrane (PEM).

Direct Use: Use of geothermal heat without first converting it to electricity, such as for space heating and cooling, food preparation, industrial processes, etc.

Displacer: This is the body moved by the waves. It might be a buoyant vessel or, as in the case of Oscillating Water Column (OWC) devices, a mass of water. If buoyant, the displacer may pierce the surface of the waves or be submerged.

Distillation Unit (Atmospheric): The primary distillation unit processes crude oil (including mixtures of other hydrocarbons) at approximately atmospheric conditions. It includes a pipe still for vaporizing the crude oil and a fractionation tower for separating the vaporized hydrocarbon components into fractions with different boiling ranges. This is done by continuously vaporizing and condensing the components to separate higher boiling-point material. The process scheme, the properties of the crude oil, and the product specifications set the selected boiling ranges.

Distributed Generation (Distributed Energy Resources): Refers to electricity provided by small, modular power generators (typically ranging in capacity from a few kilowatts to 50 MW) located at or near customer demand.

Disturbance: Any relatively discrete event in time that disrupts ecosystem, community, or population structure and changes resources, substrate availability, or the physical environment.

Diversity Exchange: A capacity, energy, or exchange between systems whose peak loads occur at different times.

Divestiture: The separation of one utility function from the others by selling (spinning-off) or, in some other way, changing the ownership of the assets related to that function. Stripping off is most commonly associated with spinning off generation assets so they are no longer owned by the shareholders that own the transmission and distribution assets.

Downwind Turbine: A turbine that does not face into the wind and whose direction is controlled directly by the wind.

Drag Bit: A drilling bit that drills by scraping or shearing the rock with fixed hard surfaces, "cutters."

Drainage Basin: The land drained by a river system.

Drawdown: The lowering of the water level of a reservoir as a result of withdrawing water.

Drilling: Boring into the Earth to access geothermal resources, usually with oil and gas drilling equipment modified to meet geothermal requirements.

E-10: A mixture of 10% ethanol and 90% gasoline by volume.

E-85: A mixture of 85% ethanol and 15% gasoline by volume.

E-95: A fuel containing a mixture of 95% ethanol and 5% gasoline.

Ecological Integrity: Native species populations in their historic variety and numbers naturally interact in naturally structured biotic communities. For communities, integrity is governed by the demographics of component species, the intactness of landscape-level ecological processes (e.g., natural fire regime), and the intactness of internal community processes (e.g., pollination).

Ecological System: Dynamic assemblages of communities that occur together on the landscape at some spatial scale of resolution, are tied together by similar ecological processes, and form a cohesive, distinguishable unit on the ground. Examples include spruce-fir forests, Great Lakes dune and swale complexes, and Mojave desert riparian shrublands.

Economy of Scale: The principle that larger production facilities have lower unit costs than smaller facilities.

Ecoregion: A territory defined by a combination of biological, social, and geographic criteria rather than geopolitical considerations; generally, a system of related, interconnected ecosystems.

Ecosystem: A natural community of organisms interacting with their physical environment, regarded as a unit.

Ecosystem Service: A benefit or service provided free by an ecosystem or the environment, such as clean water, flood mitigation, or groundwater recharge.

Efficiency: The ratio of the useful energy output of a machine or other energy-converting plant to the energy input.

Embayment: A bay or bay-like formation.

Emissions: Anthropogenic releases of gases to the atmosphere. In the context of global climate change, they consist of significant greenhouse gases (e.g., the release of carbon dioxide during fuel combustion).

Emissions Coefficient: A unique value for scaling emissions to activity data in terms of a standard rate of emissions per unit of activity (e.g., pounds of carbon dioxide emitted per Btu of fossil fuel consumed).

Emissions Factor: A rate of carbon dioxide output resulting from the consumption of electricity, natural gas, or any other fuel source.

End User: A firm or individual that purchases products for its own consumption and not for resale (i.e., an ultimate consumer).

Endothermic: A chemical reaction that absorbs or requires energy (usually in the form of heat).

Energy: The ability to do work. That is, the capacity for doing work as measured by the capability of doing work (potential energy) or the conversion of this capacity to motion (kinetic energy). Energy has several forms, some easily convertible and can be changed to another form useful for work. Most of the world's convertible energy comes from fossil fuels that are burned to produce heat, which is then used as a transfer medium to mechanical or other means to accomplish tasks. Electrical energy is usually measured in kilowatt-hours, while heat energy is usually measured in British thermal units (Btu).

Energy Crops: Crops grown specifically for their fuel value. They include food crops such as corn and sugarcane, as well as nonfood crops such as poplar trees and switchgrass. Currently, low-energy crops are under development: short-rotation woody crops, which are fast-growing hardwood trees harvested in five to eight years, and herbaceous energy crops, such as perennial grasses, which are harvested annually after taking two to three years to reach full productivity.

Energy Efficiency: A ratio of service provided to energy input (e.g., lumens to watts in the case of light bulbs). Services provided can include building-sector end uses such as lighting, refrigeration, and heating, industrial processes, or vehicle transportation. Unlike conservation, which involves some reduction of service, energy efficiency provides energy reductions without sacrificing service.

Energy Loss: Deleted because there is no need for a general term to encompass all forms of energy loss. Terms referring to losses specific to particular energy sources are defined separately.

Energy Source: Any substance or natural phenomenon that can be consumed or transformed to supply heat or power. Examples include petroleum, coal, natural gas, nuclear, biomass, electricity, wind, sunlight, geothermal, water movement, and hydrogen in fuel cells.

Enhanced Geothermal Systems (EGS): Engineered reservoirs that can extract economic amounts of heat from geothermal resources. Rock fracturing, water injection, and water circulation technologies sweep heat from unproductive areas of existing geothermal fields or new fields lacking sufficient production capacity.

Enthalpy: A thermodynamic property of a substance, defined as the sum of its internal energy plus the pressure of the substance times its volume, divided by the mechanical equivalent of heat. The total heat content, the sum of dry air and water vapor enthalpies, and the per unit weight of dry air are measured in Btu per pound (or calories per kilogram).

Environment: The total of all biological, chemical, and physical factors exposed to organisms.

Environmental Health (abiotic aspects): The composition, structure, and functioning of soil, water, air, and other abiotic features comparable with historic conditions, including the natural abiotic processes that shape the environment.

Environmental Impact Statement: A document created from a study of the expected environmental effects of a new development or installation.

Environmental Restoration: Although usually described as "cleanup," this function encompasses a wide range of activities, such as stabilizing contaminated soil, treating groundwater, decommissioning process buildings, nuclear reactors, chemical separation plants, and many other facilities, as well as exhuming sludge and buried drums of waste.

Environmental Restrictions: In reference to coal accessibility, land-use restrictions that constrain, postpone, or prohibit mining to protect an area's environmental resources, such as surface or groundwater quality, air quality affected by mining, or plants or animals or their habitats.

Evacuated Tube: In a solar thermal collector, an absorber tube contained in an evacuated glass cylinder through which collector fluids flow.

Evaluation: Examination of how an organization's plans and actions have turned out—and adjusting them for the future.

Evaporation Pond: A containment pond (preferably with an impermeable lining of clay or synthetic material such as hypalon) to hold liquid waste and concentrate it through evaporation.

Evaporative Cooler ("swamp cooler"): This air conditioner uses evaporation from a centrally located pad; no refrigeration unit is involved.

Evapotranspiration (ET): The combined evaporation from the soil surface and plant transpiration. Transpiration is the evaporation of water from the internal surfaces of living plant organics and its subsequent diffusion into the atmosphere. Evaporation is the physical process by which liquid water is converted to vapor.

Exothermic: A chemical reaction that gives off heat.

Exotic Species: A species not native to an area and introduced intentionally or unintentionally by humans; not all exotics become successfully established.

Expenditure: The incurrence of a liability to obtain an asset or service.

Externality: A cost or benefit not accounted for in the price of goods and services. Often, "externality" refers to the cost of pollution and other environmental impacts.

Extinction: Terminating any lineage of organisms, from subspecies to species and higher taxonomic categories from genera to phyla. Extinction can be local, in which one or more populations of a species or other unit vanish, but others survive elsewhere, or total (global), in which all the populations vanish.

Fahrenheit: A temperature scale on which the boiling point of water is 212° above zero, and the freezing point is 32° above zero at standard atmospheric pressure.

Failure or Hazard (electrical power distribution): Any electric power supply equipment or facility failure or other event that, in the judgment of the reporting entity, constitutes a hazard to maintaining the continuity of the bulk electric power supply system, such that load reduction action may become necessary and a reportable outage may occur. Types of abnormal conditions that should be reported include the imposition of a special operating procedure, the extended purchase of emergency power, other bulk power system actions that may be caused by a natural disaster, a significant equipment failure that would impact the bulk power supply, and an environmental and/or regulatory action requiring equipment outages.

Far Trees: Trees located more than 15 m (50 ft) from buildings influence building energy use through their aggregate effect on air temperature and wind speed at the neighborhood scale.

Fault: A fracture in rock exhibiting relative movement between the adjoining surfaces.

Fauna: All animal life associated with a given habitat, country, area, or period.

Federal Land: Public land owned by the Federal Government, including national forests, national parks, and national wildlife refuges.

Federal-Listed Species: A species listed as endangered, threatened, or at risk under the Endangered Species Act of 1973, as amended.

Fiber Products: Products derived from the fibers of herbaceous and wood plant material. Examples include pulp, composition board products, and wood chips for export.

Fine Materials: Wood residues unsuitable for chipping, such as planer shavings and sawdust.

Fire Management: All activities related to the management of wildland fires.

Fire Regime: The characteristic frequency, intensity, and spatial distribution of natural fires within a given ecoregion or habitat.

Fixing: Near-shore and offshore devices may be either bottom-mounted or floating, with the former fixed to the seabed by a static member and the latter moored to hold on to the station.

Flare: A tall stack equipped with burners used as a safety device at wellheads, refining facilities, gas processing plants, and chemical plants. Flares are used for the combustion and disposal of combustible gases. The gases are piped to a remote, usually elevated, location and burned in an open flame in the open air using a specially designed burner tip, auxiliary fuel, and steam or air. Combustible gases are often flared due to emergency relief, overpressure, process upsets, startups, shutdowns, and other operational safety reasons. Natural gas is also flared when it is uneconomical for sale due to the unavailability of a method for transporting such gas to markets.

Flexible-fuel Vehicle: A single fuel tank vehicle that runs on varying unleaded gasoline blends with either ethanol or methanol.

Flora: All the plants found in a particular place.

Floodplain: Flat or nearly flat land that floodwaters may submerge; a plain built up or in the process of being built up by stream deposition.

Flow Battery: An electrochemical energy storage device that utilizes tanks of rechargeable electrolytes to refresh the energy-producing reaction. Since its capacity is limited only by the size of its electrolyte tanks, it is useful for large-scale backup systems to supplement other intermittent generation forms.

Fluidized-bed Combustion: A method of burning a particular fuel, such as coal, in which the amount of air required for combustion far exceeds that found in conventional burners. The fuel particles are continually fed into a bed of mineral ash in the proportions of 1 part fuel to 200 parts ash while a flow of air passes through the bed, causing it to act like a turbulent fluid.

Fluorescent Lamp: A glass enclosure in which light is produced when electricity is passed through mercury vapor inside the enclosure. The electricity creates a radiation discharge that strikes a coating on the inside surface of the enclosure, causing the coating to glow. Note: Traditional fluorescent lamps are usually straight or circular, while glass tubes are used in specially designed fixtures. A newer type of fluorescent lamp, the compact fluorescent lamp, takes up much less room, comes in many differently shaped configurations, and is designed to be used in some fixtures initially intended to house incandescent lamps.

Flux Material: A substance used to promote fusion, e.g., of materials or minerals.

Fly Ash: Particulate matter mainly from coal, with a particle diameter of less than 1 × 104 m. This ash is removed from the flue gas using flue gas particulate collectors such as fabric filters and electrostatic precipitators.

Forest Land: Land at least 10% stocked by forest trees of any size, including land that formerly had such tree cover and that will be naturally or artificially regenerated. Forest land includes transition zones, such as areas between heavily forested and non-forested lands that are at least 10% stocked with forest trees and forest areas adjacent to urban and built-up lands. Also included are pinyon-juniper and scrub areas in the West and the afforested regions. The minimum area for classification of forest land is 1 acre. Roadside, streamside, and shelterbelt strips of trees must have a crown width of at least 120 ft to qualify as forest land. Unimproved roads and trails, streams, and clearings in forest areas are classified as forests if they are less than 120 ft wide.

Fossil Fuels: A general term for combustible geologic carbon deposits in reduced (organic) form and of biological origin, including coal, oil, natural gas, oil shales, and tar sands. A significant concern is that they emit carbon dioxide into the atmosphere when burnt, thus significantly contributing to the enhanced greenhouse effect.

Fractionation: The process by which saturated hydrocarbons are removed from natural gas and separated into distinct products, or "fractions," such as propane, butane, and ethane.

Fracture: A natural or induced break in the rock.

Fuel Cell: One or more cells capable of generating an electrical current by converting the chemical energy of a fuel directly into electrical energy. Fuel cells differ from conventional electrical cells in that the active materials, such as fuel and oxygen, are not contained within the cell but are supplied from outside.

Fugitive Emissions: Unintended gas leaks from fossil fuel processing, transmission, and transportation.

Full Sun: The power density in sunlight received at the Earth's surface at noon on a clear day (about 1,000 W/m^2).

Fumarole: A vent or hole in the Earth's surface, usually in a volcanic region, from which steam, gaseous vapors, or hot gases issue.

Gallon: A volumetric measure equal to 4 quarts (231 cubic inches) used to measure fuel oil. One gallon equals 3.785 L; 1 barrel equals 42 gallons.

Gas: A non-solid, non-liquid combustible energy source that includes natural gas, coke-oven gas, blast-furnace gas, and refinery gas.

Gasification: A chemical or heat process that converts solid fuel to a gaseous form.

Generally Accepted Accounting Principles (GAAP): The FASB defines GAAP as the conventions, rules, and procedures necessary to define accepted accounting practice at a particular time. GAAP includes both broad guidelines and relatively detailed practices and procedures.

Generation (Electricity): The process of producing electric energy from other forms; also, the amount of electric energy produced, expressed in watt-hours (Wh).

Geologic Hazards: A geologic hazard is one of several types of adverse geologic conditions capable of causing damage or loss of property and life. These hazards include:

- Avalanches
- Earthquakes
- Forest fires
- Geomagnetic storms
- Ice jams
- Landslides
- Mudslides
- Rock falls
- Torrents
- Volcanic eruptions
- Alluvial fans
- Geyser deposits
- Liquefaction

- Dune migration
- Thermal springs
- Stream erosion

Geographic Information Systems (GIS): A computerized system to compile, store, analyze, and display geographically referenced information.

Geothermal: Of or relating to the Earth's interior heat.

Geothermal Energy: The Earth's interior heat is made available to humans by extracting it from hot water or rocks. As used at electric power plants, hot water or steam extracted from geothermal reservoirs in the Earth's crust is supplied to steam turbines that drive generators to produce electricity.

Gigawatt (GW): One billion watts or one thousand megawatts.

Gigawatt-electric (GWe): One billion watts of electric capacity.

Gigawatthour (GWh): One billion watt-hours.

Global Positioning System (GPS): A navigation system using satellite signals to fix the location of a radio receiver on or above the Earth's surface.

Global Warming: An increase in the near-surface temperature of the Earth. Global warming has occurred in the distant past due to natural influences (a cyclical event throughout Earth's history). Still, the term is most often used to refer to the warming predicted to occur due to increased emissions of greenhouse gases from commercial or industrial sources.

Grassland Pasture and Range: All open land used primarily for pasture and grazing, including shrub and brush land types of pasture; grazing land with sagebrush and scattered mesquite; and all native grasses, legumes, and other forage used for pasture or grazing. Because of the diversity in vegetative composition, grassland pasture and range are not always clearly distinguishable from other types of pasture and range. At one extreme, permanent grassland may merge with cropland pasture, or grassland may often be found in transitional areas with forested grazing land.

Greenhouse Effect: The greenhouse effect produced by greenhouse gases allows incoming solar radiation to pass through the Earth's atmosphere but prevents most of the outgoing infrared radiation from the surface and lower atmosphere from escaping into outer space.

Greenhouse Gas: Gases that trap the sun's heat in the Earth's atmosphere, producing the greenhouse effect. The two major greenhouse gases are water vapor and carbon dioxide. Other greenhouse gases include methane, ozone, chlorofluorocarbons, and nitrous oxide.

Grid: The layout of an electrical distribution system.

Gross Domestic Product (GDP): The total value of goods and services produced by labor and property in the United States. As long as the labor and property are located in the United States, the supplier (the workers and, for property, the owners) may be either U.S. residents or residents of foreign countries.

Gross Generation: The total amount of electric energy produced by the generating units at a generation station or stations, measured at the generator terminals.

Gross Head: A dam's maximum allowed vertical distance between the upstream surface water (headwater) forebay elevation and the downstream surface water (tailwater) elevation at the tail race for reaction wheel dams or the elevation of the jet at impulse wheel dams during specified operation and water conditions.

Groundwater: Water in the ground in the saturation zone, from which wells and springs and groundwater runoff are supplied. Because groundwater is a significant source of drinking water, concern is growing over areas where leaching agricultural or industrial pollutants or substances from leaking underground storage tanks (USTs) contaminate groundwater.

Groundwater Hydrology: The branch of hydrology that deals with groundwater, its occurrence and movements, its replenishment and depletion, the properties of rocks that control groundwater movement and storage, and the methods of investigation and use of groundwater.

Groundwater Recharge: The inflow to a groundwater reservoir.

Groundwater Runoff: A portion of runoff that has passed into the ground, has become groundwater, and has been discharged into a stream channel as spring or seepage water.

HDR: Hot dry rock. These are subsurface geologic formations of abnormally high heat content that contain little or no water.

Habitat: The place or type of site where species and species assemblages are typically found and successfully reproduce.

Habitat Conservation: Protecting an animal or plant habitat to ensure that the use of that habitat by the animal or plant is not altered or reduced.

Habitat Fragmentation: The breaking up of a specific habitat into smaller, unconnected areas.

Half-life: The time it takes for an isotope to lose half of its radioactivity.

Hardwoods: Usually broad-leaved and deciduous trees.

Head: The product of the water's weight and a usable difference in elevation, giving a measurement of the potential energy water possesses.

Head-of-tide: The farthest point upstream where a river is affected by tidal fluctuations.

Heat Content: The amount of heat energy available to be released by the transformation or use of a specified physical unit of an energy form (e.g., a ton of coal, a barrel of oil, a kilowatt-hour of electricity, or a cubic foot of natural gas). The amount of heat energy is commonly expressed in British thermal units (Btu). Note: The heat content of combustible energy forms can be expressed in terms of either gross heat content (higher or upper heating value) or net heat content (lower heating value), depending on whether or not the available heat energy includes the energy used to vaporize water (contained in the original energy form or created during the combustion process). The Energy Information Administration typically uses gross heat content values.

Heat Exchanger: A device for transferring thermal energy from one fluid to another.

Heat Flow: Movement of heat from within the Earth to the surface, where it is dissipated into the atmosphere, surface water, and space by radiation.

Heat Pump: A year-round heating and air-conditioning system employing a refrigeration cycle. In a refrigeration cycle, a refrigerant is compressed (as a liquid) and expanded (as a vapor) to absorb and reject heat. The heat pump transfers heat to a space to be heated during the winter period and, by reversing the operation, extracts (absorbs) heat from the same space to be cooled during the summer period. The refrigerant within the heat pump in the heating mode absorbs the heat to be supplied to the space to be heated from an outside medium (air, ground, or groundwater) and, in the cooling mode, absorbs heat from the space to be cooled to be rejected to the outside medium.

Heating Value: The maximum amount of energy available from burning a substance.

Heliostat: A mirror that reflects solar rays onto a central receiver. A heliostat automatically adjusts its position to track daily or seasonal changes in the sun's position. The arrangement of heliostats around a central receiver is also called a solar collector field.

Herbaceous: Non-woody vegetation, usually lacking permanent strong stems, such as grasses, cereals, and canola (rape).

High-Speed Shaft: Transmits force from the gearbox to the generator.

High-Temperature Collector: A solar thermal collector designed to operate at 180°F or higher.

Historic Conditions: The composition, structure, and functioning of ecosystems resulting from natural processes that we believe, based on sound professional judgment, were present before substantial human-related changes to the landscape.

Holding Pond: A structure built to contain large volumes of liquid waste to ensure that it meets environmental requirements before release.

Horizontal Axis Wind Turbine: The most common type of wind turbine where the axis of rotation is oriented horizontally.

Horsepower: A unit for measuring the work rate (or power) equivalent to 33,000 ft-lbs/min or 746 W.

Hydraulic Fracturing: The fracturing of rock at depth with fluid pressure. It may be accomplished by pumping water into a well at very high pressures. Under natural conditions, vapor pressure may rise high enough to cause fracturing in a process known as hydrothermal brecciation.

Hydraulic Head: The distance between the elevations of the upstream water surface (headwater) above and the downstream surface water (tailwater) below a hydroelectric power plant.

Hydraulic Stimulation: A stimulation technique performed using fluid.

Hydrocarbon: An organic chemical compound of hydrogen and carbon in the gaseous, liquid, or solid phase. The molecular structure of hydrocarbon compounds varies from the simplest (methane, a constituent of natural gas) to the heavy and very complex.

Hydroelectric Power: The use of flowing water to produce electrical energy.

Hydrogen: The lightest of all gases, occurring chiefly in combination with oxygen in water; exists also in acids, bases, alcohols, petroleum, and other hydrocarbons.

Hydrokinetic Energy: The energy a body of water possesses because of its motion (Kinetic Energy = ½ mass × velocity^2).

Hydrologic or Flow Regime: Characteristic fluctuations in river flows.

Hydrology: The science of the Earth's waters: their occurrences, distributions, and circulations; their physical and chemical properties; and their reactions with the environment, including living beings.

Hydrolysis: Decomposition of a chemical compound by reaction with water.

Hydrostatic Energy: The energy a body possesses because of its position or location at an elevation or height above a reference or datum (Potential Energy = mass × gravity × height).

Hydrothermal: Pertaining to hot water.

Hydrothermal Reservoir: An aquifer or subsurface water with sufficient heat, permeability, and water to be exploited without stimulation or enhancement.

Idle Capacity: The component of operable capacity that is not in operation but not under active repair, capable of being placed within 30 days, and capacity not in operation but under active repair that can be completed within 90 days.

Idle Cropland: Land in cover and soil improvement crops, as well as cropland on which no crops were planted. Some cropland is idle each year for various physical and economic reasons. Acreage diverted from crops to soil-conserving uses (if not eligible for and used as cropland pasture) under federal farm programs is included in this component. Cropland enrolled in the Federal Conservation Reserve Program (CRP) is included in the idle cropland.

Impedance: The opposition to power flow in an AC circuit; that is, any device that introduces such opposition in the form of resistance, reactance, or both. The impedance of a circuit or device is measured as the ratio of voltage to current, where a sinusoidal voltage current of the same frequency is used for the measurement; it is measured in ohms.

Impoundment: A body of water, such as a pond, confined by a dam, dike, floodgate, or other barrier, used to collect and store water for future use.

Incandescent Lamp: A glass enclosure in which light is produced when a tungsten filament is electrically heated so that it glows. Much of the energy is converted into heat; this lamp is a relatively inefficient light source. Included in this category are the familiar screw-in light bulbs, as well as somewhat more efficient lamps, such as tungsten halogen lamps, reflector or r-lamps, parabolic aluminized reflector (PAR) lamps, and ellipsoidal reflector (ER) lamps.

Incident Light: Light that shines onto the face of a solar cell or module.

Indicator Species: A species used to gauge the condition of a particular habitat, community, or ecosystem. A characteristic or surrogate species for a community or ecosystem.

Indigenous: Native to an area.

Indigenous Species: A species that, other than as a result of an introduction, historically occurred or currently occurs in a particular ecosystem.

Industrial Wood: All commercial roundwood products except fuelwood.

Injection: The process of returning spent geothermal fluids to the subsurface. Sometimes referred to as reinjection.

Interpretive Facilities: Structures (e.g., kiosks that offer printed materials and audiovisuals, signs, and trailheads) that provide information about an event, place, or thing through various means, including printed, audiovisual, or multimedia materials.

Introduced Invasive Species: Non-native species that have been introduced into an area and, because of their aggressive growth and lack of natural predators, displace native species.

Invasive Species: An alien species whose introduction causes or is likely to cause economic or environmental harm or harm to human health.

Invertebrate: Any animal lacking a backbone or bony segment that encloses the central nerve cord.

Irradiance: The direct, diffuse, and reflected solar radiation that strikes a surface.

Ion Exchange: Reversible exchange of ions adsorbed on a mineral or synthetic polymer surface with ions in solution in contact with the surface. A chemical process used to recover uranium from solution by the interchange of ions between a solution and a solid, commonly a resin.

Isotopes: Forms of the same chemical element that differ only by the number of neutrons in their nucleus. Most elements have more than one naturally occurring isotope. Many isotopes have been produced in reactors and scientific laboratories.

Joule: This is the primary energy unit for the metric system or, in a later, more comprehensive formulation, the International System of Units (SI). It is ultimately defined in terms of the meter, kilogram, and second.

kBtu: A unit of work or energy measured as 1,000 British thermal units. One kBtu is equivalent to 0.293 kWh.

Kilowatt (kW): One thousand watts of electricity.

Kilowatt-hour (kWh): One thousand watt-hours.

Kinetic Energy: Energy available as a result of motion that varies directly in proportion to an object's mass and the square of its velocity.

Lamp: A term generally used to describe artificial light. The term is often used when referring to a bulb or tube.

Land Use: The ultimate uses to be permitted for currently contaminated lands, waters, and structures at each Department of Energy installation. Land-use decisions will strongly influence the cost of environmental management.

Landfill Gas: Gas generated by the decomposition of organic material at landfill disposal sites. Landfill gas is approximately 50% methane.

Landform: The physical shape of the land reflecting geologic structure and processes of geomorphology that have sculpted the structure.

Landscape: A heterogeneous land area composed of a cluster of interacting ecosystems repeated in similar forms throughout.

Langley (L): Unit of solar irradiance. One gram calorie per square centimeter. 1 L = 85.93 kWh/m^2.

Leachate: The liquid that has percolated through the soil or other medium.

Leeward: Away from the direction of the wind. Opposite of windward.

Life-cycle Analysis: This analysis focuses on a product's environmental impact throughout its life cycle, from resource extraction to post-consumer waste disposal. It is a comprehensive approach to examining a product or package's environmental impacts.

Lignin: Structural constituent of wood and (to a lesser extent) other plant tissues, which encrusts the cell walls and cements the cells together.

Limiting Factor: An environmental limitation that prevents further population growth.

Liner: A casing string that does not extend to the top of the wellbore but instead is anchored or suspended from inside the bottom of the previous casing string.

Load: The simultaneous demand of all customers required at any specified point in an electric power system.

Load Balancing: Keeping the amount of electricity produced (the supply) equal to the consumption (the demand). This is one of the challenges of wind energy production, which produces energy on a less predictable schedule than other methods.

Local Solar Time: A system of astronomical time in which the sun crosses the true north-south meridian at noon, which differs from local time according to longitude, time zone, and the equation of time.

Lumen: An empirical measure of the quantity of light. It is based on the spectral sensitivity of the photosensors in the human eye under high (daytime) light levels. Photometrically, it is the luminous flux emitted with a solid angle (1 steradian) by a point source that has a uniform luminous intensity of 1 candela.

Macroinvertebrates: The emphasis on aquatic insect studies has expanded exponentially in the last three decades and has been broadly ecological. Freshwater macroinvertebrates are ubiquitous; even polluted waters contain some representative of this diverse and ecologically important group of organisms. For example, benthic macroinvertebrates are aquatic organisms without backbones that spend at least part of their life cycle on the stream bottom. Examples include aquatic insects—such as stoneflies, mayflies, caddisflies, midges, beetles—and crayfish, worms, clams, and snails. Most hatch from eggs and mature from larvae to adults. Most insects spend their larval phase on the river bottom and emerge as winged adults after a few weeks to several years. The aquatic beetles, true bugs, and other groups remain in the water as adults. Macroinvertebrates typically collected from the stream substrate are either aquatic larvae or adults.

Magnetic Survey: Measurements of the Earth's magnetic field that are mapped and used to determine subsurface geology.

Magneto-Telluri: An electromagnetic method of determining structures below the Earth's surface using electrical currents and the magnetic field.

Mantle: The Earth's inner layer of molten rock, lying beneath the Earth's crust and above the Earth's core of liquid iron and nickel.

Marshlands: Areas interspersed with open water, emergent vegetation (hydrophytes), and terrestrial vegetation (phreatophytes).

Median: The middle number of a data set when the measurements are arranged in ascending (or descending) order.

Medium Pressure: This implies that valves and fittings are suitable for working pressures between 125 and 175 pounds per square inch.

Megavolt Amperes (MVA): Millions of volt-amperes, which are a measure of apparent power.

Megawatt (MW): One million watts of electricity.

Megawatt Hour (MWh): The amount of energy produced by 1 megawatt over one hour.

Methane: A colorless, flammable, odorless hydrocarbon gas (CH_4), a major component of natural gas. It is also an important source of hydrogen in various industrial processes. Methane is a greenhouse gas.

Mineral: Any naturally occurring organic substances, such as metals, salt, sand, stone, sulfur, and water, usually obtained from the Earth. Note: For reporting on the Financial Reporting System, the term includes organic non-renewable substances extracted from the Earth, such as coal, crude oil, and natural gas.

Mineral Rights: The ownership of the minerals beneath the Earth's surface with the right to remove them. Mineral rights may be conveyed separately from surface rights.

Mini-Frac: A small fracturing treatment performed before the main hydraulic fracturing treatment to acquire stress data and to test pre-simulation permeability.

Minimum Streamflow: The lowest rate of flow of water past a given point during a specified period.

Mitigation: Actions to compensate for the adverse effects of a particular project.

Moisture Content: The water content of a substance (a solid fuel) as measured under specified conditions being the "dry basis," which equals the weight of the wet sample minus the weight of a (bone) dry sample divided by the weight of the dry sample times 100 (to get percent); "wet basis," which is equal to the weight of the wet sample minus the weight of the dry sample divided by the weight of the wet sample times 100.

Mole: The quantity of a compound or element that weighs grams numerically equal to its molecular weight. Also referred to as "gram molecule" or "gram molecular weight."

MTBE: Methyl Tertiary Butyl Ether is a fuel oxygenate that reacts with methanol with isobutylene.

MSW (Municipal Solid Waste): Residential solid waste and non-hazardous commercial, institutional, and industrial wastes.

Municipal Waste: As defined in the Energy Security Act (P.L. 96-294; 1960) as "any organic matter, including sewage, sewage sludge, and industrial or commercial waste, and mixtures of such matter and inorganic refuse from any publicly or privately operated municipal waste collection or similar disposal system, or from similar waste flows (other than such flows which constitute agricultural wastes or residues, or wood wastes or residues from wood harvesting activities or production of forest products)."

Multiplier Effect: The multiplier effect is sometimes called the ripple effect because a single expenditure in an economy can have repercussions throughout the entire economy. The multiplier measures how much additional economic activity is generated from an initial expenditure.

Naphtha: A generic term applied to a petroleum fraction with an approximate boiling range between 122°F and 400°F.

Native: A species that, other than as a result of an introduction, historically occurred or currently occurs in a particular ecosystem.

Native Gas: A gaseous mixture of hydrocarbon compounds, the primary one being methane.

Native Plant: A plant that has grown in the region since the last glaciation and occurred before European settlement.

Natural Gas: A gaseous mixture of hydrocarbon compounds, the primary one being methane.

Natural Processes: A complex mix of interactions among animals, plants, and their environment that ensures the maintenance of an ecosystem's full range of biodiversity. Examples include population and predator-prey dynamics, pollination and seed dispersal, nutrient cycling, migration, and dispersal.

Natural Sinks: Regarding greenhouse gases, refers to any natural process in which these gases are absorbed from the atmosphere.

Near Tree: Trees located within approximately 15 m (50 ft) of a building that directly influence irradiance and airflow on the building envelope.

Net Head: The gross head minus all hydraulic losses except those chargeable to the turbine.

Net Metering: An arrangement that permits a facility (using a meter that reads inflows and outflows of electricity) to sell any excess power it generates over its load requirement back to the electrical grid to offset consumption.

Net Summer Capacity: The maximum output, commonly expressed in megawatts (MW), that generating equipment can supply to system load, as demonstrated by a multi-hour test during summer peak demand. This output reflects a reduction in capacity due to electricity use for station services or auxiliaries.

Niche: The specific part or smallest unit of a habitat occupied by an organism.

Nitrogen Dioxide: A compound of nitrogen and oxygen formed by the oxidation of nitric oxide (NO), which is produced by the combustion of solid fuels.

Nitrogen Oxides (NO_x): Compounds of nitrogen and oxygen produced by burning fossil fuels.

Nitrous Oxide (N_2O): A colorless gas naturally occurring in the atmosphere. Nitrous oxide has a 100-year Global Warming Potential of 310.

Non-biomass Waste: Material of non-biological origin that is a byproduct or a discarded product. "Non-biomass waste" includes municipal solid waste from non-biogenic sources, such as plastics and tire-derived fuels.

Nonforested Land: Land that has never supported forests and lands formerly forested where use of time management is precluded by development for other uses. (Note: Includes areas used for crops, improved pasture, residential areas, city parks, improved roads of any width and adjoining clearings, powerline clearings of any width, and 1- to 4.5-acre areas of water classified by the Bureau of the Census as land. If intermingled in forest areas, unimproved roads and nonforested strips must be more than 120 ft wide, and clearings, etc., must be more than 1 acre in area to qualify as nonforested land.)

Nonindustrial Private: An ownership class of private lands where the owner does not operate wood-using processing plants.

Nonpoint Source: A diffuse form of water quality degradation produced by erosion of land that causes sedimentation of streams, eutrophication from nutrients and pesticides used in agricultural and silvicultural practices, and acid rain resulting from burning fuels that contain sulfur.

Nonrenewable Fuels: Fuels that cannot be easily made or "renewed," such as oil, natural gas, and coal.

Nonutility Generation: Electric generation by nonutility power producers to supply electric power for industrial, commercial, and military operations or sales to electric utilities.

Nonutility Power Producer: A corporation, person, agency, authority, or other legal entity or instrumentality that owns electric generating capacity and is not an electrical utility. Nonutility power produces include qualifying cogenerators, qualifying small power produces, and other nonutility generators without a designated, franchised service area that do not file forms listed in the Code of Federal Regulations, Title 18, Part 141.

Occupied Space: The space within a building or structure that people normally occupy and that may be conditioned (heated, cooled, and/or ventilated).

Ocean Energy Systems: Energy conversion technologies that harness the energy in tides, waves, and thermal gradients in the oceans.

Ocean Thermal Energy Conversion (OTEC): The process or technologies for producing energy by harnessing the temperature differences (thermal gradients) between ocean surface waters and ocean depths. Warm surface water is pumped through an evaporator containing a working fluid in a closed Rankine-cycle system. The vaporized fluid drives a turbine/generator.

Other Biomass: This category of biomass energy includes agricultural byproducts/crops (agricultural byproducts, straw); other biomass gas (digester waste alcohol); other biomass liquids (fish oil liquid acetonitrile, waste, tall oil, waste alcohol); other biomass solids (medical waste, solid byproducts, sludge waste, and tires).

Other Forest Land: Forest land other than timberland and reserved forest land. It includes available forest land, which cannot produce 20 cubic feet per acre of industrial wood annually under natural conditions because of adverse site conditions such as sterile soils, dry climate, poor drainage, high elevation, steepness, or rockiness.

Other Removals: Unutilized wood volume from a cut or otherwise killed growing stock, from cultural operations such as precommercial thinning, or timberland clearing. Does not include volume removed from inventory through the reclassification of timberland to productive reserved forest land.

Other Sources: Roundwood products available from sources other than growing stock. These include dead, rough, and rotten trees, trees of noncommercial species, trees less than 5.0 inches dB, tops, and roundwood harvested from non-forest land (for example, fence rows).

Overtopping: A structure over which the waves topple into a reservoir to collect the water, and hydro turbines are installed at the bottom of the reservoir.

The head of collected water turns the turbines as it flows back out to sea and the turbines are coupled to generators to produce electricity.

Packer: A device placed in the wellbore that blocks vertical fluid flow to isolate zones.

Palustrine Forested Wetlands: Dominated by trees, including wooded swamps and low-lying hardwood forests near rivers. Sixty-eight percent of the wetlands in the Chesapeake Bay watershed are forested.

Palustrine Wetlands: Palustrine wetlands include nontidal wetlands dominated by trees, shrubs, persistent emergents, emergent mosses, or lichens, and all tidal wetlands where salinity due to ocean-derived salts is below 0.5 ppt.

Paper Pellets: Paper compressed and bound into uniform-diameter pellets to be burned in a heating stove.

Parabolic Dish: A high-temperature (above 180°F) solar thermal concentrator, generally bowl-shaped, with two-axis tracking.

Parabolic Trough: A high-temperature (above 180°F) solar thermal concentrator with the capacity for tracking the sun using one axis of rotation.

Particulate: A small, discrete mass of solid or liquid matter that remains individually dispersed in gas or liquid emissions. Particulates are aerosols, dust, fumes, mists, smoke, or sprays. Each of these forms has different properties.

Passive Solar: A system in which solar energy alone transfers thermal energy. Pumps, blowers, or other heat transfer devices that use energy other than solar are not used.

Peak Demand: The maximum load during a specified period of time.

Peak Watt: A manufacturer's unit indicating how much power a photovoltaic cell or module will produce at standard test conditions (typically 1,000 W/m^2 and 25°C).

Permeability: The ability of a rock to transmit fluid through its pores or fractures when subjected to a difference in pressure. Typically measured in darcies or millidarcies.

Petrochemical Feedstocks: Chemical feedstocks derived from petroleum, principally for manufacturing chemicals, synthetic rubber, and various plastics.

Phenology: The study of periodic plant and animal life cycle events and how seasonal and interannual variations in climate influence these events.

Photosynthesis: The process in which plants produce carbohydrates and oxygen from carbon dioxide and water in the presence of chlorophyll, with sunlight as the energy source. Carbon is sequestered, and oxygen and water vapor are released.

Photovoltaic (PV) Cell: An electronic device consisting of layers of semiconductor materials fabricated to form a junction (adjacent layers of materials with different electronic characteristics) and electrical contacts, capable of converting incident light directly into electricity (direct current).

Photovoltaic (PV) Module: An integrated assembly of interconnected photovoltaic cells designed to deliver a selected level of working voltage and current at its output terminals, packaged for protection against environmental degradation and suited for incorporation in photovoltaic power systems.

Planetary Albedo: The fraction of incident solar radiation reflected by the Earth-atmosphere system and returned to space, primarily by backscatter from atmospheric clouds.

Plate Tectonics: A theory of global-scale dynamics involving the movement of rigid plates of the Earth's crust. Tectonic activity is evident along the margins of the plates, where buckling, grinding, faulting, and vulcanism occur as the plates are propelled by the forces of deep-seated mantle convection currents. Geothermal resources are often associated with tectonic activity since groundwater can come in contact with deep subsurface heat sources.

Point Source: A source of pollution that involves waste discharge from an identifiable point, such as a drainage pipe, smokestack, or sewage treatment plant.

Pondage: The amount of water stored behind a hydroelectric dam with a relatively small storage capacity. The dam usually controls the river's flow daily or weekly.

Population: An interbreeding group of plants or animals. The entire group of organisms of one species.

Porosity: The ratio of the aggregate volume of pore spaces in rock or soil to its total volume, usually expressed as a percentage.

Power Factor: The ratio of real power (kilowatts) to apparent power (kilovolt-amperes) for any given load and time.

Powerhouse: A structure at a hydroelectric plant site that contains the turbine and generator.

Prescribed Fire: The application of fire to wildland fuels by natural or intentional ignition to achieve identified land use objectives.

Primary Wood-Using Mill: A mill that converts roundwood products into other wood products. Common examples are sawmills that convert saw logs into lumber and pulp mills that convert pulpwood roundwood into wood pulp.

Process Heating: The direct process end use in which energy is used to raise the temperature of substances involved in the manufacturing process.

Program Tree: A tree planted within the site area as a result of the shade tree program.

Propane (C3H8): A straight-chain saturated (paraffinic) hydrocarbon extracted from natural gas or refinery gas streams, which is gaseous at standard temperature and pressure.

Proppant: Sized particles mixed with fracturing fluid to hold fractures open after hydraulic stimulation.

Propylene (C_3H_6): An olefinic hydrocarbon recovered from refinery or petrochemical processes, gaseous at standard temperature and pressure. Propylene is a vital petrochemical feedstock.

Public Land: Land owned by local, state, or federal governments.

Public Utility Regulatory Policies Act of 1978 (PURPA): Part of the National Energy Act, PURPA contains measures designed to encourage energy conservation, more efficient resource use, and equitable rates. Principal among these are suggested retail rate reforms and new incentives for electricity production by cogenerators and the use of renewable resources.

Pulpwood: Roundwood, whole tree chips, or wood residues used to produce wood pulp.

Pumped-Storage Hydroelectric Plant: A plant that generates electric energy during peak load periods by using water previously pumped into an elevated storage reservoir during off-peak periods when excess generating capacity is available. When additional generating capacity is needed, the water can be released from the reservoir through a conduit to turbine generators located in a power plant at a lower level.

Pyrolysis: The thermal decomposition of biomass at high temperatures (greater than 400°F or 200°C) in the absence of air. The end product of pyrolysis is a mixture of solids (char), liquid (oxygenated oils), and gases (methane, carbon monoxide, and carbon dioxide), with the proportions determined by operating temperature, pressure, oxygen content, and other conditions

Quadrillion Btu (Quad): Equivalent to 10 to the 15th power Btu.

Qualifying Facility (QF): A cogeneration or small power production facility that meets certain ownership, operating, and efficiency criteria established by the Federal Energy Regulatory Commission (FERC) pursuant to the Public Utility Regulatory Policies Act of 1978 (PURPA).

Radiant Energy: Energy that transmits away from its source in all directions.

Reactance: A phenomenon associated with AC power characterized by a time difference between voltage and current variations.

Reactor: The body that provides a reaction to the displacer. It could be a body fixed to the seabed or the seabed itself. It could also be another structure or mass that is not fixed but moves to create reaction forces (e.g., by moving by a different amount or at different times). A degree of control over the forces acting on each body and/or acting between the bodies (particularly stiffness and damping characteristics) is often required to optimize the amount of energy captured. In some designs, the reactor is inside the displacer; in others, it is an external body.

Reclamation: The process of restoring the surface environment to acceptable pre-existing conditions. Includes surface contouring, equipment removal, well plugging, revegetation, etc.

Recovery Factor: The fraction of resources that can be extracted for productive uses.

Recycling: The process of converting materials no longer valuable as designed or intended into a new product.

Reflectivity: The ratio of the energy a wave carries after reflection from a surface to its energy before reflection.

Refuge Lands: Lands in which the National Forest Service holds full interest in fee title like an easement.

Renewable Energy: Energy produced using resources that regenerate quickly or are inexhaustible. Wind energy is considered inexhaustible because while it may blow intermittently, it will never stop.

Renewable Energy Resources: Energy resources that are naturally replenishing but flow-limited. They are virtually inexhaustible in duration but limited in the energy available per unit of time. Renewable energy resources include biomass, hydro, geothermal, solar, wind, ocean thermal, wave action, and tidal action.

Renewable Portfolio Standard (RPS): A mandate requiring that renewable energy provide a certain percentage of total energy generation or consumption.

Reservoir: A natural underground container of liquids, such as water or steam (or, in the petroleum context, oil or gas).

Residues: Bark and woody materials that are generated in primary wood-using mills when roundwood products are converted to other products. Examples are slabs, edgings, trimmings, sawdust, shavings, veneer cores and clippings, and pulp screenings. Includes bark and wood residues (coarse and fine materials) but excludes logging residues.

Resistivity Survey: The measurement of the ability of a material to resist or inhibit the flow of electrical current, measured in ohm-meters. The voltage between two electrodes measures resistivity, while an electrical current is generated between two other electrodes. Resistivity surveys can be used to delineate the boundaries of geothermal fields.

Resource Base: All of a given material in the Earth's crust, whether its existence is known or unknown, regardless of cost considerations.

Riparian: Referring to the interface between freshwater habitats and the terrestrial landscape.

River In-Stream Energy Systems: River in-stream systems tend to be smaller than their tidal counterparts; however, many possess similar designs. Unlike tidal facilities, river systems produce electricity only from flows in one direction.

Rotten Tree: A live tree of commercial species that does not contain a saw log now or prospectively, primarily because of rot (that is, when rot accounts for more than 50% of the total cull volume).

Rough Tree: (a) A live tree of commercial species that does not contain a saw log now or prospectively, primarily because of roughness (that is, when sound cull, due to such factors as poor form, splits, or cracks, accounts for more than 50% of the total cull volume) or (b) a noncommercial tree.

Run-of-River Hydroelectric Plant: A low-head plant using the flow of a stream as it occurs and having little or no reservoir capacity for storage.

Runoff: Water from rain, melted snow, or agricultural or landscape irrigation that flows over a land surface into a water body.

Salinity: A measure of the quantity or concentration of dissolved salts in water.

Saplings: Live trees 1.0 inch through 4.9 inches in diameter at breast height (dbh).

Seismic: Pertaining to the nature of, or caused by an earthquake or earth vibration, whether natural or man-made.

Seismicity: The phenomenon of earth movements, including the frequency, distribution, and intensity of earthquakes.

Seismometer: An electrical device used on the surface and within wellbores to measure the magnitude and direction of seismic events.

Self-Potential: A measure of currents induced in the subsurface due to the flow of fluids in geothermal systems.

Septic Tank: A tank where bacteria disintegrate the solid matter of continuously flowing sewage.

Sequestration: The annual net rate at which a tree removes carbon dioxide from the atmosphere through photosynthesis and respiration (kgCO_2/tree/year).

Shale Gas: Natural gas produced from wells open to shale formations. Shale is a fine-grained sedimentary rock composed of mud from flakes of clay minerals and tiny fragments (silt-sized particles) of other materials. The shale acts as both a source and a reservoir for natural gas.

Short Ton: A measure of weight equal to 2,000 (lbs) or 0.9072 metric tons.

Silicon: A semiconductor material made from silica, purified for photovoltaic applications.

Silviculture: The practice of tending and regenerating forest stands to realize sought-after benefits and sustain them over time.

Sludge: A dense, slushy, liquid-to-semi-fluid product that accumulates as the end result of an industrial or technological process designed to purify a substance. Industrial sludges are produced from the processing of energy-related raw materials, chemical products, water, mined ores, sewage, and other natural and man-made products. Sludges can also form from natural processes, such as the runoff produced by rainfall, and accumulate on the bottom of bogs, streams, lakes, and tidelands.

Species: The basic category of biological classification intended to designate a single kind of animal or plant. Any variation among the individuals may be regarded as not affecting the essential sameness that distinguishes them from all other organisms.

Stimulation: A treatment performed to restore or enhance the productivity of a well. Stimulation treatments fall into two main groups: hydraulic fracturing treatments and matrix treatments.

Stormwater: Water runoff generated when precipitation from rain and snowmelt flows over land or impervious surfaces.

Stress: The forces acting on rock. In the subsurface, the greatest force or stress is generally vertical, caused by the weight of overlying rock.

Structural Discontinuity: A discontinuity in the rock fabric, which can be a fracture, fault, intrusion, or differing adjacent rock type.

Submersible Sump: A pump with both the pumping mechanism and a driving electric motor suspended together at depth in the well.

Subsidence: The sinking of an area of the Earth's crust due to fluid withdrawal and pressure decline.

Succession: The natural, sequential change in the species composition of a community in a given area.

Sulfur Dioxide (SO_2): A toxic, irritating, colorless gas soluble in water, alcohol, and ether. It is used as a chemical intermediate, in paper pulping and ore refining, and as a solvent.

Surface Water: All waters whose surface is naturally exposed to the atmosphere or wells or other collectors directly influenced by surface water.

TDS: Total dissolved solids. Used to describe the amount of solid material in water.

Terrestrial Sequestration: Basic carbon sequestration in above- and below-ground biomass and soils.

Thermal Drawdown: The decline in formation temperature due to geothermal production.

Thermal Gradient: The rate of temperature increase as a function of depth into the Earth's crust.

Tower: Steel structures that support the turbine assembly. Higher towers allow for longer blades and the capture of faster-moving air at higher altitudes.

Tracer: A chemical injected into the flow stream of a production or injection well to determine fluid path and velocity.

Transformer: A device used to step up or step down AC voltage.

Transmission Line: Structures and conductors that carry bulk supplies of electrical energy from power-generating units.

Transmission System (Electric): An interconnected group of electric transmission lines and associated equipment that moves or transfers electric energy in bulk between points of supply and points at which it is transformed for delivery over the distribution system lines to consumers or other electrical systems.

Transportation Sector: An energy-consuming sector that consists of all vehicles whose primary purpose is transporting people and goods from one physical location to another. Included are automobiles, trucks, buses, motorcycles, trains, subways, other rail vehicles, aircraft, ships, barges, and other water-borne vehicles. Vehicles whose primary purpose is not transportation (e.g., construction cranes and bulldozers, farming vehicles, warehouse tractors, and forklifts) are classified in the sector of their primary use.

Turbidity: The extent to which light penetrates a body of water. Turbid waters do not generally support the net growth of photosynthetic organisms.

Turbine: A machine for the generation of rotary mechanical power from the energy of a stream of fluid (such as water, steam, or hot gas). Turbines convert the kinetic energy of fluids to mechanical energy through the principles of impulse and reaction or a mixture of the two.

Useful Heat: Heat stored above room temperature (in a solar heating system).

Useful Thermal Output: The thermal energy made available in any industrial or commercial process or used in any heating or cooling application, i.e., total thermal energy made available for processes and applications other than electrical generation.

Vernal Pool: Seasonal wetlands formed by isolated depressions in the landscape that hold water in the winter and spring and are usually dry by midsummer or fall.

Viewshed: The scenic characteristics of an area when referred to as a resource.

Voltage: The measure of electrical potential difference.

Watt (Electric): The electrical unit of power. The rate of energy transfer is equivalent to 1 ampere (A) of electric current flowing under a pressure of 1 V at unity power factor.

Watt (Thermal): A unit of power in the metric system, expressed in terms of energy per second, equal to the work done at a rate of 1 J/s.

Watt-hour (Wh): The electrical energy unit of measure equal to 1 W of power supplied to, or taken from an electric circuit steadily for 1 hour.

Well Log: A record of measurements of the diameter of the well and various electrical, mass, and nuclear properties of the rock, which can be correlated with the rock's physical properties. The well log is a chart of the measurements relative to depth in the well.

Windward: Into or facing the direction of the wind. Opposite of leeward.

Wood/Wood Waste: This category of biomass energy includes black liquor; wood/wood waste liquids (red liquor, sludge wood, spent sulfite liquor); wood/wood waste solids (peat, paper pellets, railroad ties, utility poles, wood/wood waste).

Wood Energy: Wood and wood products used as fuel, including roundwood (cordwood), limb wood, wood chips, bark, sawdust, forest residues, charcoal, pulp waste, and spent pulping liquor.

Wood Pellets: Sawdust compressed into uniform diameter pellets to be burned in a heating stove.

Yaw: The rotation of a horizontal-axis wind turbine around its tower or vertical axis.

Yaw Drive: A motor that keeps an upwind turbine facing into the wind.

Zonal Isolation: Various methods are used to partition portions of the wellbore for stimulation, testing, flow restriction, or other purposes.

Zone: An area within the interior space of a building, such as individual room(s), to be cooled, heated, or ventilated. A zone has its thermostat to control the flow of conditioned air into the space.

NOTE

1 Definitions retrieved 01/05/2010 @ http://www1.eere.enegy.gov/site_adminstaration/glossary.html from DOE's EERE Glossary of Energy-Related Terms.

Index

Note: **Bold** page numbers refer to tables and *italic* page numbers refer to figures.

For Product Safety Concerns and Information please contact our EU representative GPSR@taylorandfrancis.com
Taylor & Francis Verlag GmbH, Kaufingerstraße 24, 80331 München, Germany

www.ingramcontent.com/pod-product-compliance
Lightning Source LLC
LaVergne TN
LVHW011105300926
849160LV00008B/290
* 9 7 8 1 0 3 2 9 8 9 5 1 8 *